Major Constrains and Verdict of Crop Productivity

The Editor

Dr. U.N. Bhale, presently Assistant Professor, Department of Botany, Arts, Science and Commerce College Naldurg Tq. Tuljapur, Dist. Osmanabad affiliated to Dr.Babasaheb Ambedkar Marathwada University Aurangabad (MS). He has completed his Ph.D. in Botany (Plant Pathology) from Dr.Babasaheb Ambedkar Marathwada University Aurangabad (MS). He has 12 years teaching and 15 years research experience. He has contributed more than 90 papers in National and International journals of repute. He has published 3 books, guided 06 Ph.D. students and attended 37 National and International conferences. He has completed 03 Minor and 01 Major research projects. He has successfully organized a National Conference as Organizing Secretary funded by University Grants Commission (UGC). His areas of interest are Mycology, Plant pathology and VAM technology. He is an Associate Editor of International Journal of Bioscience Discovery and Life Member of Academy of Plant Sciences, IPS, ISMPP, BBRC, MSI, IMS, JBB and IJPS.

Major Constrains and Verdict of Crop Productivity

Volume 2

Editor
Dr. U.N. Bhale
Department of Botany,
Arts, Science and Commerce College,
Naldurg – 413 602
Tal. Tuljapur, Dist. Osmanabad,
Maharashtra (India)

2015
Daya Publishing House®
A Division of
Astral International Pvt. Ltd.
New Delhi – 110 002

Cataloging in Publication Data--DK
Courtesy: D.K. Agencies (P) Ltd. <docinfo@dkagencies.com>
Major constrains and verdict of crop productivity / editor, Dr. U.N. Bhale.
pages cm
Includes bibliographical references and index.
ISBN 9789351243489 (Set)
ISBN 9789351243694 (Vol. 1)
ISBN 9789351243700 (Vol. 2)
ISBN 9789351305545 (International Edition)
1. Crops and soils--India. 2. Soil productivity--India. I. Bhale, U. N., editor.
DDC 631.4954 23

Published by : **Daya Publishing House®**
A Division of
Astral International Pvt. Ltd.
– ISO 9001:2008 Certified Company –
4760-61/23, Ansari Road, Darya Ganj
New Delhi-110 002
Ph. 011-43549197, 23278134
E-mail: info@astralint.com
Website: www.astralint.com

Laser Typesetting : **Classic Computer Services**, Delhi - 110 035

Digitally Printed at : **Replika Press Pvt. Ltd.**

PRINTED IN INDIA

Prof. C. MANOHARACHARY
M.Sc,Ph.D,D.Sc.,F.N.A.Sc.,F.N.A.A.S.,F.B.S.,F.P.S.I.,F.A.P.A.S.
Professor Emeritus (CSIR)
Former Head, Chairman, Principal,
Co-ordinator (UGC SAP)
Dean, Development & UGC Affairs
NASI Senior Scientist

DPARTMENT OF BOTANY
Osmania University,
Hyderabad – 500 007, A.P. India.
Off: 91-40-2768 22 44
Mob:9391164243

Foreword

I have great pleasure in introducing the academic venture entitled " Major constrains and verdict of crop productivity" an edited volume of 54 scientific articles by Dr. U.N. Bhale, Department of Botnay, Arts, Science and Commerce College, Naldurg, Maharashtra state, India. India is an agricultutally dependent country and the growing population needs food and nutritional security.

Though India reached targeted productivity both in wheat and rice during first green revolution. Huge losses have incurred in crop prouductivity due to increase in pests and diseases, and also loss of soil fertility because of non judicious usage or overuse of pesticides, fungicides and fertilizers.

Further there has been repeated drought periods because of failure of monsoon and less precipitation. All these factors advocate that we need a second green revolution where in drought and disease resistant varieties need to be produced besides increasing the production of pulses, oil seeds and fertile land. All the 56 scientific contributions deal with biotechnology of crop production, usage of biofertilizers and biopesticides, management of pests and diseases, biodiversity, phytotoxicity, plant and microbial metabolites and related issues. The contributed authors have necessary expertise. I hope that this book will add its strengths to the existing knowledge and furtherance of new knowledge besides bringing an awareness on issues related to Agriculture. I hope that all categories of readers and libraries will find it useful. Finally I compliment the editor Dr. U.N. Bhale and authors for their concerted efforts.

1/7/2014.

Prof. C. Manohara Chary
NASI Senior Scientist

Preface

India is an Agricultural country and reached targeted productivity especially rice, wheat and others crops in green revolution. Now it is huge losses in crop productivity due to pest, diseases and loss of soil fertility because of overuse of chemical pesticides. Therefore number of constrains are found in crop productivity. Drought conditions are thoroughly increased due to failure of monsoon in time.A constraint free soil environment is very important for achieving higher food production. Abiotic stresses like nutrient-poor, degraded and often acidic, soils limit crop production in many tropical regions. Good agronomic practices help minimize the water and other physiological stresses imposed by subsoil constraints.When coupled with the high cost of inorganic fertilizer, much small-scale agriculture occurs under conditions of nutrient scarcity and/or metal ion toxicity. Drought is perhaps the biggest constraint to agricultural productivity worldwide. Biotic stresses like diseases are a huge problem worldwide for crop productivity.

This book is an outcome of to create awareness about the emerging trends in agriculture designing, developing new economic process to farmers and provide a platform for forthcoming researchers. I have attempted to compile the invaluable and carefully selected contributions of plant biologists in this consolidation of this book with all originality and alertness.The book considers more specifically the particular implications of different approaches to technical improvement of productivity in farming. The book concludes by outlining agricultural areas in which research needs to take a new approach to investigating the constraints to crop productivity improvement and the relationship of these to rural poverty. The expertise has in the fields of plant protection, plant genetic resources, forestry, soil science, and extension and is well resourced to address crop production problems in the region. To my mind, the best approach to identifying key constraints is to establish a much better dialogue with between bench scientists, researchers and those key breeders in the country who actually talk to farmers and understand local agricultural issues.

I am sincerely thanks to all contributors and those directly or indirectly have rendered their help to bring out this book. My special genuine gratitude to Prof. C. Manoharachary, Professor Emeritus (CSIR),NASI Senior Scientist, Osmania University, Hyderabad for encouraging and giving Foreword letter. Lastly, my sincere thanks to publisher Astral International (P) Limited, New Delhi who provides an opportunity to publish this book.

Dr. U.N. Bhale

Contents

Volume-2

2015, Major Constrains and Verdict of Crop Productivity *Pages 349–365*
Editor: **Dr. U.N. Bhale**
Published by: **ASTRAL INTERNATIONAL (P) LIMITED, NEW DELHI**

Chapter 26

Studies on AM Fungi Associated with Four Sugarcane Cultivars

B.P. Shinde*

P.G. Department of Botany,
Fergusson College, Pune, M.S., india

Introduction

Sugarcane, *Saccharum officinarum* L. belongs to family Poaceae of the tribe Andropogoneae. It is a tropical crop, grows most successfully in those areas where the climate is more or less tropical, but it can grow in sub-tropics also *i.e.* in North India. Under warm and humid conditions it can continue its growth, unless terminated by flowering. It has stout, jointed, fibrous stalks that are rich in sugar and measure two to six meters (six to nineteen feet) in height. Today sugarcane is cultivated in more than 110 countries. The highest production of sugarcane in the world is reported from Brazil. Brazil and India accounts about 50 per cent production of sugarcane in world. In India, sugarcane is cultivated in Maharashtra, Uttar Pradesh, Karnataka and Tamil Nadu.

The term mycorrhiza was coined by German Botanist Frank, (1885) which literally means fungus root. He used this term for the association of fungi with the roots of plants. Mycorrhiza is defined as the symbiotic (mutually beneficial) relationship between root of plants and fungi. In this mutually beneficial association the soil nutrients are made available for the plant roots and in turn plant gives sugars to the mycorrhizal fungi.

---

* E-mail: scindiab2002@gmail.com

Review of Literature

Frank, (1885) gave the name "mycorrhiza" to the peculiar association between plant roots and fungi. Earlier the word was written as Mycorhiza in this the second 'r' is added by Kelley (1931, 1950). The origin and evolution of arbuscular mycorrhizal (AM) fungi is directly related to that of land plants, yet it is crucial to understand if the AM forming fungi share a common phylogenetic origin (*i.e.* are monophyletic), or if the present day AM symbioses are more or less independent results representing the convergent evolution of plants with different fungal lineages. Trappe and Berch, (1985) cited other early observations of the symbiosis during the period 1875-1895.

Extensive surveys of host plants and sophisticated anatomical descriptions of what are most certainly arbuscular mycorrhizas are given by Janse, (1897) and Gallaud, (1905). Gallaud, (1905) made very accurate observations of the arbuscule and concluded, that it is entirely surrounded by a host membrane. He also noted that partial digestion of the arbuscule resulted in a structure called the "sporangiole" by Janse, (1897). Gallaud, (1905) further distinguished between *Arum* and *Paris* types of arbuscules (Smith and Read, 1997).

The AM fungi were classified by various researchers. The classification proposed by Morton and Benny, (1990) included one order, two sub orders, three families and six genera. After this the recent classification was published by Schüßler *et al.* (2001) with emendations of Walker and Schüßler, (2004).

Phosphorus is a major macronutrient for all organisms and serves multiple functions as a key structural element in nucleic acids, phospholipids and several enzymes and coenzymes. Enhanced uptake of phosphate is generally regarded as the most important benefit that AM provide to their host plant and plant phosphate status is often the main controlling factor in the plant fungal relationship (Thompson, 1987; Smith and Read, 1997; Graham, 2000).

The AM fungi play a significant role in crop phosphate nutrition, increasing total uptake and in some cases phosphate use efficiency (Koide and Kabir, 2000). This may be associated with increased growth and yield (Koide and Kabir, 2000).The inoculum of selected AM fungi is generally raised by 'funnel technique' using wheat as a host plant (Menge and Timmer, 1982).

Sustaining sugarcane productivity under depleting water resources has been studied by Srivastava *et al.* (2011). Whereas the arbuscular mycorrhizal fungal diversity in sugarcane rhizosphere in relation with soil properties has been reported by Datta and Kulkarni, (2012). An *in-vitro* study on sugarcane with respect to VAM fungi was carried out by Muniyamma *et al.* (2000).

Seasonal dynamics of AM fungi in sugarcane (*Saccharum officinarum* L. Cv. Spf-213) in relation to red rot (*Colletotrichum falcatum*) disease from Punjab, Pakistan has noted by Nasim *et al.* (2008). The effect of pesticides used in sugarcane cropping systems on soil organisms and biological functions associated with soil health has been reported by Pankhurst, (2006). The crop rotation biomass and arbuscular mycorrhizal fungi and its effects on sugarcane yield have been reviewed by Edmilson *et al.* (2010).

Aims and Objectives

1. To isolate and identify various indigenous AM fungi from rhizosphere of Sugarcane plants.
2. To check the correlation of AM fungi with respect to chemical properties of soil.
3. To determine and compare the effect of AM fungi on different varieties of sugarcane.
4. To study the biochemical changes in sugarcane due to inoculation of AM fungi.
5. To produce most efficient strain on large scale and make it available for sugarcane cultivation.

Materials and Methods

The rhizosphere and non rhizosphere soil samples and roots of sugarcane plants were collected from two selected sites at an interval of 30 days each. Isolation of AM fungi was carried out by wet sieving and decanting method (Gerdemann and Nicolson, 1963). Identification of AM fungi was done by using the manual of Schenck and Perez, (1990). Clearing of roots and staining of AM fungi for assessment of infections was done by Phillips and Hyman's, (1970) method. Analysis of macro and micronutrients from soil was carried out by Tandon's method (1993). Most Probable number of AM fungi was estimated by Cochran's (1950) method. Isolation and quantification of chlorophyll was carried out by Arnon's method (1949). Isolation and estimation of DNA was carried out by Dellaporta, *et al.* (1983) and Burton's method (1956) respectively. Isolation and estimation of RNA was carried out by, Brawerman, (1974) and Bial method (1902) respectively. Estimation of protein was carried out by Lowry *et al.* Method (1951). Estimation of vitamin C was carried out by Harris and Ray method (1935). The amount of sugars was determined by Hedge and Hofreiter's (1962) method.

Results and Discussion

The present work includes taxonomy of AM fungi isolated from rhizosphere soil of sugarcane plant from two selected localities, physicochemical parameters of soil like pH, electrical conductivity, organic carbon and macronutrients like phosphorus and potassium and micronutrients like Zn, Cu, Fe, and Mn. The effect of AM fungi and farm yard manure (FYM) alone, combination of both AM fungi and farm yard manure (FYM) on growth and biochemical contents in sugarcane plant was also studied. The parameters like number of tenders, leaves, height of plant, number of internodes and fresh weight of sugarcane plants were noted at regular intervals. The biochemical contents like chlorophyll, DNA and RNA, proteins, vitamin C and sugars were estimated from mycorrhizal and non mycorrhizal plants at regular intervals.

Taxonomy of AM Fungi

Isolation and Identification of AM Fungi from Rhizosphere and Non-rhizosphere Soil

The pH is a measure of the hydrogen ion concentration *i.e.* acidity or alkalinity of the soil. The pH can affect the availability of nutrients and activity of many essential micro-organisms. It may influence crops grown in the field and the types of soil micro biota. The pH of all the soil samples was ranging in between 7.65 and 7.85 which was slightly alkaline. The unit of electrical conductivity (EC) is mmhos/cm^3. It was ranging in between 0.07 and 0.2 mmhos/cm^3. There was no specific correlation found between EC and spore density. The response of VAM fungi is variable according to the strains present in that soil. The AM fungi were positively correlated with alkaline pH value (Sharrif and Moawad, 2006). Maximum organic carbon (OC) percentage was recorded at VSI Pune locality (0.92 per cent) and minimum at Sangamner locality (0.52 per cent). The OC content has positive effect on the number of *Glomus* species. At VSI Pune locality nine species of *Glomus* were recorded and at Sangamner eight species of *Glomus* were recorded. Our results match with the results of Kliromonos *et al.* (1998).

The nutrient content of the soil is one of the important factors related with association of AM fungi with the roots of plants. Phosphorus is one of the key macronutrients required for plant growth and metabolism. Inorganic phosphate supplied to the soil as a fertilizer is rapidly converted into unavailable form. Soluble P converted into insoluble phosphate involves microorganisms. Mycorrhizal plants can take up more phosphorus than non mycorrhizal plants (Bhat *et al.*, 2011). The spore density was more at Sangamner locality than VSI Pune and the phosphate content of VSI Pune locality was more than Sangamner. This view was supported by Janaki Rani and Manoharchary (1994), Shinde and Nair (1994), Manoharachary *et al.* (1996).

The potassium content of the soil at Sangamner locality was 15 kg/ac and VSI Pune locality was 13 kg/ac. The potassium content has positive effect on the mycorrhizal colonization and number of propagules. Similar findings were recorded by Joshi and Singh (1995).

Mycorrhizal response is greater in soil with lower amount of Cu and Fe (Sreenivasa and Bagyaraj, 1988). The amount of Cu and Fe contents was lower at Sangamner locality but the root colonization and number of propagules was more. Our results corroborates with the results of Radhika and Rodrigues (2010).

Percentage Root Colonization and Number of AM Propagules per 100 g of Rhizosphere Soil of Sugarcane

It is well established fact that soil pH, soil P, available P, cropping season has an influence on AM population (Bharadwaj *et al.*, 1997). The AM fungi are often colonizing the plant roots in soil with limited phosphorus content (Allen, 1991). The AM fungi use some of the root exudates and modify root functions. Microbial communities in the rhizosphere soil differ from those in the non rhizosphere community because of AM fungi influences on the nature of root exudates (Perry *et al.*, 1989, Kumar *et al.*, 1999). The hypha of AM fungi cause the aggregation of the soil particles (Linderman 1988; Hooker and Black 1995; Filion *et al.*, 1999), which increase the rate of movement of water and nitrates from the bulk soil to the root surface (George *et al.*, 1995). Thus, the number of AM propagules and the percentage root colonization in the rhizosphere was more than the non rhizosphere soil. Soil phosphate content of Sangamner locality was low as compared with the VSI Pune locality but the numbers of AM propagules were more at Sangamner locality than the VSI Pune locality. The pH near to the neutral range, high organic content might have created favorables condition for the sporulation of AM fungi. Our results are in agreement with results reported by Plenchette *et al.*, 1983; Schubert *et al.*, 1990, Delowara *et al.*, 2006. Most of the AM fungal species detected were belonging to Acaulospora, *Glomus* and *Scutellospora*.

The percentage root colonization was recorded in between zero per cent and 70 per cent. At the beginning *i.e.* in the month of June it was zero per cent at both the localities and it was maximum (70 per cent) in the month of May of next year.

The number of AM propagules was ranging in between 316 and 450. The number of AM propagules recorded in the month of June was 316 at Sangamner and 320 at VSI Pune. It went on increasing with the time and at the time of harvesting it was recorded 450 and 432 at Sangamner and VSI Pune localities respectively.

Number of AM Propagules per 100 g of Non-rhizosphere Soil of Sugarcane

The number of AM propagules from the non rhizosphere soil was less than the rhizosphere soil. This may be due to lack of soil nutrients, organic contents and root exudates. The number of AM propagules was ranging in between 147 and 306. It was recorded minimum 147 at VSI Pune locality and 171 at Sangamner locality at the beginning of season *i.e.* in June. On the other hand it was recorded maximum 280 and 306 at VSI Pune and Sangamner locality respectively at the end of season *i.e.* in May.

Soil Analysis

The analysis of soil from the rhizosphere of sugarcane plant showed slightly alkaline (7.65 to 7.85) pH. It is a very important property of soil as it determines the availability of nutrients, microbial activity and physical condition of soil. The acidic range of pH decreases the sporulation of AM propagules. The acidic range of pH harbours the species of *Entrophospora* and *Gigaspora*. The range of the pH of the selected localities was slightly alkaline thus the number of the species of AM fungi from the selected localities was less. It is known that most AM fungal species occur in soils around and above a neutral pH (Schubert and Cravero, 1995). Thus, AM fungi depends upon edaphic factors such as pH, moisture content, and phosphorus (P) and nitrogen (N) availability (Ahulu *et al.*, 2006; Radhika and Rodrigues, 2010).

An electrical conductivity (EC) of the selected localities was extremely low (0.07 to 0.2 mmhos/cm^3). It was reported very low as compared to standard values. There was no correlation between electrical conductivity of soil and root colonization and the number of AM propagules/100 g soil. Higher OC percentage was recorded at VSI Pune locality (0.92 per cent) and Lower at Sangamner locality (0.52 per cent). The OC content has an effect on the species of *Glomus*. Generally the higher OC content harbours more species of *Glomus* and similar findings were recorded in present investigation.

The phosphorus and potassium are also very important major nutrients essential for the growth of plants. The phosphorus is one of the important resources that limit plant growth in tropical soils (Jordan and Herrera, 1981). The AM fungi do not add P to the soil. They only improve its availability to the plants. There is evidence that phosphatase activity is higher in the rhizosphere around AM than in non-mycorrhizal roots (Dodd *et al.*, 1987). Plants rely more on AM fungi when growing in soils which are deficient in phosphorus (Augé *et al.*, 1994; Bationo *et al.*, 2000; Merryweather and Fitter 1995). The symbiotic association with AM fungi allows the plant to access phosphorus beyond the depletion zone through the extraradical fungal hyphae, in addition to the root uptake (Pearson and Jakobsen, 1993). The AM fungi play a major role in soil fertility, nutrient acquisition, (Bagyaraj, 1986) especially uptake of phosphorus from the soil and thereby enhancing plant vigor (Mosse, 1972).

The phosphorus content of soil was ranging in between 180 and 207 kg/ac. This amount of phosphorus was moderate. The phosphate content of the soil is correlated with number of AM propagules and percentage root colonization. At low P concentrations there is high colonization and at high concentration there is low colonization. The amount of phosphorus at Sangamner locality after 270 days was 180 kg/ac whereas at VSI Pune locality it was 205 kg/ac.

The amount of iron in the rhizosphere soil of sugarcane was in between 2.3 and 3.2 ppm; which were extremely low. It was utilized by the plants and thus it was lowered further with the time. But iron content not affected the number of AM propagules and per cent root colonization in sugarcane.

The potassium content of the soil was ranging between 12 and 15 kg/acre which were extremely low. But there was no any relationship between potassium content and number of AM propagules and per cent root colonization.

The amount of copper content was high after 150 days (2.6 to 4.2 ppm) at Sangamner and VSI Pune localities respectively but it was reduced after 270 days (1.6 and 3.1) at Sangamner and VSI Pune respectively. The amount of the available copper was utilized by the plant.

The amount of zinc was high (5.94 and 2.99 ppm) after 150 days in rhizosphere soil of sugarcane at Sangamner and VSI Pune localities respectively. It went on decreasing with time span. The extent of manganese was very low (1.8 and 1.68 ppm) at Sangamner and VSI Pune respectively. Our results match with the results of Chen *et al.* (2003).

Manganese content in rhizosphere soil of sugarcane was also reported very low (1.8 and 1.68 ppm) at Sangamner and VSI Pune localities respectively. It was also reduced with the time at both the selected localities.

Effect of AM Fungi on Chlorophyll Content in Four Varieties of Sugarcane

The plant colonized with AM fungi leads to better functioning of chloroplast and photosynthetic efficiency (Al-Karaki, 2000; Rabie and Almadini, 2005). The AM fungi provide soil nutrients to plant roots. The roots thus absorb the nutrients from the soil and due to which plants growth was vigorous. The AM fungi facilitate the uptake of P from the soil in the plants. Therefore, mycorrhizal fungi can enhance pigment content by increasing P uptake (Rahmaty and Khara, 2011). In present studies the mycorrhizal sugarcane plants showed more amount of chlorophyll a, chlorophyll b and total chlorophylls than non mycorrhizal plants. Thus it can be stated that AM fungi enhanced the photosynthetic efficiency of sugarcane plants. Demir (2004) and Selvaraj and Chellappan (2006) indicated an increase in chlorophyll content by mycorrhizal colonization in pepper plants and *Prosopis juliflora*, respectively. Our results corroborates with the results of Allen *et al.*, 1981; Ganesan and Mahadevan, 1994; Gemma *et al.*, 1997; Giri and Mukerji, 2004; Sannazzaro *et al.*, 2006; Colla *et al.*, 2008 and Sheng *et al.*, 2008.

After 90 days the amount of chlorophyll a, chlorophyll b and total chlorophylls was more in mycorrhizal plants than control plants. It was reported maximum (2.38 mg/g) in CoVSI 9805 and VSI 434 and minimum in (1.96 mg/g) in CoC. 671. After 150 days the amount of chlorophyll a, chlorophyll b and total chlorophylls went on increasing in all the varieties. There was similar trend of results after 210 days in all the four varieties. This might be because of availability and mobility of the nutrients by AM fungi. At the time of maturity *i.e.* after 270 days it was reported that the amount of chlorophyll was less as compared to 150 and 210 days old sugarcane plants. The amount of chlorophyll a, b and total chlorophylls after 90 and 150 days was significant at $P \leq 0.05$ level. Whereas, the amount of chlorophyll a, b (except variety CoC 671) and total chlorophyll after 210 days was significant at $P \leq 0.05$ level. After 270 days chlorophyll a content in CoVSI 9805, Co 86032 and VSI 434 were significant at $P \leq 0.05$ level. However the chlorophyll b was significant in CoVSI 9805 and VSI 434. The total chlorophylls were significant at $P \leq 0.05$ level in CoVSI 9805, CoC 671 and CoC 434.

Effect of AM Fungi on DNA and RNA Content in Four Varieties of Sugarcane

It is well established fact that behind the synthesis of proteins there is involvement of nucleic acids. The protein content in any plant is based upon its nucleic acid content.

The amount of DNA and RNA was more in control plants than the experimental plants in all the varieties of sugarcane. After 150 days an amount of DNA and RNA went on increasing. After 210 days it again increased in all the varieties. At the time of maturity *i.e.* after 270 days it again went on increasing and recorded maximum among all. As the plant grows there is chlorosis in the plants which results in yellowing of the plants and leads to the shrinkage. The shrinkage of cells decreases the size of cell. The decreased size of cell increases number of cells per gram of tissue. Hence, DNA content increases; which also increase the RNA content per gram of plant tissue. The DNA and RNA content were significant in all the varieties of sugarcane after 90, 150, 210 and 270 at $P \leq 0.05$ level.

Effect of AM Fungi on Proteins, Vitamin C and Sugar Content in Four Varieties of Sugarcane

It is suggested that AM hyphae have a potential for supplying nitrogen to mycorrhizal plants through atmospheric nitrogen fixation (Rabie *et al.*, 2005) and ultimately results in increase in the percentage of N_2 and P in plants. The organic acids added to soils increase the plant uptake of P from a water soluble P (Bolan *et al.*, 1994). Increase in N2 content is normally related to increased protein content and the humus is believed to stimulate plant nutrient uptake and metabolism, have influence on protein synthesis (Bago *et al.*, 1996). The phosphorus with in the soil is taken up via a phosphate transporter located in the extra-radical hyphae of this fungus (Harrison and Van Buuren, 1995). The phosphate is then condensed into polyphosphate (Poly-P) and translocate back towards the plant by cytoplasmic streaming into the intradical hyphae (Cox and Tinker, 1976). Finally the Poly-P may be hydrolyzed and released as phosphate to the plant across the fungal membrane probably at the arbuscule. Other possible mechanisms might be in the potential of AM *i.e.*, inducing changes in phytohormone levels in the host plant, such as cytokinins and gibberellins (Allen *et al.*, 1982). Increased levels of protein in the inoculated plants could be attributed to either the presence of fungal proteins or post inflectional stimulation of protein synthesis in the host plant (Krishna and Bagyaraj, 1984). However, if the plants are growing in little stressed conditions then AM fungi helps the plants by avoiding the shrinkage of the plants by availing the soil nutrients to the plants. Due to this reason the amounts of nucleic acids were recorded less as compared with the non mycorrhizal fungi so that the protein content may decrease in mycorrhizal plants.

After 90 days of cultivation the amount of proteins was more in experimental plants than control plants except variety CoVSI 9805. This was because of mycorrhizal inoculation and due to availability of ample water in soil. In later stages of growth *i.e.* after 150, 210 and 270 days the protein content in control plants was more than the experimental plants. This might be because of little scarcity of water and little stressed conditions. Stress has negative impact on nucleic acid content and this result in less protein content in experimental plants. Our results are in agreement with Bakalova *et al.* (2008). The results were significant at $P \leq 0.05$ level.

Plants inoculated with AM fungi increases the vitamin C content (Dey *et al.*, 2005). The AM fungi are capable of promoting phosphorus availability by acidifying the soil and consequently, exploiting the phosphorus in nutrient patches and by facilitating the growth and development of host plants (Shi *et al.*, 2011). After 90, 150, 210 and 270 days of growth the vitamin C content were more in experimental plants than that of control plants in all the varieties of sugarcane. This might be due to availability of all the nutrients to the plants due to presence of AM fungi. The AM fungi may improve P nutrition, enhance N_2 uptake, in their host plants or adaptation to various environmental stresses. The agriculturally important symbiotic microorganisms play a remarkable role in nutrients namely nitrogen, phosphorus, potassium and microelements acquisition for plants (Rabie *et al.*, 2005). There was significant difference in vitamin C content at $P \leq 0.05$ level. Our results match with the results of Sivakumar and Thamizhiniyan (2012), Nedorost and Pokluda (2012) and Albrechtova *et al.* (2012).

The amount of sugar in experimental plants was more than control plants. After 90, 150, 210 and 270 days all the varieties of sugarcane showed more amount of sugar in experimental plants than control plants. The accumulations of carbohydrates and sugars compounds in plants inoculated with AM fungi are due to the greater photosynthetic capacity (Ahmed *et al.,* 2008). Phosphorus is very effective in the synthesis of starch from glucose (Demir, 2004). As AM fungi increase the uptake of phosphorus, they may also increase the synthesis of carbon compounds (Graham, 2000). Thus, it is seen that photosynthesis activity, increasing as a result of this symbiotic association, has a close relationship with the increase in the function of AM fungi. The greater sugars and concentration in experimental plants may be due to the sink effect of AM fungus demanding sugar from the leaves. However, the net accumulation of sugar suggested that AM colonization enhanced host plant water stress tolerance (Porcel and Ruiz-Lozano, 2004). Sorial, (2001) also observed that AM inoculation recorded highly significant induction in chemical constituents *e.g.* chlorophyll a, b; total chlorophylls; N, P, K uptakes; total sugars; total amino acids as well as proline concentration under water stress conditions, which increased the osmoregulation of wheat plants exposed to water stress. Higher levels of sugar may improve the ability of the plants to withstand water stress and recover after the condition is restored (Kameli and Lösel, 1996). Similar observations have been found by Subramanian and Charest (1998), Mathur and Vyas (2000), Sorial (2001), Wu *et al.* (2008), Thangavel *et al.* (2009). There was significant difference in sugar content at $P \leq 0.05$ level.

Effect of FYM and Mycorrhiza on Number of Tenders, Leaves and Height of Four Varieties of Sugarcane

The effect of farm yard manure (FYM), mycorrhiza and combination of both FYM and mycorrhiza have been studied on the growth parameters of sugarcane. These parameters included number of tenders, leaves, height, number of internodes and fresh weight of sugarcane plants. In present investigation it has been recorded that the effect of combination of FYM and mycorrhiza was more effective than FYM and mycorrhiza alone. As compared to control plants the plants treated with FYM showed better results. In comparison with FYM the plants treated with mycorrhiza showed improved performance with respect to number of tenders, leaves, height, number of internodes and fresh weight of sugarcane. However the best results were recorded from the plants treated with the combination of FYM and mycorrhiza both.

These results indicated that inoculation of sugarcane with AM fungi alone and in combination with FYM improved number of tenders, leaves, height, number of internodes and fresh weight. This might be because of nutrient accumulation by the roots of the sugarcane plants due to presence of mycorrhizal fungi. Root infection and spores concentrations increased with the inoculation of AM fungi with different organic fertilizers. The AM fungal symbiosis is recognized for its multiple positive effects on plant growth and for its contribution toward improving and maintenance of soil quality. The enhancement in soil organic matter (SOM) content might be associated with the increased roots dry weight in these treatments. Addition of organic materials along with chemical fertilizers improves the SOM status and decreases pH values slightly (Subramanian and Kumaraswamy, 1989). Jadhav and Patil (1996) reported that inoculation of plants with AM fungi has potential to increase shoot dry weight and crops grain yield along with improved N, P and K accumulation by plants. It was further concluded that AM fungi can be established in previously undisturbed and un-sterilized soil to the benefits of crops. Arihara and Karasawa (2000) found that inoculation of wheat, maize, potato and rape with AM fungi alone and in combination with different organic fertilizers in the previous season promoted root colonization of succeeding crops, which in turn enhanced the P uptake, crops growth and grains yield. Jeffries (1987), Millner (1990), and Demir (2004), investigated that inoculation of AM fungi play a major role in plants P accumulation of most agronomically important crops. Wang *et al.* (1997) reported that the infection of host plant with AM fungi markedly increased plants accumulation of P, Cu, Zn and Mn. Ammani *et al.* (1994) examined roots of 21 grass species and concluded that these species formed multiple mycorrhizal associations with 30 to 70 per cent root infection rates. Bagyaraj *et al.* (1989) and Lakshmipathy *et al.* (2001) reported that different strains of AM fungi differ in the extent they increase nutrient uptake, physiological parameters and plant growth.

After 90, 150, 210 and 270 days the number of tenders, leaves, height of plant, number of internodes and fresh weight of sugarcane plants was more in the plants treated with combination of FYM and mycorrhiza in all the varieties of sugarcane. After 90 days variety CoVSI 9805 showed lowest number of leaves, internodes, fresh weight whereas variety VSI 434 showed highest number of leaves, internodes, fresh weight as compared to other varieties. After 150, 210 and 270 variety VSI 434 showed the highest number of tenders, leaves, height of plant, number of internodes and fresh weight. Our results match with the results of Thangavel *et al.* (2009), Mohammad *et al.* (2012) and Fuqiang *et al.* (2012). All the growth parameters after 90, 150, 210 and 270 days were significant at $P \leq 0.05$ level.

References

Ahmed A. Khalafallah and Hoda H. Abo-Ghalia (2008). Effect of Arbuscular Mycorrhizal Fungi on the Metabolic Products and Activity of Antioxidant System in Wheat Plants Subjected to Short-term Water Stress, Followed by Recovery at Different Growth Stages. *Journal of Applied Sciences Research*, **4**(5): 559-569.

Ahulu, M. E., A. Gollotte, V. Gianinazzi–Pearson and Nonaka, M.(2006). Co occurring plants forming distinct arbuscular mycorrhizal morphologies harbor similar AM fungal species. *Mycorrhiza*, **17**: 37"49.

Albrechtova, J., L. Ales L. Nedorost, R. Pokluda, K. Posta, and Vosatka, M.(2012). Dual Inoculation with Mycorrhizal and Saprotrophic Fungi Applicable in Sustainable Cultivation Improves the Yield and Nutritive Value of Onion. *The Scientific World Journal*. **1**: 1-8.

Al-Karaki, G. N. (2000). Growth and mineral acquisition by mycorrhizal tomato grown under salt stress. *Mycorrhiza*, **10**: 51-54.

Allen, M. F. (1991). The ecology of mycorrhizae. *Cambridge University Press*, Cambridge.

Allen, M. F., T. S., Moore and Christensen, M.(1982). Phytohormone changes in *Boutelouga gracilis* infected by vesicualr-arbuscular mycorrhizae I. Cytokinin increase in the host plant. *Can. J. Bot.*, **58**, 371-374.

Allen, M. F., W. K. Smith, T. S. Moore, and Christensen, M.(1981). Comparative water relations and photosynthesis of mycorrhizal and non-mycorrhizal *Bouteloua gracilis* H.B.K. Lag ex Steud. *New phytol.,* **88:** 683-643.

Ammani, K., K. Venkateswarlu and Rao, A.S. (1994). Arbuscular mycorrhizal fungi in grasses; their occurrence, identity and development. *Phytomorphology,* **44**(3): 159-168.

Arihara, J. and T. Karasawa (2000). Effect of previous crops on AM fungal formation and growth of succeeding maize. *Soil Sci. and Pl. Nutrition.,* **46**(1): 43-51.

Arnon, D. T. (1949). Copper enzymes in isolated Chloroplast polyphenol oxidase in *Beta vulgaris. Plant Physiology,* **24**: 1-15.

Augé, R. M., X. Duan, R. C. Ebel and Stodola, A. J. W. (1994). Non hydraulic signalling of soil drying in mycorrhizal maize. *Planta.,* **193**: 74-82.

Bago B, H., Vierheilig, Y. Piche, C. Azc´on-Aguilar (1996). Nitrate depletion and pH changes induced by the extraradical mycelium of the arbuscular mycorrhizal fungus *Glomus intraradices* grown in monoxenic culture. *New Phytologist.,* **133**: 273–280.

Bagyaraj, D. J. (1986). Mycorrhizal association in crop plants and their utilization in agriculture. In: Nair, M. C. and Balakrishnan, S. Eds., *Beneficial fungi and their utilization,* Scientific Publication, Jodhpur, pp. 59-72.

Bagyaraj, D. J., M. S. Byra Reddy, and Nalini, P. A. (1989). Selection of efficient inoculants of VAM fungus for *Leucaena. Forest Ecol. Manage.,* **27**, 791-801.

Bakalova, S., D. Nedeva and Mckee, J.(2008). Protein profiles in wheat seedlings subjected to dehydration stress. *Applied Ecology and Environmental Research,* **6** (2): 37-48.

Bationo, A., S. P. Wani, C. L. Bielders, P. L. G. Vlek and Mokwanye, A. U. (2000). Crop residue and fertilizer management to improve soil organic carbon content, soil quality and productivity in the desert margins of West Africa. pp. 117-145. *In*: R. Lal, J.M. Kimble and B.A. Steward (eds.) *Advances in Soil Science. Global Climate Change and Tropical Ecosystems*. CRC Press, LLC Washington DC, USA.

Bharadwaj, J., S. S. Dudeja, and Khurana, A. L. (1997). Distribution of vesicular arbuscular mycorrhizal fungi in the natural ecosystem. *Floria Microbiol.,* **42:** 589-594.

Bhat S. H., A. B. Darzi, M. S. Dar, M. M. Ganaie and Bakhshi, S. H. (2011). Correlation of soil physico-chemical factors with VAM fungi distribution under different agroecological conditions. *International Journal of Pharma and Bio Sciences,* **2** (2): B98-B107.

Bial, M. (1902). Estimation of RNA nucleic acid. *Den Med Woch.* **28**: p. 253.

Bolan, N.S., R. Naidu, S. Mahimairaja and Baskaran, S.(1994). Influence of low molecular weight organic acids on the solublization of phosphates. *Biol. Fert. Soils.,* **18**, 311-319.

Brawerman, G. (1974). *In: Methods in Enzymol.* 30 (*Eds.* Moldave and Grossman, L.) Academic Press. New York, p. 605.

Burton, K. (1956). A study of the Conditions and mechanism of the diphenylamine reaction of the colorimetric estimation of Deoxyribonucleic acid *Biochem., J.* **62**: 315.

Chen, B. D., X. L. Li., H. Q. Tao, P. Christie and Wong, M. H. (2003). The role of mycorrhiza in zinc uptake by red clover growing in a calcareous soil spiked with various quantities of zinc. *Chemosphere,* **50**(6): 839-846.

Cochran, W. G. (1950). Estimation of bacterial densities by means of the most probable number. *Biometric, s* **6**: 105-116.

Colla, G., Y. Rouphael, M. Cardarelli, M. Tullio, C. M. Rivera, and Rea, E.(2008). Alleviation of salt stress by arbuscular mycorrhizal in zucchini plants grown at low and high phosphorus concentration. *Biol. and Fert. of Soil,* **44**: 501-509.

Cox, G. and Tinker, P.B.(1976). Translocation and transfer of nutrients in vesicular arbuscular mycorhizas. I. The arbuscular and phosphorus transfer: a quantitative ultra structural study. *New Phytol.* **77:** 371-378.

Datta P. and Kulkarni, M.(2012). Arbuscular Mycorrhizal Fungal Diversity in Sugarcane Rhizosphere in Relation with Soil Properties. *Not Sci Biol,* **4**(1): pp. 66-74.

Dellaporta, S., L. J. Wood and J. B. Hicks 1983. A plant DNA mini preparation: version II. *Plant Molecular Biology Reporter,* **1**(4): 19-21.

Delowara, K. M., A.U. Mridha, A. R. M. Solaiman and Hossain, T.(2006). Effect of edaphic factors on root colonization and spore population of arbuscular mycorrhizal fungi *Bull. Inst.trop. Agr. Kyushu. Univ.,* **29**: 97-104.

Demir S. (2004). Influence of arbuscular mycorrhiza on some physiological growth parameters of pepper. *Turk. J. Biol.,* **28**: 85- 90.

Dey, P., M. Rai, S. Kumar, V. Nath, B. Das, N. and Reddy, N.(2005). Effect of biofertilizer on physico-chemical characteristics of guava (*Psidium guajava*) fruit. *Indian Journal of Agricultural Sciences,* **75** (2): 95"96.

Dodd, J. C., C. C. Burton, R. G. Burns and Jeffries, P.(1987). Phosphate activity associated with the roots and rhizosphere of plants infected with vesicular arbuscular mycorrhizal fungi. *New Phytologist,* **107**: 163- 172.

Edmilson, J. A., R. Azcón, H. Cantarella, G. Maria, B. Ambrosano, E. A.Schammass, T. Muraoka, P. C. O. Trivelin, F. Ross, N. Guirado, M. R. G. Ungaro and Teramoto, J. R. S. (2010). Crop rotation biomass and arbuscular mycorrhizal fungi effects on sugarcane yield *Sci. Agric.* (*Piracicaba, Braz.*), **67** (6): 692-701.

Filion, M., M. S. Arnaud and Fortin, J. A. (1999). Direct interaction between the arbuscular mycorrhizal fungus *Glomus intraradices* and different rhizosphere microorganisms. *New Phytologist,* **141**: 525-533.

Frank, A. B. (1885). Ueber die auf Wurzelsymbiose beruhende Ern hrung gewisser Ba_me durch unterirdische Pilze. Ber Dtsch Bot Ges., **3**: 128–145

Fuqiang S., X. Kong, A. Dong and Liu, X.(2012). Impact of arbuscular mycorrhizal fungi on the growth and related physiological indexes of *Amorpha fruticosa*. *Journal of Medicinal Plants Research.*, **6**(2): 3648-365.

Gallaud, J. (1905). Etude sur les mycorrhizes endotrophes. *Rev. Gen. Bot.*, **17**: 5-48.

Ganesan, V. and Mahadevan, A.(1994). Effect of mycorrhizal inoculation of cassava, elephant foot yarn and taro. *J. Root Crops*, **20**(1): 1-14.

Gemma, J. N., R. E. Koske, E. M. Roberts, N. Jackson and K. De Antonis (1997). Mycorrhizal fungi improve drought resistance in creeping bentgrass. *J. Turfrass. Sci.*, **73:** 15-29.

George, E., H. Marschner and Jakobsen, I. (1995). Role for arbuscular mycorrhizal fungi in uptake of phosphorus and nitrogen from soil. *Critical Reviews in Biotechnology*, **15**: 257-270.

Gerdemann, J.W. and Nicolson, T. H. (1963). Spores of mycorrhizal *Endogone* species extracted from soil by wet sieving and decanting.*Trans. Br. Mycol. Soc.*, **46:** 235–244.

Giri, B. and Mukerji, K. G. (2004). Mycorrhizal inoculants alleviates salt stress in *Sesbania aegyptiaca* and *Sesbania grandiflora* under field conditions: evidence for reduced sodium and improved magnesium uptake. *Mycorrhiza*, **14**: 307-12.

Graham, J. H. (2000). Assessing cost of arbuscular mycorrhizal symbiosis in agroecosystems", in "Current Advances in Mycorrhizae Research (Ed. G. K. Podila and D. D. Douds), APS Press, *St Paul*, pp.127-140.

Harris, L. J. and Ray, S. N. (1935). Diagnosis of Vitamin C in Subnutrition by Urine Analysis. *Lancet*, **1**: 462.

Harrison, M. J. and Van Buuren, M.L. (1995). A phosphate transporter from the mycorrhizal fungus *Glomus versiforme*. *Nature*, **376**: 626-629.

Hedge, J. E. and Hofreiter, B. T. (1962). *In Carbohydrate Chemistry*, 17 (Eds. Whistler R.L. and Be Miller, J.N.) Academic press, New York.

Hooker, J. E. and Black, K. E. (1995). Arbuscular mycorrhizal fungi as a component of sustainable soil plant systems. *Critical Reviews in Biotechnology*, **15**: 201- 212.

Jadhav, D. N. and Patil, M. S. (1996). Effect of AM inoculation on growth and NPK uptake of groundnut. *Acta Agronomica Hungarica.*, **44**(2): 153-159.

Janaki Rani, S. and Manoharachary, C.(1994). Occurrence and distribution of VAM fungi associated with safflower. *Indian Phytopathol.*, **47:** 263-265.

Janse, J. M. (1897). Les endophytes radicaux de quelques plantes Javanaises. *Ann. Jardin. Bot. Buitenzorg*, **14**: 53-201.

Jeffries, P. (1987). Use of mycorrhiza in agriculture", *Crit. Rev. Biotechnol.*, **5**: 319-357.

Jordan, C. F. and Herrera, R.(1981). Tropical rain forest: are nutrients really critical? *Am. Nat.*, **117**: 167-180.

Joshi, K. C. and Singh, H. P. (1995). Interrelationships among vesicular arbuscular mycorrhizae population; Soil properties and root colonization capacity of soil. *Journal of the Indian Society of soil science*, **43** (2): 204-207.

Kameli, A. and Lösel, D. M. (1996). Growth and sugar accumulation in durum wheat plants under water stress. *New Phytologist*, **132**: 57-62.

Kelley, A. P. (1931). The concept of mycorrhiza. *Mycologia*, **23**: 147–151.

Kelley, A. P. (1950). Mycotrophy in plants. Chronica Botanica, Waltham, Mass.

Kliromonos, J. N., M. Ursic, M. C. Rillig, Allen, M. F. (1998). Interspecific difference in the response of arbuscular mycorrhizal fungi to *Artemisia tridenta* grown under elevated atmospheric CO_2. *New Phytologist*, **138**: 99-605.

Koide, R. T. and Kabir, Z.(2000). Extraradical hyphae of the Mycorrhizal fungus *Glomus intraradices* can hydrolyse organic phosphate. *New Phytologist*., **148:** 511–517.

Krishna, K. R. and Bagyaraj, D. J. (1984). Growth and nutrient uptake of peanut inoculated with the mycorrhizal fungus *Glomus fasciculatum* compared with own inoculated ones. *Plant Soil*, **77**: 405-408.

Kumar, A., Nivedita and Upadhyaya, R.S.(1999). VA mycorrhizae and revegetation of coal mine soils: a review. *Tropical Ecology*, **40**: 1-10.

Lakshmipathy, R. K. Chandrika, B. Gowda, A. N. Balakrishna, and Bagyaraj, D. J.(2001). Response of *Saraca asoca* (Roxb.) de Wilde to inoculation with *Glomus mosseae*, *Bacillus coagulans* and *Trichoderma harzianum*. *J. Soil Biol. Ecol.*, **21**, 76-80.

Linderman, R.G. (1988). Mycorrhizal interactions with the rhizosphere microflora, the mycorrhizophere effect. *Phytopathology*, **78**: 366-371.

Lowry, O. H., N. J. Rosebrough, A. L Farr, and Randall, R. J.(1951). Protein measurement with the Folin-Phenol reagents. *Journal of Biological Chemistry*, **193**: 265.

Manoharachary, C., P. Wani, P. Satyaprakash and Prakash, P.(1996). vesicular arbuscular mycorrhizal symbiosis with special reference to castor production. In: *Persp. Biol. Sci.* 197-216.

Mathur, N. and Vyas, A.(2000). Influence of arbuscular mycorrhizae on biomass production, nutrient uptake and physiological changes in *Zizyphus mauritiana* Lam. under water stress. *J. of Arid Environ.*, **45**: 191-195.

Menge, J. A and Timmer, L. M. (1982). Procedure for inoculation of plants with VAM in the laboratory, greenhouse and field. In: Methods and principles of mycorrhizal research, (Ed.) Schenck N. C., *Am. Phytopath. Soc. St. Paul*, pp. 59-68.

Merryweather, J. and Fitter, A.(1995). Arbuscular mycorrhiza and phosphorus as controlling factors in the life-history of *Hyacinthoides nonscripta* (L.) Chourad ex Rothm. *New Phytologist*, **129**: 629-636.

Millner, P. D. (1990). A minor review of biotechnology as applied to AM fungi. *Soil Microbe System Lab. USA.*, **29**: 149- 158.

Mohammad S. S. saud, T. Burni, M. Afzal, F. Khan, M. J. Khan and Wahid, F. (2012). Effect of arbuscular mycorrhizal fungal inoculation in Combination with different organic fertilizers On maize crop in eroded soils. *Pak. J. Bot.*, **44**(4): 1427-1432.

Morton, J. B. and Benny, G. L. (1990). Revised classification of arbuscular mycorrhizal fungi (Zygomycetes): a new order, Glomales, two new suborders, Glomineae and Gigasporineae, and two new families, Acaulosporaceae and Gigasporaceae, with an emendation of Glomaceae. *Mycotaxon,* **37**: 471-491.

Mosse, B. (1972). The influence of soil type and *Endogone* strain on growth of mycorrhizal plants in phosphate deficient soils. *Review of Ecological Biology Society,* **9**: 529.

Muniyamma, M., B. K. Bharati and C. N. Reddy 2000. Effect of VAM on root induction in in vitro sugarcane (*Saccharum officinarum* L.) seedlings - a new technique *Mycorrhiza News,* 12(1): pp. 13-16.

Nasim, G., A. Ali, A. Munawar And Bajwa, R.(2008). Seasonal dynamics of AM fungi in sugarcane (*Saccharum officinarum* L. Cv. Spf-213) in relation To red rot (*Colletotrichum falcatum*) Disease from punjab, Pakistan *Pak. J. Bot.,* **40**(6): pp. 2587-2600.

Nedorost, L. and Pokluda, R.(2012). Effects of arbuscular mycorrhizal fungi on tomato yield and nutrient uptake under different fertilization levels. *Acta Universitatis Agriculturae et Silviculturae Mendelianae Brunensis.,* **60** (21): 181-186.

Pankhurst, C. (2006). Effects of pesticides used in sugarcane cropping systems on soil organisms and biological functions associated with soil health. A report prepared for the Sugar Yield Decline Joint Venture.

Pearson, J. N. and Jakobsen, I.(1993). The relative contribution of hyphae and roots to phosphorus uptake by arbuscular mycorrhizal plants, measured by dual labeling with 32P and 33P. *New Phytologist,* **124**: 489-494.

Perry, D. A., M. P. Amaranthus, J. G. Borchers, S. L. Borchers and Brainerd, R. E. (1989). Bootstrapping in ecosystems. *Bioscience* **39**: 230-237.

Phillips, J. M. and Hayman, D. S. (1970). Improved procedures for clearing roots and staining parasitic and vesicular-arbuscular mycorrhizal fungi for rapid assessment of infection. *Trans. Br. Mycol. Soc.,* **55**: 158–160.

Plenchette, C., V. Furlan, and Fortin, J. A. (1983). Responses of endomycorrhizal plants grown in calcined montmorillonite clay to different levels of soluble phosphorus. I. Effect on growth and mycorrhizal development. *Can. J. Bot.,* **61**: 1377-1383.

Porcel, R. and Ruiz-Lozano, J. M. (2004). Arbuscular mycorrhizal influence on leaf water potential, solute accumulation, and oxidative stress in soybean plants subjected to drought stress. *J. of Exp. Bot.,* **55**: 1743-1750.

Rabie, G. H. and Almadini, A. M. (2005). Role of bioinoculants in development of salt-tolerance of *Vicia faba* plants under salinity stress. *Afri. J. of Biotech.,* **4**: 210-220.

Rabie, G. H., M. B. Aboul-Nasar and Al-Huminary, A.(2005). Increased salinity tolerance of cowpea plants by dual inoculation of AM fungus *Glomus clarum* and nitrogen-fixer *Azospirillum brasilense. Mycobiology,* **33**: 51-61.

Radhika, K. P. and Rodrigues, B. F.(2010). Arbuscular mycorrhizal fungal diversity in some commonly occurring medicinal plants of Western Ghats, Goa region *Journal of Forestry Research*, **21** (1): 45–52.

Rahmaty, R. and Khara, J.(2011). Effects of vesicular arbuscular mycorrhiza *Glomus intraradices* on photosynthetic pigments, antioxidant enzymes, lipid peroxidation and chromium accumulation in maize plants treated with chromium Turk *J. Biol.*, **35**: 51-58.

Sannazzaro, A. I., R. Oscar, A. Edgardo and Ana, M.(2006). Alleviation of salt stress in *Lotus glaber* by *Glomus intraradices*. *Plant Soil.*, **285**: 279-287.

Schenck, N. C. and Y. Perez 1990. Manual for the identification of VA Mycorrhizal fungi, 3rd ed. Synergistic Publications, Gainesville, Florida.

Schubert, A. and Cravero, M. C. (1995). Occurrence and infectivity of Vesicular mycorrhizal fungi in north western Italy. *Vitis*, **24:** 129-138.

Schubert, A., M. Mazzitelli, O. Ariusso and Eynard, I.(1990). Effects of vesicular arbuscular mycorrhizal fungi on micropropagated grapevines: Influence of endophyte strain, P fertilization and growth medium. *Vitis*, **29**, 5-13.

Schüßler, A., Schwarzott, D., and Walker C. (2001). A new fungal phylum, the Glomeromycota: phylogeny and evolution. *Mycol Res.*, **105**: 1413–1421

Selvaraj, T. and Chellappan, P.(2006). Arbuscular mycorrhizae: A diverse personality. Review Paper. *Central European Agriculture Journal*, **7**: 349-358.

Sharrif, M. and Moawad, A. M.(2006). Arbuscular Mycorrhizal Incidence and Infectivity of crops in North West Frontier of Pakistan. *World Journal of Agricultural Sciences*, **2** (2): 123-132.

Sheng, M., T. Ming, C. Hui, Y. Baowei, Z. Fengfeng and Yanhui, H.(2008). Influence of arbuscular mycorrhizae on photosynthesis and water status of maize plants under salt stress. *Mycorrhiza*, **18**: 287-296.

Shi, Z., F. Wang, C. Zhang, and Yang, Z.(2011). Exploitation of phosphorus patches with different phosphorus enrichment by three arbuscular mycorrhizal fungi. *Journal of Plant Nutrition*, **34**(8): 1096–1106.

Shinde, B. P. and Nair, L. N.(1994). Studies on VAM fungi and *Rhizobium* of *Phaseolus* sp. from Baglan area Nashik district, Ph.D thesis.

Sivakumar, P.V. and Thamizhiniyan, P.(2012). Influence of AM fungi and *Azospirillum* on fruit nutritional values of *Lycopersicon esculentum*, Mill. *International Journal of Environmental Biology*., **2**(3): 133-136.

Smith, S. E. and Reid, D. J.(1997). Mycorrhizal symbiosis, 2nd ed. San Diego, CA: Academic Press.

Sorial, M. E. (2001). Growth, Phosphorus uptake and water relations of wheat infected with an arbuscular mycorrhizal fungus under water stress. *Ann. of Agric. Sci. – Moshtohor.*, **39**(2): 909-931.

Sreenivasa, M. N. and Bagyaraj, D. J. (1988). *Chloris gayana* (Rhodes grass) a better host for mass production of *Glomus fasciculatum* inoculum. *Plant and Soil*, pp.289-290.

Srivastava Ashok. K., Arun K. Srivastava and Solomon, S.(2011). Sustaining sugarcane productivity under depleting water resources *Current Science*, **101**(6): 748-754.

Subramanian, K. S. and Charest, C.(1998). Arbuscular mycorrhizae and nitrogen assimilation in maize after drought and recovery. *Physiol. Plant.*, **102**: 285-296.

Subramanian, N. S. and Kumaraswamy, N.(1989). Effect of continuous cropping and fertilization on chemical properties of soil. *J. Ind. Soc. Soil Sci.*, **37**: 171-173.

Tandon, H.L.S. (1993). Methods of Analysis of Soils, Fertilizers and Waters. FDCO, New Delhi.

Thangavel S., M. C. Nisha and Rajeshkumar, S.(2009). Effect of indigenous arbuscular mycorrhizal fungi on some growth parameters and phytochemical constituents of *Pogostemon patchouli* Pellet *Maejo Int. J. Sci. Technol.*, **3**(01): 222-234.

Thompson, J. P. (1987). Decline of vesicular-arbuscular mycorrhizae in long fallow disorder of field crops and its expression in phosphorus deficiency of sunflower. *Australian Journal of Agricultural Research*; **38** (5): 847-867.

Trappe, J. M, and Berch, S. M. (1985). The prehistory of mycorrhizae: A.B. Frank's predecessors. In: Proceedings of the 6th North American conference on mycorrhizae. Forest Research Laboratory, Oregon State University, Corvallis, Ore., pp. 2-11

Walker, C. and Schußler, A.(2004). Nomenclatural clarifications and new taxa in the Glomeromycota. *Mycol. Res.*, **108**: 981-982.

Wang, S., D., D. F. Wang, and Fang, D. H.(1997). Effect of AM fungi on vegetative growth and physiology of tea trees and tea quality. *Acta Pedologica Sinica.*, **34**(1): 97-102.

Wu, Q. S., R.X. Xia and Zou, Y. N. (2008). Improved soil structure and citrus growth after inoculation of three arbuscular mycorrhizal fungi under drought stress. *Eur. J. of Soil Biol.* **44**: 122-128.

2015, Major Constrains and Verdict of Crop Productivity *Pages 366–383*
Editor: Dr. U.N. Bhale
Published by: ASTRAL INTERNATIONAL (P) LIMITED, NEW DELHI

Chapter 27

Diseases: Major Constraints in Sugarcane Cultivation and Production

Seweta Srivastava*

Center for Sugarcane Biotechnology,
Sugarcane Research Institute, Shahjahanpur – 242 001, U.P., India

Introduction

Sugarcane belongs to the grass family (Poaceae), an economically important seed plant family that includes maize, wheat, rice, and sorghum and many forage crops. The main product of sugarcane is sucrose, which accumulates in the stalk internodes. Sucrose, extracted and purified in specialized mill factories, is used as raw material in human food industries or is fermented to produce ethanol. Ethanol is produced on a large scale by the Brazilian sugarcane industry.

Sugarcane is the world's largest crop by production quantity (FAO, 2014). In 2012, FAO estimates it was cultivated on about 26.0 million hectares, in more than 90 countries, with a worldwide harvest of 1.83 billion tons. Brazil was the largest producer of sugar cane in the world. The next five major producers, in decreasing amounts of production, were India, China, Thailand, Pakistan and Mexico.

The world demand for sugar is the primary driver of sugarcane agriculture. Sugarcane contributes about 60 per cent of the total world sugar requirement while the remaining 40 per cent came from sugar beet, (Onwueme and Sinha, 1999).

---

* E-mail: shalu.bhu2008@gmail.com

Sugarcane predominantly grows in the tropical and subtropical regions, and sugar beet predominantly grows in colder temperate regions of the world. Other than sugar, products derived from sugarcane include falernum, molasses, rum, *cachaça* (a traditional spirit from Brazil), bagasse and ethanol. In some regions, people use sugarcane reeds to make pens, mats, screens, and thatch. The young unexpanded inflorescence of *tebu telor* is eaten raw, steamed or toasted, and prepared in various ways in certain island communities of Indonesia (Dahlia *et al.*, 2010).

In India, between the sixth and fourth centuries BC, the Persians, followed by the Greeks, discovered the famous "reeds that produce honey without bees". They adopted and then spread sugar and sugarcane agriculture (FAO, 2009). A few merchants began to trade in sugar–a luxury and an expensive spice until the 18th century. Before the 18th century, cultivation of sugar cane was largely confined to India. Sugarcane plantations, like cotton farms, were a major driver of large human migrations in the 19th and early 20th century, influencing the ethnic mix, political conflicts and cultural evolution of various Caribbean, South American, Indian Ocean and Pacific island nations (Mintz, 1986).

Numerous pathogens infect sugarcane, such as sugarcane grassy shoot disease caused by *Phytoplasma*, whiptail disease or sugarcane smut, *pokkah boeng* caused by *Fusarium moniliforme*, *Xanthomonas axonopodis* bacteria causes Gumming Disease, and red rot disease caused by *Colletotrichum falcatum*. Viral diseases affecting sugarcane include sugarcane mosaic virus, maize streak virus, and sugarcane yellow leaf virus.

List of Sugarcane Diseases

This article mainly focuses on the diseases of sugarcane (*Saccharum* spp. hybrids).

Bacterial Diseases

1) Gumming disease	*Xanthomonas axonopodis* pv. *vasculorum*
2) Leaf scald	*Xanthomonas albilineans*
3) Mottled stripe	*Herbaspirillum rubrisubalbicans*
4) Ratoon stunting disease	*Leifsonia xyli* subsp. *xyli*
5) Red stripe (top rot)	*Acidovorax avenae* subsp. *avenae*

Fungal Diseases

1) Banded sclerotial (leaf) disease	*Thanatephorus cucumeris* = *Pellicularia sasakii* *Rhizoctonia solani* [anamorph]
2) Black rot	*Ceratocystis adiposa* *Chalara* sp. [anamorph]
3) Black stripe	*Cercospora atrofiliformis*
4) Brown spot	*Cercospora longipes*
5) Brown stripe	*Cochliobolus stenospilus* *Bipolaris stenospila* [anamorph]

Disease	Pathogen
6) Downy mildew	*Peronosclerospora sacchari* = *Sclerospora sacchari*
7) Downy mildew, leaf splitting form	*Peronosclerospora miscanthi* = *Sclerospora mischanthi* *Mycosphaerella striatiformans*
8) Eye spot	*Bipolaris sacchari* = *Helminthosporium sacchari*
9) Fusarium sett and stem rot	*Gibberella fujikuroi* *Fusarium moniliforme* [anamorph] *Gibberella subglutinans*
10) Iliau	*Clypeoporthe iliau* = *Gnomonia iliau* *Phaeocytostroma iliau* [anamorph]
11) Leaf blast	*Didymosphaeria taiwanensis*
12) Leaf blight	*Leptosphaeria taiwanensis* *Stagonospora tainanensis* [anamorph]
13) Leaf scorch	*Stagonospora sacchari*
14) Marasmius sheath and shoot blight	*Marasmiellus stenophyllus* = *Marasmius stenophyllus*
15) Myriogenospora leaf binding (tangle top)	*Myriogenospora aciculispora*
16) Phyllosticta leaf spot	*Phyllosticta hawaiiensis*
17) Phytophthora rot of cuttings	*Phytophthora* spp. *Phytophthora megasperma*
18) Pineapple disease	*Ceratocystis paradoxa* *Chalara paradoxa* = *Thielaviopsis paradoxa* [anamorph]
19) Pokkah boeng (that may have knife cut symptoms)	*Gibberella fujikuroi* *Fusarium moniliforme* [anamorph] *Gibberella subglutinans*
20) Red leaf spot (purple spot)	*Dimeriella sacchari*
21) Red rot	*Glomerella tucumanensis* = *Physalospora tucumanensis* *Colletotrichum falcatum* [anamorph]
22) Red rot of leaf sheath and sprout rot	*Athelia rolfsii* = *Pellicularia rolfsii* *Sclerotium rolfsii* [anamorph]

23) Red spot of leaf sheath	*Mycovellosiella vaginae*
	= *Cercospora vaginae*
24) Rhizoctonia sheath	*Rhizoctonia solani*
	and shoot rot
25) Rind disease (sour rot)	*Phaeocytostroma sacchari*
	= *Pleocyta sacchari*
	= *Melanconium sacchari*
26) Ring spot	*Leptosphaeria sacchari*
	Phyllosticta sp. [anamorph]
27) Root rots	*Marasmius sacchari*
	Pythium arrhenomanes
	Pythium graminicola
	Rhizoctonia sp.
	Unidentified Oomycete
28) Rust, common	*Puccinia melanocephala*
	= *Puccinia erianthi*
29) Rust, orange	*Puccinia kuehnii*
30) Schizophyllum rot	*Schizophyllum commune*
31) Sclerophthora disease	*Sclerophthora macrospora*
32) Seedling blight	*Alternaria alternata*
	Bipolaris sacchari
	Cochliobolus hawaiiensis
	Bipolaris hawaiiensis [anamorph]
	Cochliobolus lunatus
	Curvularia lunata [anamorph]
	Curvularia senegalensis
	Setosphaeria rostrata
	Exserohilum rostratum [anamorph]
	= *Drechslera halodes*
33) Sheath rot	*Cytospora sacchari*
34) Smut, culmicolous	*Ustilago scitaminea*
35) Target blotch	*Helminthosporium* sp.
36) Veneer blotch	*Deightoniella papuana*
37) White rash	*Elsinoë sacchari*
	Sphaceloma sacchari [anamorph]
38) Wilt	*Fusarium sacchari*
	= *Cephalosporium sacchari*

39) Yellow spot	*Mycovellosiella koepkei* = *Cercospora koepkei*
40) Zonate leaf spot	*Gloeocercospora sorghi*

Nematodes and Parasitic Diseases

1) Lesion	*Pratylenchus* spp.
2) Root-knot	*Meloidogyne* spp.
3) Spiral	*Helicotylenchus* spp. *Rotylenchus* spp. *Scutellonema* spp.

Viral Diseases

1) Chlorotic streak	Virus (assumed)
2) Dwarf	Sugarcane dwarf virus
3) Fiji disease	Sugarcane Fiji disease virus
4) Mosaic	Sugarcane mosaic virus
5) Sereh	Virus (assumed)
6) Streak disease	Maize streak virus, sugarcane strain
7) Yellow Leaf	Sugarcane Yellow Leaf Virus

Phytoplasma Diseases

Phytoplasmas were previously known as 'mycoplasma-like organisms' (MLOs) (Rao and Ford, 2000; Nasare *et al.*, 2007).

Grassy Shoot (SCGS), Leaf Chlorosis, Early Bud Sprouting	Sugarcane Grassy Shoot Phytoplasma

Various biological agents including bacteria, fungi and viruses cause diseases of sugarcane. Important diseases of sugarcane that have been identified are listed in Appendix 2 and the major pathogens are discussed below.

Disease control in sugarcane is based on an integration of legislative control, resistant cultivars and other management procedures. Short term spraying options are available, but their economic viability may not be sustained. Hygiene is important to disease management strategies, particularly for diseases transmitted through cuttings such as ratoon stunting disease (RSD) and leaf scald. Cutting one infected stalk may lead to significant infection to the next 100 cuttings which are subsequently cut by the same blade (Croft *et al.*, 2000). Machine harvesters can also transmit disease. Many sugarcane diseases are also managed through the use of disease-free planting material supplied through Cane Protection and Productivity Boards. Hot-water treatments are used to disinfect planting material. Long hot-water treatment (three hours at 50°C) is used to control RSD. Soaking in ambient temperature running water for ~40 hours followed by three hours at 50°C is used to control leaf scald bacteria.

Short hot-water treatment (50°C for 30 minutes) is used to control chlorotic streak and some insect pests (Croft *et al.*, 2000).

I. Bacterial Diseases

Bacterial diseases ratoon stunting disease (RSD) are probably the most important disease of sugarcane. It is a highly infectious disease caused by *Leifsonia xyli*, (formerly named *Clavibacter xyli* subsp *xyli*) which infects vascular tissues of sugarcane. It was first reported in QLD in 1944–45 and has been identified in most countries that grow sugarcane (Bailey, 2004). It was estimated in the early 1990's that it affected approximately 30 per cent of farms in NSW (Roach *et al.*, 1992; McLeod *et al.*, 1999) and the estimated loss from this disease Australia-wide was $6.3 million in 1996 (McLeod *et al.*, 1999).

The symptoms are poor growth and stunted shoots, which might not be obvious if most plants in the field are infected. The visual symptoms of red-orange dots in the vascular tissues can be seen only when the stalks are cut and sliced (Croft *et al.*, 2000). The disease is transmitted through healthy plants coming in contact with diseased plant material or contaminated cutting implements. Yield loss is higher in dry weather and often becomes more severe in subsequent ratoon crops (Frison and Putter, 1993). The incidence of the disease in parts of NSW has been reduced due to an extension campaign promoting clean seed cane (McGuire *et al.*, 2009).

Leaf scald is caused by the bacterium *Xanthomonas albilineans* which infects the vascular tissues of sugarcane. It is found in most sugarcane districts in QLD (Croft *et al.*, 2000), although it is hard to identify and the disease often has a latent period after infection (Bailey, 2004). Leaf scald is characterized by a long white to cream streak on the leaves. Severely infected leaves appear scalded and roll inwards, with the top of the shoots becoming chlorotic. Yield loss occurs through the death of infected cane stalks and poor ratooning (BSES Ltd, 2005b). Leaf scald can spread by wind-blown rain, plant material and contaminated cutting equipment such as planters and harvesters (Croft *et al.*, 2000). Leaf scald can infect many other grasses which are alternate hosts and act as a reservoir for the disease. Extremes of moisture and temperature favor disease transmission. Resistant cultivars are used to curb the spread of the disease and susceptible plants are not used in breeding programs (BSES Ltd, 2005b).

II. Fungal Diseases

The two major rusts in sugarcane are orange and common sugarcane rusts (Braithwaite *et al.*, 2009).

1. Orange Rust

It is caused by *Puccinia kuehnii* and is not as economically important as the common rust, caused by *P. melanocephala*. These are both obligate parasitic fungi spread by windblown spores. The disease symptoms of the two rusts are distinct. Pustules of the orange rust are orange and tend to be grouped in clusters, while those of sugarcane rust are reddish brown and are distributed evenly on leaves. Pustules rupture the leaves and allow water to escape from the plant, leading to moisture stress (Croft *et al.*, 2000). Both diseases are most severe in humid environments with temperatures below 25°C (Walker, 1987). In 1999–2000, sugarcane crops in Australia were affected by an outbreak of orange rust, which severely damaged the most widely grown commercial cultivar, Q124 (Croft *et al.*, 2000).

Highly susceptible parents are no longer used in any breeding programs. More recently, cultivars such as Q173 and Q182 have also been found to be affected by the disease (NSW Sugar, 2005). Yield loss from sugarcane rust depends on environmental conditions and was estimated to cause an economic loss of $3.5 million in 1996 (McLeod *et al.*, 1999).

2. Sugarcane Smut

It is caused by *Ustilago scitaminea*, is a serious disease of sugarcane that can reduce yields by 30–100 per cent (Watson, 2007). Infection occurs through the sugarcane buds from wind-blown spores (Walker, 1987). The disease causes severe stunting and multiple thin stalks.

It is characterized by black, whip-like structures that form at the growing points of sugarcane plants (Croft *et al.*, 2000). These whips replace the spindle leaves and are formed in the shoots developing from infected cane cuttings (Frison and Putter, 1993). The whips break open to release the mature spores which are spread by wind (BSES Ltd, 2006). Sugarcane smut is now seen as being widely spread and established (Croft *et al.*, 2008).

Estimated losses in susceptible cultivars are up to 62 per cent in the Herbert region (Magarey *et al.*, 2010). The spread and occurrence of the disease is being controlled through planting of resistant cultivars, using uninfected seed canes and removing infected crops (BSES Ltd 2009b). However, there has not been a complete conversion to resistant cultivars as the resistant cultivars can have a lower yield than the non-resistant cultivars. Yield of resistant cultivars is also dependant on soil type in the region (Watson, 2007). In order to reduce the spread of sugarcane smut, the movement of sugarcane and sugarcane machinery is restricted in QLD by the Plant Protection Regulation 2002-3. Provisions under the Plant Protection Act 1989 (QLD) allow for inspectors to order the destruction of diseased cane and practical guidelines have been developed to control the spread of the disease (BSES Ltd, 2005a; BSES Ltd 2007a; BSES Ltd 2007b; Queensland Government, 2009).

3. Red Rot

It is one of the oldest known diseases of sugarcane. It occurs in most cane-growing countries. Red rot occurs in various parts of the cane plant but it is usually considered a stalk and a seed-piece disease. Its symptoms are highly variable depending upon the susceptibility of the sugarcane variety and the environment. Symptoms may not be readily apparent in the field, especially in the early stages of the disease. In the later stages of the disease, red rot may cause standing cane to "break down". Diagnostic symptoms can best be observed by splitting the stalk lengthwise. The infected tissues have a dull red color interrupted by occasional whitish patches across the stalk. These white patches are specific to the disease and are of significance in distinguishing red rot from other stalk rots. Reddened vascular bundles may also pass through to the healthy tissues. In susceptible varieties the red color, sometimes along with some gray color, may be seen throughout the length of the stalk. The infection is largely confined to the internodes in resistant varieties. On the leaves, the pathogen may produce elongated red lesions on the midribs, reddish patches on the leaf sheaths, and, infrequently, small dark spots on the leaf blades. The lesions may eventually develop a straw color in the center. In seed pieces, the entire seed piece may become rotted and the internal tissues turn various shades of red, brown or gray.

Red rot disease is caused by the fungus *Glomerella tucumanensis*. An older name, *Colletotrichum falcatum*, is still preferred by some pathologists. The red rot fungus can be readily isolated from infected tissues. At least two races have been identified. Fungus growth is affected by temperature, pH, nutrition and environmental conditions.

Midrib lesions are probably the major source of inoculum during the growing season. Diseased stalks generate a great deal of inoculum. Dissemination of the inoculum takes place by wind, rain, heavy dews and irrigation water. Infected plant material can readily spread or cause secondary infections. Crop debris or stubble may also provide inoculum to infect a new crop. Although the fungus is not a true soil-borne organism, spores washed into the soil may produce infection in planted seed pieces. Hosts other than sugarcane are not considered important inoculum sources. Climatic factors affect both the spread and severity of red rot. In newly-planted cane, the disease is favored by excessive soil moisture, drought conditions, and low temperatures.

The use of resistant varieties is the most effective method of prevention and control. The factors determining resistance to red rot are not fully understood. There are two kinds of resistance: (1) morphological, which may prevent or retard the infection process, and (2) physiological, in which the living cells of the plant suppress or prevent pathogen growth. Physiological resistance is considered to be of greater importance.

The incidence of red rot can be reduced through good cultural practices, such as clearing fields of excessive trash and efficient drainage. Agronomic practices that hasten germination are important in reducing seed rotting and obtaining good stands. The avoidance of planting susceptible cultivars during excessively cool and wet weather has been effective in several countries. Regular rouging of diseased plants, burning of trash, plowing out badly affected fields, maintenance of proper soil moisture, and prompt harvesting of infected or susceptible crops are other management practices recommended for red rot control. Foliar fungicides have not been effective in the control of red rot. However, better crop stands have been achieved from enhanced germination obtained by treating seed pieces with a fungicide before planting. This treatment reduces the incidence of red rot infection in the treated seed piece. Heat treating of seed cane has also been effective in controlling seed piece infection of red rot.

4. Pineapple Disease

It is an economically important sugarcane disease that occurs in almost all countries where sugarcane is grown. The disease is caused by the fungus *Ceratocystis paradoxa* which induces seed piece decay following planting. Shortly after infection, the internal tissue of the seed piece turns red and eventually black. The black coloration results from the production of fungal spores within the seed piece. Nodes act as partial barriers to the spread of rotting, but with susceptible varieties, entire seed pieces may become colonized by the fungus. The disease severely retards bud germination, shoot development and early shoot vigor. Pineapple disease can result in young plant-cane crops having a patchy, uneven appearance. When severe, the disease may reduce germination over large areas. Although pineapple disease is not considered important in standing cane, infection may occur if the stalks are physically damaged or stressed.

The pineapple disease fungus, *Ceratocystis paradoxa,* grows readily in culture and can easily be isolated from diseased tissue. At least some variability has been shown to occur within *Ceratocystis paradoxa,* but evidence for multiple races of the pathogen is lacking. *Ceratocystis paradoxa* also causes diseases of pineapple, banana, cacao, coconut and oil palm. *Ceratocystis paradoxa* produces several types of spores which are important in its survival and spread. These spores, produced by the millions on the internal tissues of infected seed pieces, are released into the soil upon seed piece decay. There, the spores may survive for several years, serving as a source of inoculum for the next crop.

Pineapple disease is essentially soil-borne, being transmitted by the fungal spores present in the soil. The fungus is found mainly in the top 25 cm of the soil profile. Any factor that delays germination of the buds on the seed piece increases the likelihood of infection. Excessively deep planting, wet or dry soil conditions and low temperature are all conducive to the development of the disease. Long hot-water treatments used to control other diseases may have a detrimental effect on germination and thus may actually increase susceptibility to pineapple disease. Even short hot-water treatments, which usually stimulate germination, may result in increased susceptibility. In general, rapid germination decreases the impact of the disease. Infection of standing cane occurs through wind-blown or rain-splashed spores gaining entry through damaged tissue.

The use of resistant cultivars is often the easiest, most economical method for controlling plant diseases. If susceptible cultivars are planted, there are a number of ways to minimize the effects of the disease. If possible, planting should take place when conditions favor rapid germination. For example, the use of seed cane that is relatively young improves the prospect of rapid germination, thus hindering development of the disease. Since pineapple disease is a soil-borne disease, crop rotation or a fallow period between cane crops may prove to be of some benefit in reducing its impact. Site selection and site preparation may be important tools in limiting the impact of pineapple disease. Susceptible cultivars should be planted on dry, well-drained soils. Cool, wet conditions inhibit seed piece germination more severely than they inhibit fungal growth. Preventing standing water on planted soils, therefore, will dramatically reduce the effects of pineapple disease.

Seed piece infection by pineapple disease frequently proceeds from the exposed cut ends to the center of the seed piece. Therefore, the use of seed pieces containing at least three nodes increases the likelihood that buds closer to the center will germinate. Similarly, care should be taken to minimize the amount of cracking and wounding of seed pieces, since the pathogen also enters through wounds. Since mechanically-harvested seed piece billets are typically shorter and have more damage than whole stalks planted by hand, growers may want to compensate by planting mechanically-harvested seed cane at higher densities.

Propiconazole (Tilt) is registered for use as a hot or cold water dip or as a seed piece topical spray to be applied as the seed is being transported by mechanical conveyor to the furrow. The former method of application is the more efficacious, but treating large quantities of seed in this manner is frequently infeasible.

5. Pokkah Boeng

It was originally described in Java and the name is a Javanese term denoting a malformed or distorted top. It is caused by the fungus *Fusarium moniliforme* and/or *Fusarium moniliforme* var. *subglutinans*. The disease is present in most, if not all, sugarcane producing areas of the world. Pokkah boeng may cause serious yield losses in commercial plantings. However, there have been many reported outbreaks of the disease which have looked spectacular, but have caused little economic loss.

The earliest symptoms of the disease are chlorotic areas at the base of young leaves, distortion (wrinkling and twisting) and shortening of affected leaves, and in severe cases, death of the stalk. The base of affected leaves is often narrower than that of normal leaves. As leaves mature, irregular reddish stripes and specks develop within the chlorotic parts. The reddish areas sometimes develop into lens-shaped holes which have no definite arrangement. This reddish tissue may form ladder-like lesions, often with dark edges. Leaf sheaths may also become chlorotic and develop irregular necrotic areas of reddish colour.

The infection in the spindle sometimes continues downward into the stalk and dark reddish streaks may be found extending through several internodes. Also, in the internodes, the infection may form long lesions with cross depressions that give them a ladder-like appearance. These lesions sometimes break through the surface of the rind causing curvature and distortion of the stalk. Exaggerated versions of these depressions may look like neatly made "knife-cuts" in the stalk. In the most advanced stage of pokkah boeng, the entire top (growing point) of the plant dies (referred to as "top rot"). Symptom severity varies with the susceptibility of the cultivar and the environmental conditions governing the development of the organism. In the stem, the fungus causes a dark-brown discoloration of the infected tissues. The ladder-like lesions are due to rupturing of the diseased cells which cannot keep up with growth of the healthy tissue.

The fungi *Fusarium moniliforme* and *Fusarium moniliforme* var. *subglutinans* can persist on decaying plant residues and isolates can be collected on specialized media. The pathogens can infect a wide range of hosts. Rice, corn, sorghum and many other grasses are susceptible to infection. These fungi also cause a wide range of diseases such as seedling blight, scorch, stalk rot, root rot, and stunting in different crops.

The spread of pokkah boeng is mainly by airborne spores. Dissemination of the disease by seed pieces may occur but is considered of little importance. Infection usually occurs through the spindle along the margin of a partially unfolded leaf. Spores which enter the spindle germinate and grow into the inner tissue of the spindle leaves. The fungus reaches the immature portion of the stem by way of the vascular bundles. It may pass through the vascular bundles of the leaf sheath without entering the surrounding tissue, but ladder-like lesions are often found in the sheath. Pokkah boeng appears to be favored by dry climatic conditions being followed by a wet season. Cane that is three to seven months old and growing vigorously appears to be most susceptible.

Pokkah boeng is seldom serious enough to warrant concern or control. However, if control is desired, the only satisfactory control measure for pokkah boeng is the use of resistant varieties. Fortunately, sugarcane resistance to pokkah boeng has been shown to be highly heritable. In Florida, while pokkah boeng resistance is not actively screened for, the relative susceptibility of prospective clones to pokkah boeng is noted during breeding program field trials. Susceptible clones are typically not released for commercial production, thereby limiting the impact of this disease.

III. Viral Diseases

1. Sugarcane Yellow Leaf Disease

It was first recognized as Yellow Leaf Syndrome during the 1980s in Hawaii. Later it was reportedly associated with yield losses of 25 per cent or more in the cultivar SP 71-6163 in Brazil. Subsequently the virus sugarcane Yellow Leaf Virus (SCYLV) was discovered to be associated with the disease. Yellow leaf has been found in numerous countries of the world. However, somewhat confusedly, another disease, "leaf yellows", has been found in other areas. Although it has similar symptoms, it is caused by a phytoplasm rather than a virus. To date, this phytoplasm pathogen has not been identified on sugarcane in Florida.

Symptoms of SCYLV are a yellowing of the leaf midrib on the underside of the leaf. The yellowing first appears on leaves 3 to 6 counting down from the top expanding spindle leaf (Figure 27.1). Yellowing is most prevalent and noticeable in mature cane from October until the end of harvest in March. The yellowing expands out from the leaf midrib into the leaf blade as the season progresses until a general yellowing of the leaves can be observed from a distance. Eventually, almost all leaves of the plant turn yellowish. Cold and nutrient stress appears to intensify the symptoms.

Taxonomically, SCYLV is a member of the Luteoviridae family that contains barley yellow dwarf virus, soybean dwarf virus, potato leaf roll virus, and pea enation mosaic virus-1. Its classification within the family has not been fully resolved. The virus is transmitted by aphids, *Melanaphis sacchari* and *Rhopalosiphum maidis*, in a semi-persistent manner. The virus is also spread by planting infected seed cane. The virus is localized within the phloem cells of the plant and can be diagnosed by either reverse transcription-polymerase chain reaction (RT-PCR) or tissue blot immunoassays using an antibody specific for the virus. There is no thermal treatment effective for eliminating SCYLV in infected sugarcane.

Figure 27.1: Initial Symptoms of Yellow Leaf, with a Yellowing of the Lower Surface of the Leaf Midrib of Leaves 3 to 6 Counting from the Top Expanding Spindle Leaf.

SCYLV reduces both cane and sugar yield. In a yield trial conducted on a grower's field in Florida, hot-water-treated seed cane had 95 per cent SCYLV incidence compared to an initial 0 per cent incidence on commercial tissue-culture based seed cane. There was a 7 and 4 per cent overall increase in cane and sugar yield respectively in the tissue-culture based over the heat-treated seed cane. There was no difference in sucrose content in plants between these two seed cane sources. Another yield test was conducted using seed cane derived from tissue-culture and long hot-water-treated seed cane planted in one-meter plots. The numbers of stalks, cane weight and sucrose were significantly higher in tissue-culture seed cane plots in plant cane for all five cultivars (CP 72-1210, CP 80-1827, CP 84-1198, CP 85-1382 and CP 89-2143). In addition, the tissue-culture plots were SCYLV-free. In the first ratoon crop, the results were less dramatic and not all the differences were statistically significant. Generally yields were higher in the tissue-culture plots free of SCYLV. Cultivar CP 80-1827 was the exception, with higher yields in the heat-treated plots that were SCYLV infected. In a third experiment, seed cane was obtained from plots of tissue-culture derived seed cane that was either SCYLV-free or had become infected during the growing season. The plot weight and sucrose per plot of the SCYLV-free plots of all cultivars combined were higher than the plots with infected plants. The overall results indicate a yield loss due to SCYLV in Florida. Losses have also been reported Louisiana and in Brazil.

SCYLV is widespread in the CP cultivars grown commercially in Florida. Commercial fields planted with regular seed cane usually have at least 85 per cent of the plants infected with SCYLV. Currently, no CP cultivar grown commercially in Florida is free of the virus. Unfortunately, the heat therapy treatments used to control ratoon stunt do not eliminate SCYLV from sugarcane stalks. Thus, using infected seed cane transmits the virus to the emerging plants. The only way to eliminate the virus from the plant is to use meristem tissue culture techniques. Even this technique is not 100 per cent effective and plants derived using this technique must be tested to ensure the virus was eliminated. In Florida, a commercial seed cane company sells disease-free seed cane where SCYLV along with the ratoon stunt and other pathogens have been eliminated. Fields established with SCYLV-free seed cane will initially be virus-free. However in time, plants in these fields will become infected with SCYLV because the virus is aphid-transmitted. The rate at which the plants become infected is cultivar dependent. However, in most cultivars the incidence of SCYLV infection is typically 30-40 per cent after three years. An exception is the former commercial cultivar CP 72-1210 that becomes infected very rapidly.

Since SCYLV-infected plants have a 4 to 10 per cent yield loss of cane weight and sucrose, yellow leaf is a major economic threat. Actual field losses will depend on the loss per plant and the incidence of yellow leaf in the field. Unfortunately, the incidence of yellow leaf is 85 per cent or more in almost all commercial fields planted with regular seed cane of CP cultivars in Florida. Assuming a yield loss of 10 per cent per plant and an 85 per cent incidence level, there would be an 8.5 per cent loss for the field in each crop. Those fields planted with tissue-culture based seed cane have zero or low incidences of yellow leaf, depending on the number of years since it was purchased as clean seed cane. Data indicates that the incidence of SCYLV in plants in the second ratoon in fields established with tissue culture based seed cane is around 30 per cent depending on cultivar, location and other factors. Thus, using these figures there would be only a 3 per cent loss for the tissue-culture field in the second ratoon crop. The use of the tissue-culture based seed cane would also be free of the ratoon stunt pathogen and would prevent losses from this disease too. In summary, control of yellow leaf in the field is extremely important for Florida growers because its high incidence leads to greater economic losses to the industry than many other diseases.

Until resistant cultivars can be developed, only the use of SCYLV-free seed cane can moderate SCYLV yield losses.

2. Chlorotic Streak

It is thought to be caused by a virus. The disease occurs in all eastern sugarcane districts, especially in wet and poorly drained fields. Lower incidence of the disease is generally found in drier regions (Croft *et al.*, 2000). The symptoms are yellow to white streaks on the leaf, midrib and leaf sheath. Older streaks change to yellow and are more visible than younger streaks. This is followed by the appearance of chlorosis in the middle of the leaves. Internal vascular bundle tissues may be reddish in color. (Croft *et al.*, 2000). The disease is transmitted by soil water and diseased seed cane. Yield losses may be up to 40 per cent, with water-logging compounding the losses. Ratooning may also be poor (BSES Ltd, 2009a).

3. Fiji Leaf Gall (Previously called Fiji leaf disease)

It is caused by Fiji disease virus (FDV) and can lead to stunting and death of infected plants (Ridley *et al.*, 2006). The initial symptoms are whitish galls raised on the underside of the leaf blade and midrib. Galls are produced due to the disorder of cell proliferation in the phloem and xylem. Galls can vary from white to green and the surface is usually smooth. When the gall is old, the epidermis may be ruptured and appear brown. At an advanced stage of infection, stem development slows down. Successive leaves become smaller and stiffer with the whole top part of the stem developing a fan-like appearance (Croft *et al.*, 2000). Fiji disease can be transmitted by infected cuttings and plant hoppers (*Perkinsiella saccharicidae*) are a known vector for the disease. Significant yield loss was recorded in 1970s in QLD (Croft *et al.*, 2000) but due to the intensive management program put in place, there have been no reports of disease incidence since the 1980s.

However, FDV is present in southern cane growing areas and plant hoppers are present in all cane growing areas of QLD and NSW (Ridley *et al.*, 2006). Worldwide, sugarcane mosaic is caused by a number of potyviruses such as the Sugarcane mosaic virus (SCMV) (BSES Ltd, 2008b). The mosaic symptom pattern appears in young growing leaves. Once the leaves are older, infected leaves may appear relatively normal as the mosaic becomes green. Aphids transmit the disease, as can seed produced by infected cane.

IV. Other Biotic Interactions

Sugarcane may have symbiotic relationships with a number of bacteria that fix nitrogen. In Brazil, sugarcane is grown with low nitrogen inputs (50 kg ha^{-1}) compared to other countries who use >200 kg ha^{-1} (Boddey *et al.*, 1991). The cane is commonly burnt before harvesting in Brazil so little nitrogen is returned to the field. This low nitrogen requirement has led to the suggestion that some cultivars of sugarcane can obtain nitrogen via biological nitrogen fixation (BNF). The occurrence of BNF has been suggested in several pot studies where some cultivars of sugarcane have thrived for several generations without the addition of nitrogen (Boddey *et al.*, 1991; Urquiaga *et al.*, 1992). Differences were seen between plant genotypes, but it was estimated that BNF could account for 25–60 per cent of the nitrogen assimilated in one study (Boddey *et al.*, 2001) and up to 70 per cent in another study (Urquiaga *et al.*, 1992). The organisms responsible for this have not been unequivocally determined. Studies have focused on endophytic bacteria such as *Gluconacetobacter diazotrophicus* (previously called *Acetobacter diazotrophicus*), however these bacteria were not shown to be producing nitrogenase in planta (James *et al.*, 2001). Despite this, a study using *G. diazotrophicus* -innoculated plants found large increases in nitrogen fixation under nitrogen deficient conditions. This nitrogen fixation did not occur after inoculation with a mutated nitrogenase deficient form of the bacterium (Sevilla *et al.*, 2001). *G. diazotrophicus* may also play a role in defence against sugarcane pathogens. It inhibited *in vitro* growth of *Colletotrichum falcatum* (red-rot) (Muthukumarasamy *et al.*, 2000) and *Xanthomonas albilineans* (leaf scald) (Piñón *et al.*, 2002; Blanco *et al.*, 2005). Additionally *G. diazotrophicus*-inoculated sugarcane stems were resistant to infection by *X. albilineans* (Arencibia *et al.*, 2006). There is also some evidence that it may promote sugarcane growth by production of a growth promoting factor (Sevilla *et al.*, 2001), such as auxin (IAA; indole-3-acetic acid) or by solubilisation of mineral nutrients as reviewed (Saravanan *et al.*, 2008). Other bacterial species have been isolated from sugarcane that may play a role in nitrogen fixation including *Agrobacterium diazotrophicus* (Xing *et al.*, 2006), *Herbaspirillum* spp. (Reis *et al.*, 2007) and *Azospirillum* spp. (Baldani *et al.*, 1997). Experiments have shown that co-inoculation of *G. diazotrophicus* and *Herbaspirillum* spp. gave enhanced sugarcane biomass compared to inoculation with either the single species, or to uninoculated controls (Muthukumarasamy *et al.*, 2006). A field-based experiment and surveys of sugarcane fields in NSW and QLD showed no evidence of biological nitrogen fixation as a source of nitrogen (Biggs *et al.*, 2000).

V. Vesicular-Arbuscular Mycorrhizal Fungi (VAM)

VAM have been found in north QLD sugarcane fields in association with sugarcane roots. These fungi are known to colonize plant roots and may supply the plant with mineral nutrients, especially phosphorous. Pot experiments, using soil and mycorrhial spores from cane fields, showed that the addition of VAM increased the yield of soybean and maize plants. However, no effects have been seen on sugarcane growth from addition of the VAM *Glomus clarum* at various phosphorus levels in pot experiments (Kelly *et al.*, 2001; Kelly *et al.*, 2005). Similar experiments in wheat have shown that although there is no increased yield following root colonization with VAM, 50 per cent of the phosphorus in the plants had been absorbed via VAM (Li *et al.*, 2006).

Conclusion

Variety resistance, together with sugar yield, cane quality and suitable agronomic characteristics, is one of the main factors affecting the choice of varieties for planting. In areas where diseases are limiting factors to production, a high level of resistance may be a prerequisite for a successful sugarcane enterprise. A system for evaluating the reactions of varieties to prevalent diseases is therefore necessary so that sound decisions on which varieties are to be extensively propagated can be made. Given adherence to the control principles that are summarized below, the economic impact of diseases on production can be minimized.

Aspects of control that must be incorporated into every sugarcane enterprise include:

1. Variety resistance. 2) Seed cane health, achieved through a planned, well-managed system of seed cane production. 3) Effective destruction of the old crop before replanting. This is particularly important if the previous crop in the field to be planted was infected by a systemic disease.4) Monitoring the incidence of diseases to provide timely warning of any developing disease problem. Survey data can prompt enhanced attention to basic control measures or signal a need to change from a susceptible to a more resistant variety.

References

Arencibia, A.D., Vinagre, F., Estevez, Y., Bernal, A., Perez, J., Cavalcanti, J., Santana, I., Hemerly, A.S. (2006). *Gluconacetobacter diazotrophicus* elicits a sugarcane defense response against a pathogenic bacteria *Xanthomonas albilineans*. *Plant Signal Behav*, **1:** 265-273.

Bailey, R.A. (2004). Diseases. *In*: G James, ed. Sugarcane, Edition Second. Blackwell Science Ltd. p. 159.

Baldani, J., Caruso, L., Baldani, V.L.D., Goi, S.R., Döbereiner, J. (1997). Recent advances in BNF with non-legume plants. *Soil Biology and Biochemistry*, **29:** 911-922.

Biggs, I. M., Wilson, J. R., Keating, B. A. and Critchley, C. (2000). Does biological N2 fixation contribute to nitrogen requirements in Australian sugarcane? *In*: *"Proceedings of the Australian Society of Sugarcane Technologists"*, **22:** 133-138.

Blanco, Y., Blanch, M., Piñón, D., Legaz, M., Vicente, C. (2005). Antagonism of *Gluconacetobacter diazotrophicus* (a sugarcane endosymbiont) against *Xanthomonas albilineans* (pathogen) studied in alginate-immobilized sugarcane stalk tissues. *Journal of Bioscience and Bioengineering*, **99:** 366-371.

Boddey, R., Urquiaga, S., Reis, V., Döbereiner, J. (1991). Biological nitrogen fixation associated with sugar cane. *Plant and Soil*, **137:** 111-117.

Boddey, R.M., Polidoro, J., Resende, A.S., Alves, B.J.R., Urquiaga, S. (2001). Use of the N natural abundance technique for the quantification of the contribution of N_2 fixation to sugarcane and other grasses. *Functional Plant Biology*, **28:** 889-895.

Braithwaite, K.S., Croft, B.J., Magarey, R.C., Scharaschkin, T. (2009). Phylogenetic placement of the sugarcane orange rust pathogen *Puccinia kuehnii* in a historical and regional context. *Australasian Plant Pathology*, **38:** 380-388.

BSES Ltd (2005a). Guidelines for the Movement of farm Machinery from a Smut-infected Property to a Non-infected Property. IS06056 BSES

BSES Ltd (2005b). Leaf Scald. IS05052 BSES

BSES Ltd (2006). Sugarcane Smut. ISO6006 BSES

BSES Ltd (2007a). Guidelines for the Management of Smut-infected Sugarcane. IS06054 BSES

BSES Ltd (2007b). Guidelines for the Movement of an Appliance from a Non-infested Property to Another Non-infested or Infested Property Within a Quarantine Area. IS06059 BSES

BSES Ltd (2008b). Sugarcane mosaic virus. ISO8001 BSES

BSES Ltd (2009a). Chlorotic streak. S9004 BSES

BSES Ltd (2009b). Control of sugarcane smut - the three Rs. IS09003 BSES

Croft, B., Magarey, R., Whittle, P. (2000). Disease Management. Chapter 13. *In*: M Hogarth, P Allsopp, eds. *Manual of cane growing*. Bureau of Sugar Experimental Stations Indooroopilly, Australia. pp. 263-289.

Croft, B.J., Magarey, R.C., Allsopp, P.G., Cox, M.C., Willcox, T.G., Milford, B.J., Wallis, E.S. (2008). Sugarcane smut in Queensland: arrival and emergency response. *Australasian Plant Pathology*, **37:** 26-34.

Dahlia, L., Kurniawan, I., Anggakusuma, D. and Roshetko, J.M. (2010). "Consumer Preference for Indigenous Vegetables". World Agroforestry Centre.

FAO. (2009). "Agribusiness Handbook: Sugar beet white sugar". Food and Agriculture Organization, United Nations.

FAO. (2014). "Crop production". Food and Agriculture Organization of the United Nations. Retrieved 2014-04-26.

Frison, E.A. and Putter, C.A.J. (1993). FAO/IBPGR Technical guidelines for the safe movement of sugarcane germplasm. Food and Agriculture Organization of the United Nations/International Board for Plant Genetic Resources Rome.

James, E.K., Olivares, F.L., de Oliveira, A.L.M., dos Reis, F.B., Jr., da Silva, L.G., Reis, V.M. (2001). Further observations on the interaction between sugar cane and *Gluconacetobacter diazotrophicus* under laboratory and greenhouse conditions. *Journal of Experimental Botany*, **52:** 747-760.

Kelly, R.M., Edwards, D.G., Thompson, J.P., Magarey, R.C. (2001). Responses of sugarcane, maize, and soybean to phosphorus and vesicular-arbuscular mycorrhizal fungi. *Australian Journal of Agricultural Research*, **52:** 731-743.

Kelly, R.M., Edwards, D.G., Thompson, J.P., Magarey, R.C. (2005). Growth responses of sugarcane to mycorrhizal spore density and phosphorus rate. *Australian Journal of Agricultural Research*, **56:** 1405-1413.

Li, H., Smith, S.E., Holloway, R.E., Zhu, Y., Smith, F.A. (2006). Arbuscular mycorrhizal fungi contribute to phosphorus uptake by wheat grown in a phosphorus-fixing soil even in the absence of positive growth responses. *New Phytol*, **172:** 536-543.

Magarey, R. C., Bull, J. I., Sheahan, T., Denney, D. (2010). Yield losses caused by sugarcane smut in several crops in Queensland. *In:* Proceedings of the Association of Sugarcane Technologists, 32 pp. 347-354.

McGuire, P., Bambach, G., Aitken, R., Beattie, R., Lokes, S. (2009). RSD control in the NSW sugar industry. *In*: "Proceedings of the Australian Society of Sugarcane Technologists".

McLeod, R.S., McMahon, G.G., Allsopp, P.G. (1999). Costs of major pests and diseases to the Australian sugar industry. *Plant Protection Quarterly*, **14:** 42-46.

Mintz, S.W. (1986). *Sweetness and Power: The Place of Sugar in Modern History*. Penguin. ISBN 978-0-14-009233-2.

Muthukumarasamy, R., Govindarajan, M., Vadivelu, M., Revathi, G. (2006). N-fertilizer saving by the inoculation of *Gluconacetobacter diazotrophicus* and *Herbaspirillum* sp. in micropropagated sugarcane plants. *Microbiological Research*, **161:** 238-245.

Muthukumarasamy, R., Revathi, G., Vadivelu, M. (2000). Antagonistic potential of N2 fixing *Acetobacter diazotrophicus* against *Colletotrichum falcatum* Went., a causal organism of red rot of sugarcane. *Current Science*, **78:** 1063-1066.

Nasare, K., Yadav, Amit., Singh, A. K., Shivasharanappa, K. B., Nerkar, Y. S., and Reddy, V. S. (2007). Molecular and symptom analysis reveal the presence of new phytoplasmas associated with sugarcane grassy shoot disease in India. *Plant Disease*, **91:** 1413-1418.

NSW Sugar (2005). Diseases in Sugarcane. NSW Sugar Milling Cooperative, http: // www.nswsugar.com.au/index.php?option=com_content and task=view and id=48 and Itemid=85.

Onwueme, I.C. and Sinha, T.D. (1999). CTA Field Crop Production in Tropical Africa, CTA, Wageningen, Netherland, pp: 401-411.

Piñón, D., Casas, M., Blanch, M., Fontaniella, B., Blanco, Y., Vicente, C., Solas, M.T., Legaz, M.E. (2002). *Gluconacetobacter diazotrophicus*, a sugar cane endosymbiont, produces a bacteriocin against *Xanthomonas albilineans*, a sugar cane pathogen. *Research in Microbiology*, **153:** 345-351.

Rao, G. P. and Ford, R. E. (2000). Vectors of virus and Phytoplasma diseases of Sugarcane: An Overview. In: Sugarcane Pathology, Vol. III. Virus and Phytoplasma diseases, G.P. Rao, R.E. Ford, M. Tosic and D.S. Teakle (Eds) Science Publishers, Hamshere, USA, pp. 265-314.

Reis, V., Lee, S., Kennedy, C. (2007). Biological Nitrogen Fixation in Sugarcane. Chapter 10. *In:* C Elmerich, WE Newton, eds. Associative and Endophytic Nitrogen-fixing Bacteria and Cyanobacterial Associations, Volume 5. Springer Netherlands, pp. 213-232.

Ridley, A.W., Dhileepan, K., Johnson, K.N., Allsopp, P.G., Nutt, K.A., Walter, G.H., Croft, B.J. (2006). Is the distribution of Fiji leaf gall in Australian sugarcane explained by variation in the vector Perkinsiella saccharicida? *Australasian Plant Pathology*, **35:** 103-112.

Roach, B. T., Parsons, D. H., Nielsen, P. J. (1992) Incidence and control of ratoon stunting disease in sugarcane in New South Wales. In "Proceedings of Australian Society of Sugarcane Technologists", 14.

Saravanan, V.S., Madhaiyan, M., Osborne, J., Thangaraju, M., Sa, T.M. (2008). Ecological occurrence of *Gluconacetobacter diazotrophicus* and nitrogen-fixing Acetobacteraceae members: their possible role in plant growth promotion. *Microbial Ecology*, **55:** 130-140.

Sevilla, M., Burris, R.H., Gunapala, N., Kennedy, C. (2001). Comparison of benefit to sugarcane plant growth and incorporation following inoculation of sterile plants with *Acetobacter diazotrophicus* wild-type and Nif- mutant strains. *Molecular Plant-Microbe Interactions*, **14:** 358-366.

Urquiaga, S., Cruz, K.H.S., Boddey, R.M. (1992). Contribution of Nitrogen Fixation to Sugar Cane: Nitrogen-15 and Nitrogen-Balance Estimates. *Soil Science Society of America Journal*, **56:** 105-114.

Walker, D.I.T. (1987). Breeding for disease resistance. Chapter 12. In: DJ Heinz, ed. Sugarcane improvement through breeding. *Elsevier Amsterdam*. pp. 455-502.

Watson, D. (2007). The economic impact of sugarcane smut on the Queensland sugarcane industry. Report No. report to Queensland Government.

Xing, Y.X., Yang, L.T., Huang, S.L., Li, Y.R. (2006). Identification of new nitrogen fixing endo-bacterium strain isolated from sugarcane stalk. *Sugar Tech*, **8:** 49-53.

2015, Major Constrains and Verdict of Crop Productivity *Pages 384–389*
Editor: Dr. U.N. Bhale
Published by: ASTRAL INTERNATIONAL (P) LIMITED, NEW DELHI

Chapter 28

The Study of Mineral Content and Shelf Life of *Kunapajala*: A Liquid Biofertilizer from *Vrikshayurveda*

Rupali S. Deshmukh*

D-11, Shangrilla Gardens,
Vivekanad Nagar, Baramati – 413 102, India

ABSTRACT

Liquid biofertilizers play a vital role in agriculture leading to green food production, which is safer, healthier and tastier. A better option for the liquid biofertilizers is described in ancient Indian literature with more scientific formulation under the generic name "Kunapajala" The kunapajala is a fermentation product using easily available ingredients like Sesamum indicum L. (Tilaceae), bone marrow, flesh (sheep, goat, fish etc), milk, black gram, ghee, honey etc. So, the present investigation is made to standardize the procedure for preparation of kunapajala and to study the mineral content in kunapajala along with its shelf life. The present investigation showed that the kunapajala is a fermented product. The pH of the solution was acidic and the electrical conductivity 4.20 micro Siemens per cm (μS/cm). After a year, there is no considerable difference in pH (4.5) and electrical conductivity (4.27μS/cm) of kunapajala. So, to apply kunapajala on the plant, dilution is needed. The mineral analysis of liquid biofertilizer kunapajala showed that in initial stage, the percentage of total nitrogen, total phosphorus, total potash, total calcium and total magnesium was 0.42, 0.08, 0.30, 0.18 and 0.46 respectively. The content of total zinc, total copper, total iron and total manganese were 56, 3, 78 and 131 ppm respectively. After one year, the results obtained showed more or less the similar mineral composition in total calcium, total magnesium and total zinc respectively while the total nitrogen, total phosphorus, total potash and total manganese contents slightly increased compared to the mineral content of kunapajala in initial stage. In the present investigation the mineral analysis of liquid bio-

* E-mail: drupali77@yahoo.com

fertilizer kunapajala showed that it contains all the macro and micro nutrients which are essential for the growth of the plant. After one year the composition of mineral content was more or less the same compared to that in the initial stage. This shows that shelf life of kunapajala is at least up to one year from the date of preparation.

Keywords: *Kunapajala, Kunapajala ingredients, Liquid biofertilizers, Mineral analysis, Shelf life.*

Introduction

Foliar nutrition and liquid fertilizer are the concepts of modern agriculture. Liquid fertilizers provide plants with high concentrations of easily absorbed, soluble nutrients, thereby enhancing their health and productivity. Israel is a leading country manufacturing various types of liquid chemical fertilizers containing water soluble nutrients. Some Indian companies are producing metal chelates and these are used on a commercial basis. The supply of micronutrients in the form of chelates is very specific, as per the requirements of plants. So these can be used under specific conditions only.

Liquid biofertilizers play a vital role in agriculture leading to green food production, which is safer, healthier and tastier. Nutritional spray on plants can decrease the delay between absorption and consumption of elements by the plants, which is very important for the fast growth of plant. Hiltbrunner *et al.* (2005) stated that liquid manure increases the yield in organically grown wheat. Maragoni *et al.* (2004) claimed that organic farming systems are more sustainable than conventional agricultural techniques and they represent a commercially viable alternative in stone-fruits production. They further stated that application of vegetable waste materials and cow manure may improve plant nutrition and influence growth of peach and other fruit tree species. The growth boosters or fertilizers when sprayed on the plants enter the metabolic pathway of plants and alter them, which is against the concept of organic farming. This is avoided by soil application method, where the plant absorbs required nutrients in required quantities in an organic way. A better option for the liquid biofertilizers is described in ancient Indian literature with more scientific formulation under the generic name "*Kunapajala*" by Surapala (1000 AD) in '*Vrikshayurveda*', Sarangadhara (1283–1301 AD) in '*Upavan Vinoda*' and Chakrapani Mishra (1577 AD) in '*Vishvavallabha*'. The *kunapajala* is a fermentation product using easily available ingredients like *Sesamum indicum* L. (Tila), bone marrow, flesh (sheep, goat, fish etc), milk, black gram, ghee, honey etc.

Kunapajala induces healthiness in plants, so that they become tolerant to abiotic stresses, pests and diseases. By this, the yield obtained is of high quality (Nene, 2006). Action of *kunapajala* increases the vegetative and reproductive growth of plants. It overcomes the effect of a season also (Sadhale, 2004). Nene (2006) stated that *kunapajala* is a liquid manure of antiquity. It gives flexibility to farmers according to their convenience. So far, the Indian agricultural scientists have shown least interest in *kunapajala*. Only social workers have taken interest and have demonstrated the immense fertilizer value that the *kunapajala* offers. The components of *kunapajala* mentioned by the scholars may not be easily available. Every farmer may not handle them as this *kunapa* is a non vegetarian *kunapa.* So, the methods of preparation of herbal kunapa were mentioned without experimenting with it on any plant (Saxena *et al.*, 2009). Mishra (2007) pointed out that *kunapajala* can be a good substitute for synthetic fertilizers. However there is no any documentation material for the mineral content study of *kunapajala* and its shelf-life. So, the attempt is made to prepare a *kunapajala* and to study the mineral content in *kunapajala* along with its shelf life.

Materials and Methods

The *kunapajala* formulation was based on the work of Surapala (1000 AD) in *Vrikshayuraveda* verses 101 to 106 (Sadhale, 1996) and Chakrapani Mishra (1577 AD) in *'Vishvavallabha'* verses 1, 2 and 3 respectively (Sadhale, 2004) with slightly modifications.

Flesh of goat and pieces of fish (each 250 g) were cooked in earthen pot with 1000 ml water. 500 ml milk was added along with 500 g each of seasum oil cake, ghee, cooked black gram and honey. It was stirred well and the fire was stopped. The lid was placed and the pot was kept for fermentation for 15 days in a warm place. The fermented liquid was filtered through cotton and the final volume of filtrate was made 1000 ml. The solution was stored in refrigerator. 0.5 per cent solution was used for treatments (working solution) (Figure 28.1).

The physical properties of *kunapajala* were studied by routine laboratory methods. Mineral contents in *kunapajala* were analyzed in laboratory of Draksha Bagayatdar Sangh, Hadapsur, Pune (2012-2013) using following instruments and model:

Parameter	*Instruments and Model No.*
Phosphorus and Sulphur	Spectrophotometer – Thermo Scientific Model : Spectra scan UV 2700
Iron, Manganese, Zinc, Copper, Calcium and Magnesium	Atomic Absorption Spectrophotometer Model : Chemito AA 203
Potassium and sodium	Flame Photometer Model : Systronics 121
Nitrogen	Gerhardt Nitrogen Analyser Model : Vepodest 30s

Results and Discussion

Physical Properties of *Kunapajala*

Data depicted in Table 28.1 that the physical properties of *kunapajala.* It was pale yellow in colour and had a foul smell. pH of the solution was acidic (4.57) and the electrical conductivity 4.20 micro Siemens per cm (µS/cm). After a year, there is no considerable difference in pH (4.5) and electrical conductivity (4.27µS/cm) of *kunapajala*. The percentage of total soluble salts was the same as before and after a year of storage of *kunapajala* (0.6 per cent).

Mineral Analysis of *Kunapajala*

The mineral analysis of *kunapajala* in the initial stage and after one year is illustrated in Table 28.2. The results showed that in initial stage, the percentage of total nitrogen, total phosphorus, total potash, total calcium and total magnesium was 0.42, 0.08, 0.30, 0.18 and 0.46 respectively. The content of total zinc, total copper, total iron and total manganese were 56, 3, 78 and 131 ppm respectively. After one year, the results obtained showed more or less the similar mineral composition in total calcium, total magnesium and total zinc respectively while the total nitrogen, total phosphorus, total potash and total manganese contents slightly increased compared to the mineral content of *kunapajala* in initial stage.

Table 28.1: Physical Properties of *Kunapajala* after the Formulation Procedure

Sl.No.	*Parameter*	*Properties of Kunapajala*	
		Initial Stage	*After One Year*
1.	Colour	Pale yellow	Pale yellow
2.	pH	4.57	4.5
3.	EC (at 25°C) (μS/cm)	4.20	4.27
4.	Total soluble salts (per cent)	0.6	0.6

Table 28.2: Mineral Content of *Kunapajala* at Initial Stage and after One Year of Preservation in Refrigerator

Sl.No.	*Parameter*	*Mineral Content of Kunapajala*	
		Initial Stage	*After One Year*
1	Total Nitrogen (per cent)	0.42	0.43
2	Total Phosphorus (per cent)	0.08	0.10
3	Total Potash (per cent)	0.30	0.31
4	Total Calcium (per cent)	0.18	0.18
5	Total Magnesium (per cent)	0.46	0.46
6	Total Zinc (ppm)	56	56
7	Total Copper (ppm)	3	2
8	Total Iron (ppm)	78	76
9	Total Manganese (ppm)	131	135
	SEM	0.68	0.68
	CD@1 per cent	1.2	1.2
	CD@5 per cent	1.4	1.4

Kunapjala is the liquid biofertilizer. It is a fermentation product of easily available ingredients. During the fermentation the simple proteins, fats, carbohydrates etc are broken into simple, low molecular weight products. Therefore nutrients from *kunapajala* become available to the plants easily. The process takes place under partially anaerobic condition leading to accumulation of intermediate products of decomposition. Various organic acids, so liberated, combine to form ammonium salts, preventing the loss of ammonia by volatilization. The acids also lower the pH value of manure.

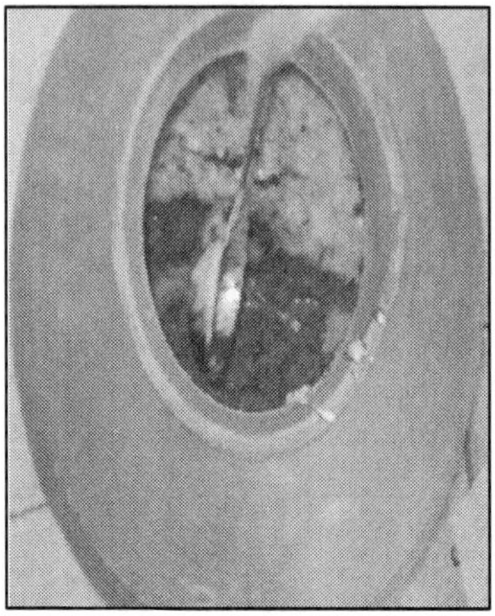

Kunapajala

Filtration of *Kunapajala* (I)

Filtration of *Kunapajala* (II)

Figure 28.1: Steps of *Kunapajala* Preparation.

There is formation of nitrate under such condition. The loss of organic matter is considerably reduced. Hence, there is greater conservation of organic matter and nitrogen. The danger of passing on dormant pathogens to fields with plant based compost is avoided by *kunapajala* because the ingredients are cooked and fermented.

The prepared *kunapajala* solution is acidic in nature. To apply it on the plant, dilution is needed. 0.5 per cent working solution was used for pilot study. Valmiki (2004) reported excellent results of *kunapajala* on mango, coconut and chilli plant but he did not study chemical analysis of *kunapajala*. Present study is the pioneer work on the analysis of *kunapajala* and it is significant.

In the present investigation the mineral analysis of liquid bio-fertilizer *kunapajala* showed that it contains all the macro and micro nutrients which are essential for the growth of the plant. The presence of these nutrients is due to the ingredients used for *kunapajala* preparation. After one year the composition of mineral content was more or less the same compared to that in the initial stage. This shows that shelf-life of *kunapajala* is at least up to one year from the date of preparation.

Conclusion

Overall, most of the macro (N, P, K, Ca and Mg) and micro (Zn, Cu, Fe and Mn) nutrients essential for the plant growth were found to be present in *kunapajala*. After one year the composition of mineral content was more or less the same compared to that in the initial stage. This shows that shelf life of *kunapajala* was up to 12 months in refrigerator.

References

Hiltbrunner, J., Liedgens, M., Stamp, P. and Streit, B. (2005). Effects of row spacing and liquid manure on directly drilled winter wheat in organic farming. *European J. Agronomy*, **22** (4): 441-447.

Maragoni, B., Rombola, A. D. and Sorrenti, G. (2004). Use of natural compounds for plant nutrition and protection in organic farmed orchards. (Dijon and France eds), *Bulletin (OIBC/SROP)*, **27**(5): 129-136.

Mishra, P. K. (2007). Effect of kunapajalam vrikshyurveda on growth of paddy. *Indian Journal of Traditional Knowledge*, **6** (2): 307-310.

Nene, Y. L. (2006). Utilization of traditional knowledge in agriculture. Asian Agri History Foundation, Secunderabad, India. *Journal of Traditional Knowledge System of India and Sri Lanka*, pp. 32-38.

Sadhale, N. (1996) (tr) Surapala's *Vrikshayurveda* (The Science of Plant Life by Surapala). Agri-History Bulletin No. 1. Asian Agri-History Foundation, Secunderabad 500 009, India, pp. 104.

Sadhale, N. (2004) (tr) *Vishvavallabha* (Dear to The World: The Science of Plant Life). Agri-History Bulletin No 5. Asian Agri- History Foundation, Secunderabad, India, pp. 76-80.

Saxena, R.C., Choudhary, S. L. and Nene, Y. L. (2009) Kunapajala: a Liquid Manure. In: A text book on ancient history of Indian agriculture. Asian Agri History Foundation (AAHF), Secunderabad and Rajasthan, Udaipur, India, pp.106-109.

Valmiki, S. A. (2004 a) Harbal *Kunapa*. Asian Agri History Bulletin- 8. Asian Agri History Foundation Secunderabad, Andhra Pradesh, India, pp. 315-317.

2015, Major Constrains and Verdict of Crop Productivity *Pages 390–413*
Editor: Dr. U.N. Bhale
Published by: ASTRAL INTERNATIONAL (P) LIMITED, NEW DELHI

Chapter 29
Micropropagation

T. Ugandhar[1]*, M. Venkateshwarlu[2] and K. Jaganmohan Reddy[2]

[1]Department of Botany,
S.R.R. Govt. Arts and Science College, Karimnagar – 505 001, A.P., India
[2]Department of Botany,
Kakatiya University, Warangal – 506 009, A.P., India

Introduction

Micropropagation is the first major and widely accepted practical application of plant tissue culture and is commonly used for clonal propagation of plants. The word 'clone' was first used by Webber for apply to cultivated plants that were propagated vegetatively. The word is derived from Greek (Clon = twig, spray or a slip, like those broken off as propagules for multiplication). Initially, the technique of micropropagation was employed basically to ornamental plants only (Ammirato *et al.*, 1983), but gradually it is extended to various vegetable and fruit crops such as potato, strawberry, oil palm, banana, etc. medicinal and aromatic plants and trees (Bajaj, 1986 & 1988). Presently, the micropropagation technique is especially being used not only for those plants which are difficult to be propagated through conventional practices, but also for the mass multiplication of existing stocks of germ plasm for biomass energy production and conservation of important, elite and rare plant species that are threatened or on the verge of extinction (Normah *et al.*, 1997; Wawrosch *et al.*, 2001). Now micropropagation has gained the status of a multibillion dollar industry through out the world. The beginning of this industry goes back to the mid of 20th century, when for the first time cultured shoot meristem were used to get virus-free plants of Dahlia and Cymbidium (Morel and Martin, 1952).

* *Corresponding Author.* E-mail: t.ugandharbiotech@gmail.com

Crop breeding and genetic engineering technology require the widespread use of reliable true to type propagation and better regeneration methods. In this context, the types of propagation required will vary immensely. Micropropagation or clonal propagation is the multiplication of genetically identical individuals by asexual means. With the current techniques of single cell and protoplast culture, thousands of plants can be derived from a single cell in a comparatively short span of time. The significant advantage of micropropagation over the conventional methods of asexual clonal propagation is that in a relatively short span of time and space, a large number of true to the type plants can be produced. For example an estimate suggests that as many as 1,000,000 propagules or plants starting from a single explants in 6 months. Murashige (1974) outlined four major stages involved in the process of micropropagation. However, Deberg and Maene (1981) introduced Stage 0, making micropropagation a five stage process:

Stage 0: Preparative Stage

Selection and maintenance of mother or stock plants for culture initiation.

Stage 1: Initiation of Cultures

Initiation and establishment of aseptic cultures (Isolation of explants, their sterilization and transfer to nutrient media for establishment, *i.e.* initiation of a sterile culture of the explants).

Stage 2: Multiplication

Proliferation or multiplication of shoots or Somatic embryo formation using a defined culture medium.

Stage 3: Rooting

Transfer of shoots/emryoids to a medium for rooting.

Stage 4: Transplantation

Transfer of *in vitro* regenerated plantlets to sterile soil for acclimatization under greenhouse environment. However, in a few cases step III is bypassed, as rhizogenesis is induced *ex vitro*.

These stages have been designed simply to describe micropropagation processes and to facilitate comparison between two or more systems. Plant regeneration can be achieved by culturing tissue sections with a preformed meristem (adventitious origin *viz.* axillary bud) or arise *de novo* from callus and cell cultures derived from non-meristematic or permanent tissues *e.g.* leaf blade, cotyledon, internode, petiole, etc. Thus adventitious or *de novo* plant regeneration can occur by organogenesis and embryogenesis.

Micropropagation and plant regeneration can be grouped into the following categories (Figure 29.1):

1. **Apical/Axillary bud**: Meristem and shoot tips or axillary shoot buds or single node can grow into several plantlets.

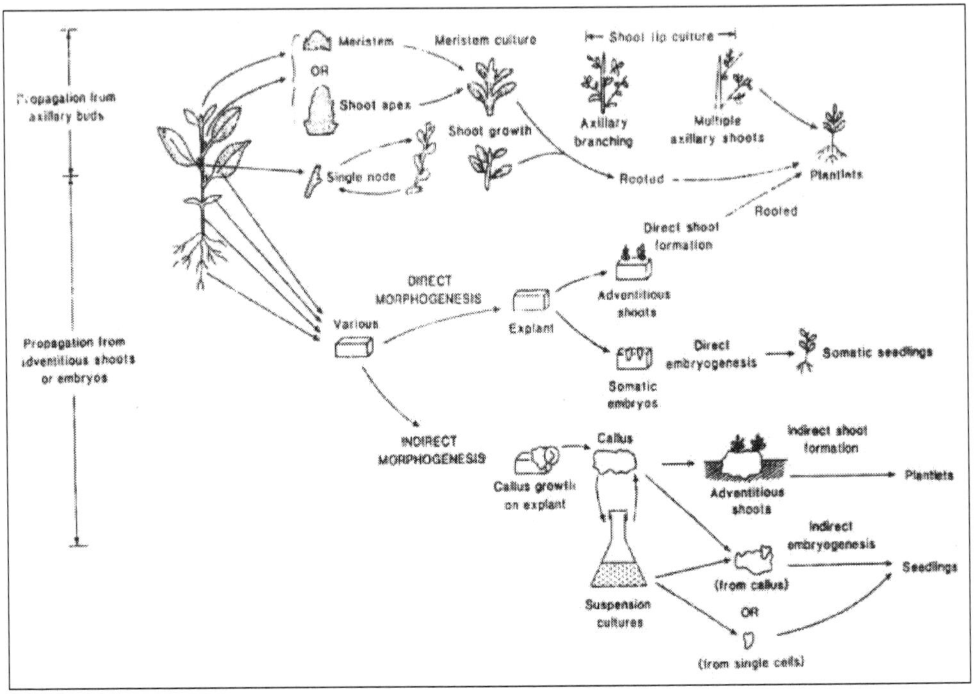

Figure 29.1: A Diagrammatic Representation of different Source Material and Methods Used for Micropropagation.

2. **Organogenesis:** The formation of individual organs such as shoots and roots either directly on the explant where a preformed meristem is lacking or *de novo* origin from callus and cell culture induced from the explant.
3. **Somatic embryogenesis**: Somatic embryogenesis is the formation of a bipolar structure containing both shoot and root meristems either directly from the explant (adventitive origin) or *de novo* origin from callus and cell culture induced from the explant.

Stage-0: Mother or Stock Plant Selection and Preparation

The success of micropropagation greatly depends upon the selection of the stock or mother plant and it should be disease free. Field grown plants normally harbor large number of microbial contaminants on their surface and since these microbes grow much faster than the explant tissue on the tissue culture medium, it is essential to eliminate them. It is not possible to kill many of these microorganisms with known sterilizing agents. Therefore, the mother plant is either sprayed with fungicides or the growing branches are covered with polythene bags. Viruses have to be considered separately from other pathogens because these are transmitted in a propagative way and pesticide treatments have no effect on them, and thus their mode of elimination is different. In order to propagate virus-free plants, one has to select a virus-free donor plant or perform donor plant treatments such as thermotherapy. One of the methods of elimination of viruses is shoot-tip meristem culture as shown by Bryan and Dodds in 1992 wherein healthy virus free potato plants were produced by meristem culture. *In vitro* culture techniques with thermotherapy have been used for elimination of viruses from fruit trees (Knapp *et al.*, 1995), *Prunus* species (Gella and Errea, 1998), pear (Cieslinska 2002) and raspberry (Karesova *et al.*, 2002).

Stage-1: Initiation of Aseptic Cultures

The plant organ used to initiate a culture is called 'explant'. All plant organs *viz.* nodal segments, internodal segments, shoot tip, root pieces etc. can give rise to complete plants. The source and nature of the explant determines the stringency of the surface-sterilization treatment, for example, explants derived directly from field grown plants or underground plant parts such as rhizomes and tubers require a stronger disinfection treatments than those obtained from plants maintained under controlled conditions in the greenhouse. In several plants such as *Saussurea lappa* (Joshi and Dhar, 2003), *Alpinia galanga* (Borthakur *et al.*, 1998), *Bacopa monnieri* (Tiwari *et al.*, 1998), *Curcuma* species (Balachandran *et al.*, 1990) etc., heavy infection has been reported at the initiation stage. At the same time, harsher treatments would damage certain explants such as in the case of *Bacopa monnieri* (Tiwari *et al.*, 1998) and *Trichopus zeylanicus* (Krishnan *et al.*, 1995) wherein explants turned brown when treated with 0.01 – 0.1 per cent mercuric chloride. Under such circumstances, alternative sources such as *in vitro* seedling derived explants can be used as explants as in the cases of *Saussurea obvallata* (Joshi and Dhar, 2003) and *Lilium nepalense* (Wawrosch *et al.*, 2001).

Explants with preformed meristems are shoot tip (terminal/apical meristem) and nodal explants (with axillary meristems) are ideal choice for micropropagation. However, if the mother plant is infested with viruses, it is obligatory to initiate cultures from the extreme meristem. While this calls for tedious dissections coupled with poor survival, once a plant is obtained it is convenient to produce multiple copies under aseptic conditions.

Stage-2: Multiplication

I. Meristem and Shoot Tip Culture

Morel and Martin (1952) developed the technique of meristem culture for *in vivo* virus eradication of *Dahlia.* Georges Morel (1965) was the pioneer in applying shoot tip culture for micropropagation of orchid *Cymbidium. In vitro* clonal propagation gained momentum in the 1970s when Murashige (1974) gave the concept of developmental stages by defining establishment, proliferation, and rooting and hardening stages. This concept stimulated awareness that a single medium is not sufficient for *in vitro* multiplication, but the propagule needs a series of transfers in specially designed physical and chemical environments to reach regeneration. This method has been more successful in herbaceous plants because of weak apical dominance and strong root regenerating capacities as compared to woody species.

Many crops are being currently propagated commercially using these *in vitro* procedures. Shoots of all angiosperms and gymnosperms grow by virtue of their apical meristems. The apical meristem is usually a dome of tissue located at the extreme tip of a shoot and measures -0.1 mm in diameter and -0.25 to 0.3 mm in length. The apical meristem together with one to three young leaf primordia measuring 0.1 to 0.5 mm constitutes the shoot apex. The explant of meristem tip culture may either be the apical meristematic dome or the apical dome with few subjacent leaf primordia. Meristem or shoot tip is cut or isolated from stem by applying a V-shaped cut with a sterilized knife. For meristem tip culture, the cut is applied 0.3 to 0.5 mm below the tip of the dome and the excised tissue (explant) is removed along with portions of procambial tissue and is immediately planted on the media. Size of the tip explant is the decisive factor which governs the success of culture. When very small explants are used, the presence of leaf primordia appears to determine the capability of an explant to develop. In general, the larger the explant, the better are the chances for survival. It is generally agreed that large explants such as shoot tips measuring up to 2 cm in length should be selected for *in vitro* micropropagation. However, when disease-free plants are the objective, meristem tips should be cultured.

Like all asexual propagation methods, success of meristem and shoot tip cultures is affected by the season during which the explants are obtained. For plant species with a dormant period, best results may be expected when the explants are dissected at the end of their dormancy period. Further, actively growing shoot tips are recommended for their culture because of their strong growth potential and low virus concentration. A general protocol of shoot tip culture has been given in Figure 29.2. MS medium salts have been very satisfactory for such cultures though White's

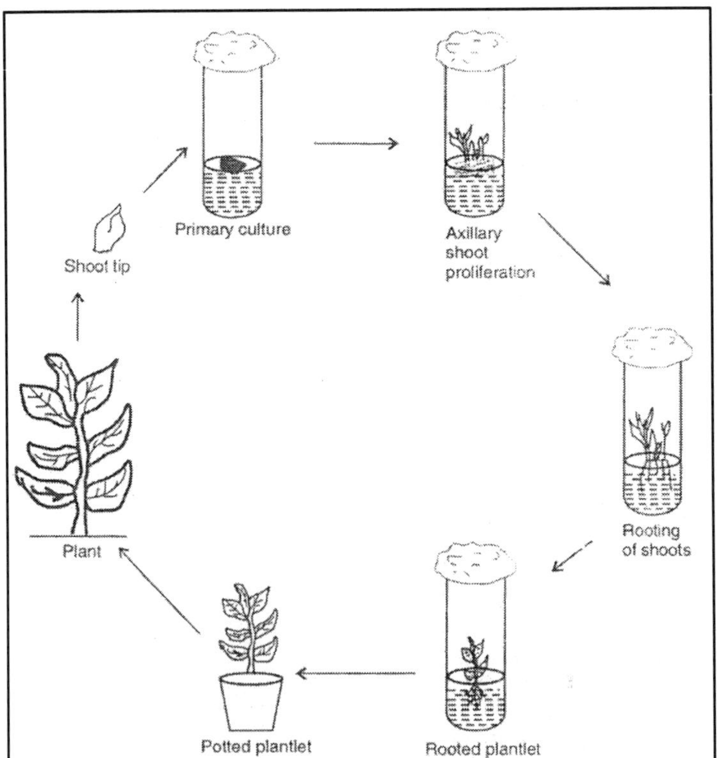

Figure 29.2: Schematic Representation of Shoot Tip Culture.

medium and the Gautheret's (1959) were the most widely used media during the early days of meristem culture. The growth additives vary from species to species and in some woody species culture one needs to add certain polyphenol antioxidant compounds. Some general conclusions from the published reports have been given.

Murashige reported that there are three stages of culture. Stage I is the culture establishment stage when explants may develop either into single shoots or multiple shoots. It is unlikely that meristem, shoot tip or bud explants have sufficient endogenous cytokinin to support growth and development. Thus at this stage media are supplemented with cytokinin like BA, kinetin, and 2iP. Auxin is another hormone required for shoot growth. Since the young shoot apex is an active site for auxin biosynthesis, exogenous auxin is not always needed in stage 1 medium, especially when relatively large shoot tip explants from actively growing plants are used. If an auxin is to be added, it should be a weak one like NAA or IBA rather than 2, 4-D. The usual range of concentration is between 0.45 and 10.0 μM.

Many plants are rich in polyphenolic compounds. After tissue injury during dissection, such compounds will be oxidized by polyphenol oxidases and the tissue will turn brown or black. The oxidation products are known to inhibit enzyme activity, kill the explants, and darken the tissues and culture media which severely affect the establishment of explant. Some of the procedures used by various workers to combat this problem are:

1. Adding antioxidants to culture medium *viz.* ascorbic acid, citric acid, Poly Vinyl Pyrrolidone (PVP) and dithiothreitol etc.
2. Presoaking explants in antioxidant before inoculating into the culture medium.
3. Incubating the initial period of primary cultures in reduced light or darkess because it is known that phenolic oxidation products are formed under illumination.
4. Frequent transfer of explants into fresh medium whenever browning of the medium is observed.

In Stage II the objective is to multiply the propagule and for this axillary shoot proliferation is followed as it maintains higher genetic stability and is more easily achievable by most plant species than organogenic or embryogenic path. In axillary shoot proliferation, high levels of cytokinin are utilized to overcome the apical dominance of shoots and to enhance the branching of lateral buds from leaf axils.

The cytokinin concentration varies but, 4.5 to 25.0 μM or higher concentrations has been used. In general BA is the most effective cytokinin followed by kinetin and 2iP. Exogenous auxin does not promote axillary shoot proliferation; however, if it is used in low concentration to nullify the suppressive effect of high cytokinin concentrations on axillary shoot elongation and to restore normal shoot growth.

II. Bud/Nodal Culture

Buds contain quiescent or active meristems, depending upon the physiological state of the plant, each of which is capable of growing into a shoot that is identical to the main axis. In this method a shoot bud is isolated, from which the axillary buds in the axils of leaves develop under the influence of a relatively high cytokinin concentration. A schematic diagram of the steps in axillary bud culture is shown in Figure 29.3. A single node explant is often used in combination with the axillary bud method. The axillary bud present in the axil of each leaf either develops into a single shoot or forms a cluster of shoots. In culture, the rate of shoot multiplication by axillary branching can be substantially enhanced by culturing shoots in a medium containing suitable cytokinin/s with or without an auxin at appropriate concentrations. Initially, the multiplication rates achieved by this method may be low but with each passage the number of shoots increases logarithmically. These shoots could be further multiplied by sub-culturing on the medium of similar composition indefinitely. The type of explant used is one of the major factors affecting the multiplication rate.

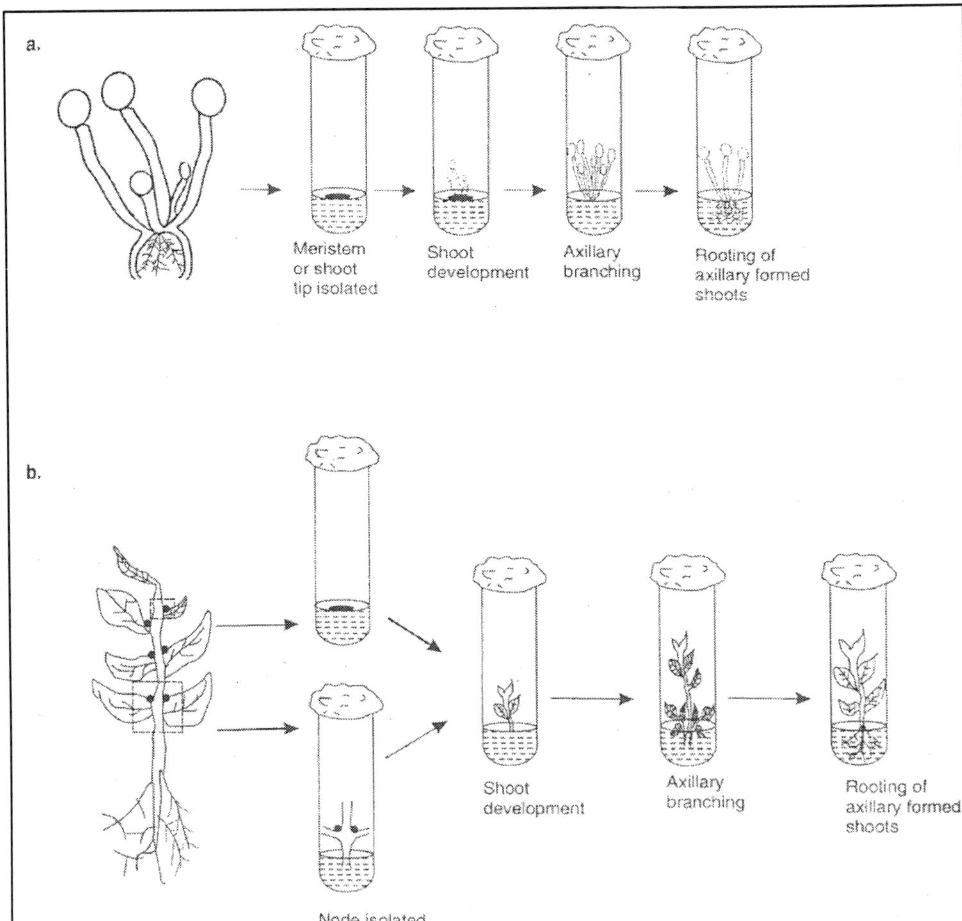

Figure 29.3: Schematic Representation of Axillary Bud Method of Vegetatively Propagating Plants. (a) Rosette plants; (b) Elongate plants showing bud culture and single node culture.

Nodal explants were found to be superior to apical shoot buds in *Melia azadiarach* (Thakur *et al.*, 1998) and *Rauwolfia micrantha* (Sudha and Seeni, 1996). Explant orientation is another factor that was found to be important for achieving optimal shoot multiplication in certain species. In literature, benzyl amino purine (BAP) has been reported to be most effective hormone in inducing multiplication and hence has been adopted in the micropropagation of several species such as *Holarrhena antidysenterica* (Kumar *et al.*, 2005), *Chlorophytum borivilianum* (Dave *et al.*, 2003), *Ocimum americanum* and *O. canum* (Pattnaik and Chand, 1996) and Indian Pandan (*Pandanus amaryllifolius* Roxb.) (Gangopadhyay *et al.*, 2004) In certain cases, the synergistic effect of two or more cytokinins was found to be necessary for optimal multiplication rates as in the case of *Eclipta alba* wherein BAP, KN and 2ip together were found to be essential for shoot proliferation (Baskaran and Jayabalan, 2005). In order to improve the shoot multiplication rate, the use of additives has been attempted in several studies. Kaur *et al.* (1998) found adenine sulphate ($AdSO_4$), ascorbic acid and glutamine to be essential for maximum shoot bud induction in *Acacia catechu*. In addition to duration, the quality and intensity of light too influences shoot multiplication. The nature of gelling agent too plays an important role in shoot proliferation. Dave *et al.* (2003) observed that Phytagel™ at (0.2 per cent) gave a marginally better multiplication response in *Chlorophytum borivilianum* than BDH™ agar. In certain plant species, liquid medium has been found to be more responsive for shoot multiplication as compared to semi-solid medium. Use of floating membrane rafts for micropropagation of *Aconitum napellus* resulted in 45 per cent higher shoot proliferation than on semi-solid medium (Watad *et al.*, 1995).

Holostemma ada-kodien, an endangered medicinal shrub indigenous to India, was conventionally propagated through seeds and stem cuttings – systems that faced problems such as low seed viability, low germination frequencies and additionally the progenies were not true-to-type due to cross pollination. Therefore, micropropagation *via* axillary bud multiplication was adopted. Martin 2003 established efficient protocols for axillary bud multiplication and indirect organogenesis in *Holostemma ada-kodien* (Asclepiadaceae) using Murashige and Skoog (MS) medium supplemented with BAP and Indole butyric acid (IBA). Misra (1996) described a protocol for micropropagation of an industrially important aromatic patchouli (*Pogostemon cablin* Benth.) plant using apical meristems and nodal segments as well as through organogenesis.

Cephaelis ipecacuanha, a shrub, was conventionally propagated by seeds and cuttings and suffered from the common problems of decreased seed viability after collection, lengthy growth time, susceptibility to diseases etc. To overcome these issues, several efficient methods of micropropagation for this plant were developed, among which Isogai *et al.* (2008) described a method for rapid propagation of *C. ipecacuanha* by utilizing Gibberellic acid (GA_3). They reported internodal elongation of shoots cultured on medium supplemented with 0.5 or 1mg/L GA_3 and this method using GA_3 propagated numerous plants at a rate of more than 100 times compared to the method without GA_3.

Karthikeyan *et al.* (2009) performed axillary shoot proliferation of *Centella asiatica* Linn. a medicinal herb, through nodal explants for rapid clonal multiplication and reported that among the two cytokinins used, BAP was superior and more effective than KN in inducing shoot development as well as shoot multiplication from nodal explants.

Micropropagation and callus culture in *Cicer arietinum* L. and Lathyrus *sativus* L. was described by Misra *et al.* (1994a, b). This metod was extended by Misra and Misra (1993) for genetic transformation of lathyrus using Ri plasmids.

The single node method has had a good success with many different plants such as potato, *Vitis,* peas, rose, tomato, cucumber, aubergine, etc. Recently, this method has also been applied to *Eucalyptus grandis* and *Araucaria cunninghammii.*

The following points should be kept in mind when applying this method:

It is sometimes recommended that the shoot tip should be allowed to develop before the cytokinin concentration is increased to induce axillary shoot formation.

1. If axillary shoot development is not satisfactory then apical meristem is either removeded or killed.
2. Isolation by this method is practically impossible when dealing with rosette plants such as gerbera, and when the chances of infection are high.
3. To reduce the chances of infection it is best to isolate closed buds.
4. The rate of propagation is dependent on the number of buds available. The more leaves that are initiated with time, the more buds became available, and more rapid is the propagation.
5. Dormancy, especially of woody plants from temperate climates often presents problems.
6. Starting material, position, age, season of the year at the time of isolation, and physical factors are of great importance as is the composition of nutrient medium.
7. The cytokinin requirement is extremely variable and should be adjusted according to the plant species and particular cultivar in use. Most plants react very well to BA and often less to kinetin.
8. It has been demonstrated in wide range of species that thidiazuron and several substituted pyridylphenylurea compounds rather than classical cytokinins stimulate more axillary branching.
9. Liquid medium sometimes promotes axillary shoot formation.

III. Organogenesis

In plant tissue culture, organogenesis is a process of differentiation by which plant organs *viz.* roots, shoots, bud flowers, stem, etc. are formed. Simultaneously adventitious development of organs (roots, buds, shoots, flowers, etc.) or embryos (embryo-like structures) from explants lacking a preformed meristem. Whole plant regeneration via organogenesis is a monopolar structure. It develops procambial strands which establish a connection with the pre-existing vascular tissue dispersed within the callus or cultured explant.

This organogenesis process is much more common than somatic embryogenesis and has far more potential for mass clonal propagation of plants. Plant production through organogenesis can be achieved by two modes (Figure 29.4):

a) Organogenesis through callus formation with *de novo* origin.

b) Emergence of adventitious organs directly from the explant.

a) Organogenesis via Callus Formation

Plant regeneration from cultured explant involves the initiation of callus and then shoots bud differentiation. Meristematic tissues or organs like shoot tip, lateral bud, inflorescence, rachis, leaf, petiole, root should be selected in preference to other tissues because of their clonal properties, culture survival, growth rates and totipotency. Establishment of callus growth with subsequent organogenesis has been obtained from many species of plants and from explants *viz.* cotyledons, hypocotyl,

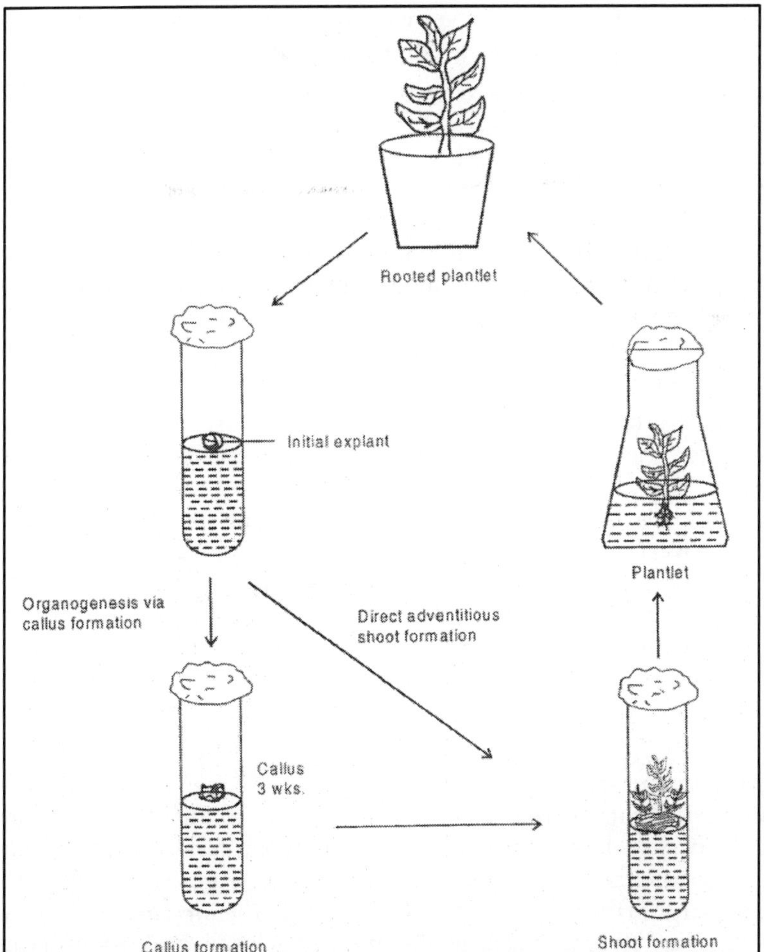

Figure 29.4: Protocol for Plant Regeneration via Organogenesis.

stem, leaf, shoot apex, root, young inflorescence, flower petals, petioles, embryos, etc. cultured *in vitro*. Explants from both mature and immature organs can be induced to form callus and then plant regeneration. The explants with mitotically active cells are generally good for callus initiation. Immature tissues and organs are invariably more morphogenetically plastic *in vitro* than mature tissues and organs. The size and shape of the explant is also crucial. Only a small proportion of the cells in a given explant contribute to the formation of callus.

The site of initiation of callus is generally at the peripheral surfaces of the inoculum or at the excised surface. Callus is produced on explants *in vitro* as a result of wounding and in response to hormones, either endogenous or exogenously supplied to the medium. Meristems shoot tips, axillary buds, immature leaf, and immature embryos in cereals are particularly suited as explants. The explants like mature leaves, roots, stems, petioles, and flower parts from herbaceous species can often be successfully cultured to initiate plantlets through organogenesis. The season of the year, donor conditions of the plant, the age and physiological state of the parent plant contribute to the success of organogenesis in cell cultures.

Various culture media are used for organogenesis which include MS, B_5, White's medium (White, 1963) and SH (Schenk and Hildebrandt, 1972). MS medium contains a high concentration of nitrogen as ammonium unlike other media. It may be difficult to transfer cells from White's, B_5 or SH medium to MS medium. Two modes of cell culture are generally used for organogenic path: (i) The cultivation of cell clusters on a semi-solid medium and (ii) The cultivation of cell suspensions in liquid medium. A suspension cell culture is usually initiated by placing friable callus into liquid culture medium. The suspension usually consists of free cells and aggregates of 2-100 cells. Suspension cultures should be subcultured at least once a week while callus culture should be subcultured after 3-4 weeks.

Growth regulator concentration in the culture medium is critical for morphogenesis. This was first indicated by Skoog and Miller (1957). Auxin at a moderate to high concentration is the hormone used to produce callus. Very often 2, 4-D a potent auxin is used alone to initiate callus. In some species a high concentration of auxin and a low concentration of cytokinin in the medium promote cell proliferation with the formation of callus. Cytokinins, if supplied are kinetin or benzyladenine. Callus tissue comprises a wide range of cell types and characteristically consists of irregularly differentiated, vacuolated cells interspersed with smaller more meristematic cells. The nature of any callus will depend on the explant tissue or tissues from which it has arisen and also on the composition of medium used to induce and maintain it.

Callus may be serially subcultured and grown for extended periods, but its composition and structure may change with time as certain cells are favored by the medium and come to dominate the culture. Under conditions favouring unorganized growth, the meristems in a callus are random and scattered. Transfer of callus pieces to conditions supporting organized growth leads to the formation of meristemoids, also termed as growth centers. The meristemoids are localized clusters of cambium—like cells which may become vascularized due to the appearance of tracheidal cells in the center. These are the sites of organ formation in the callus and can form shoots or roots. By varying the growth regulator levels and types, one can determine the route of morphogenesis *in vitro.*

Lowering the auxin and increasing the cytokinin concentration is traditionally performed to induce shoot organogenesis from callus. In tobacco, the presence of adenine or KN in the medium leads to the promotion of bud differentiation and development. Auxins inhibit bud formation. The next phase involves the induction of roots from the shoots developed. IAA or IBA auxins alone or in combination with a low concentration of cytokinin can induce root primordia. Organogenesis can be induced from either cell suspension or callus cultures. One needs to transfer only cells from callus medium to regeneration medium and then continued subculturing on regeneration medium. Thus organ formation is determined by quantitative interaction, *i.e.* ratios rather than absolute concentrations of substances participating in growth and development.

b) Direct or Adventitious Organ Formation

The somatic tissues of higher plants are capable, under certain conditions, of regenerating adventitious plants. Adventitious buds are those which arise directly from a plant organ or a piece thereof without an intervening callus phase. Induction of adventitious shoots directly on roots (blackberry, raspberry), leaves (*Begonia, Crassula, Peoeromia, and Saintpaulia*), bulb scales and other organs of intact plants is a common method of propagation. This method is suitable for herbaceous species. In *Begonia,* buds normally originate along the leaf. But in a medium containing BAP, the bud formation from the cut end of a leaf is so profuse that the entire surface of the cutting becomes covered with shoot buds (Reuther and Bhandari, 1981). Micropropagation techniques are available for large number of bulbous species *e.g.* Lilies (Capellades *et al.,* 1991). Propagation by adventitious bulblets formation directly from aseptically cultured bud-scale segments was described by Hackett (1969) Harikrishnan and Hariharan (1996) and Gupta *et al.* (1978). In the culture of the whole scale about 10 bulblets were formed within 15 days. As many as 100 bulblets could beproduced from a single scale. Promotion of bud formation is promoted by cytokinins. However, the requirement for exogenous auxin and cytokinin varies with the tissue system, apparently depending on the endogenous levels of the hormones in the tissue, and depending upon the explant taken, age of the plant and growing conditions.

A number of woody legumes have been successfully micropropagated using juvenile as well as mature explants. *Dalbergia sissoo* (Sissoo), an important timber yielding forest tree legume species of India, is conventionally propagated through seeds or root suckers. But these methods were not suitable for mass propagation because of low seed germination frequency and root suckers production depends on age of the trees. Pradhan *et al.* (1998) presented a protocol for micropropagation of *Dalbergia sissoo* (Sissoo), through *in vitro* shoot proliferation from cotyledonary nodes. A protocol to induce adventitious shoot organogenesis from semi-mature and mature cotyledons lacking the embryonic axis of *Dalbergia sissoo* is reported by Singh *et al.* (2002).

Hasbullah *et al.* (2008) studied the regeneration potential of *Gerbera jamesonii* through tissue culture techniques using leaf, petiole and root explants. They obtained adventitious shoots from petiole explants when cultured on MS medium containing BAP and NAA, and callus induction was best from leaf explants grown on MS medium supplemented with BAP and 2, 4-D. They reported petiole segments as the best explant for regeneration of this species.

IV. Somatic Embryogenesis

In somatic embryogenesis the embryos regenerate from somatic cells, tissues or organs either *de novo* or directly from the tissues (adventive origin), which is the opposite of zygotic or sexual embryogenesis. Various forms of somatic embryos are reported *viz.* adventive embryos (somatic embryos arising directly from other organs or embryos), parthenogenetic embryos (those formed by the unfertilized egg), and androgenetic embryos (formed by the male gametophyte). In sexual embryogenesis, the act of fertilization triggers the egg cell to develop into an embryo. However, it is not the monopoly of the egg to form an embryo. Any cell of the gametophytic (embryo-sac) or sporophytic tissue around the embryo-sac may give rise to an embryo. Cells of the nucellus or inner integument of members of Rutaceae family (*e.g.*, *Citrus*) may develop into embryos. The occurrence of asexual embryogenesis is generally restricted to intraovular tissues. However, in genera, somatic embryos are those which are formed from the somatic tissues like epidermis, parenchymatous cells of petioles or secondary root phloem *in vitro*. Somatic embryogenesis differs from organogenesis as the embryo being a bipolar structure with a closed radicular end rather than a monopolar structure. The embryo arises from a single cell and has no vascular connection with the maternal callus tissue or the cultured explant.

The initiation and development of embryos from somatic tissues of *Daucus carota* was first reported by Steward *et al.* (1958) and Reinert (1958, 1959). Embryos have also been obtained from generative cells of *Datura innoxia* microspores (Guha and Maheshwari (1964). The list of species from which somatic embryogenesis is reported is exhaustive. The members of Umbellifereae and Solanaceae, Leguminoseae and a range of other dicots along with monocots families' *viz.* Gramineae show the ability to generate somatic embryos. In this chapter embryogenesis will be restricted to sporophytic tissue; the discusssion on androgenesis is dealt separately. Sharp *et al.* (1980) described two routes to somatic embryogenesis.

a. Direct Embryogenesis

The embryos initiate directly from the explant tissue without the intervening callus. This occurs through 'Pre-Embryogenic Determined Cells' (PEDC) where the cells are committed to develop embryos and needs potentiation. Such PEDCs are found in embryonic tissues (*e.g.*, scutellum of cereals), and in certain tissues of young *in vitro* grown plantlets (*e.g.*, hypocotyl in *Daucus carota, Ranunculus scleratus, Linum usitatissimum, Brassica napus*), nucellus and embryo-sac (within ovules of mature plants). Thormatie embryo development is triggered by transferring the PEDCs to a low auxin containing medium or only to the basal medium. Subsequent conversion of somatic embryos into plantlets occurred mostly in hormone free media or in specific plants on media supplemented with GA_3, BAP and IBA in *Psoralea corylifolia, Eryngium foetidum* and *Aconitum carmichaeli* (Chand and Sahrawat, 2002; Ignacimuthu *et al.*, 1999; Hatano *et al.*, 1987).

b. Indirect Embryogenesis

Indirect embryogenesis involves an intermittent callus formation process from which the somatic embryos develop. The embryogenically determined cells in the callus form embryos and as these are induced to do so, they are also called 'Induced Embryogenic Determined Cells' (IEDC), *e.g.*, secondary phloem of carrot, leaf tissues of coffee, *Petunia, Asparagus,* etc. Specific growth regulator concentrations and/or cultural conditions are required for these stages in somatic embryogenesis. A diagrammatic representation of somatic embryogenesis in carrot is shown in Figure 29.5. Somatic embryos arise from single cells located within clusters of meristematic cells either in the callus mass or in suspension. Such cells develop into proembryos, with polarity following a pattern that tends to mimic the general pattern associated with the development of embryos in the ovule. Proembryo initials may be single cells or multicellular groups.

When the conditions are suitable these embryos germinate to produce plantlets. For some species any part of the plant body serves as an explant for embryogenesis (*e.g.* carrot), whereas in some species only certain regions of the plant body may respond in culture *e.g.*, Cereals. Floral or reproductive tissue in general has proven to be an excellent source of embryogenic material. The physiological state of the plant

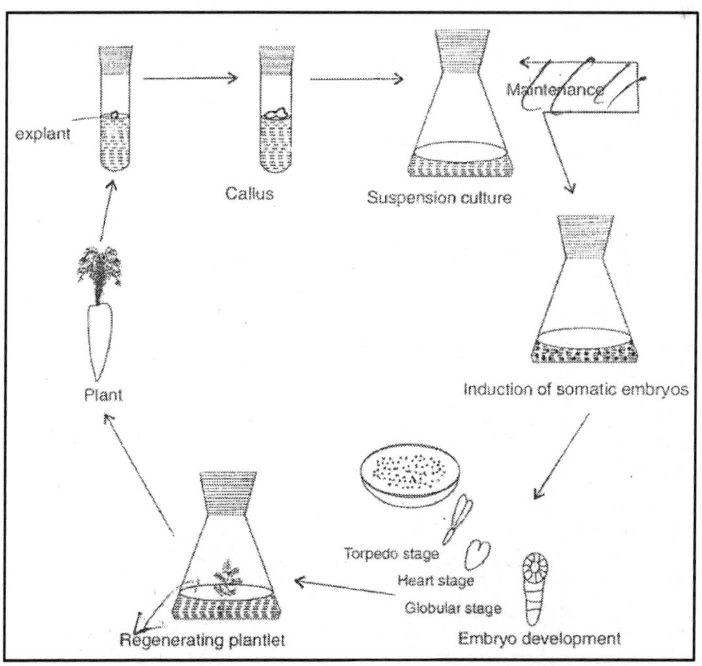

Figure 29.5: Protocol for Somatic Embryogenesis in Carrot.

from which the explant is taken is also extremely important, as is the season during which the material is obtained. Equally important is the sequence of media and especially the growth regulators.

Somatic embryos have been grown on a range of media from the dilute White's medium to very high salt MS medium, but the latter has been extensively employed. The addition of reduced nitrogen in the medium helps in both embryo initiation and maturation. The sources of reduced nitrogen have already been explained. Of all the amino acids, L-glutamine seems to play a special role. Another factor is the chelated form of iron in the media. In the absence of iron, embryo development fails to pass from the globular to the heart-shaped stage.

Growth regulators in the medium, especially auxin or auxin in combination with cytokinin appear essential for the onset of growth and the induction of embryogenesis. Of all the auxins, 2, 4-D followed by NAA has proven to be extremely useful. Effective concentration ranges are 0.5-27.6 µM for 2, 4-D and 0.5-10.7 µM for NAA. The auxin for the primary (callus initiation) and secondary media (embryo development) may be same or different. One auxin or several may be used in the same medium. Cytokinins have been important in a number of species. Cytokinins have been used in the primary medium invariably during embryogenesis of crop plants. The effective concentration range for kinetin is 0.5-5.0 µM. Cytokinins are important in fostering somatic embryo maturation and especially cotyledon development. Cytokinins are sometimes required for growth of embryos into plantlets. Gibberellins are rarely incorporated in primary culture media.

However, they have proven useful in fostering embryo maturation or in stimulation of rooting and subsequent growth of plants in a number of cases. The role of growth inhibitor ABA in somatic embryogenesis has emerged when added at non-inhibitory levels (0.1-1.0 µM). ABA promotes somatic embryo development and maturation and at the same time inhibits abnormal proliferation and initiation of accessory adventitious embryos. ABA, antiauxins (5-hydroxynitrobenzyl bromide, 7-azaindoie, 2, 4, 6-trichlorophenoxy acetic acid, p-chlorophenoisobutyric aoid) and other growth inhibitors may serve to promote somatic embryo maturation by countering the effects of growth promoters. Their addition to culture media may permit somatic maturation to proceed under conditions when it normally would not occur. The addition of activated charcoal to the medium has proved to be useful for somatic embryo development. Charcoal media shows lower levels of phenylacetic acid and p-OH benzoic acid compounds, which inhibit somatic embryogenesis. Also, it absorbs 5-hydroxymethyl furfural, an inhibitor formed by sucrose degradation during autoclaving. Environmental conditions of light, temperature, and density of embryogenic cells in medium are important. Regarding culture vessel, the position of embryos (floating or submerged) and the physical state of the medium (semisolid or liquid) have little effect.

Somatic embryogenesis as a means of propagation is seldom used because of:

i) High probability of mutation.

ii) Methodological difficulty.

iii) Loss of regenerative capacity with repeated subcultures.

iv) Induction of embryogenesis is often very difficult or impossible with many recalcitrant plants.

v) Occurrence of somatic embryo dormancy.

Stage-3: Rooting of *In vitro* Formed Shoots

Numerous studies have indicated that NAA followed by IBA, IAA, 2, 4-D and other auxins are used for induction of root regeneration. In many species addition of auxin is unnecessary because of its production in young shoots. Also, high level of auxin will induce callus formation. Sometimes if roots are unable to initiate in such high salt media irrespective of the hormone present, then the salt concentration in the medium should be lowered to one-half, one-third, or one-fourth of the standard strength. Riboflavin is reported to improve the quality of root system in *Eucalyptus ficifolia*. Similarly, phloroglucinol has been observed to promote rooting in a number of rosaceous trees, but its role is yet to be established equivocally. IBA in *Artemisia judaica* (Liu *et al.*, 2003), NAA in *Rotula aquatica* (Martin, 2003), and, IAA in *Tylophora indica* (Chaudhuri *et al.*, 2004) is reported to be beneficial for *in vitro* rooting of regenerated shoots,. It is a common practice to transfer shoots from a high to low strength media in order to induce rooting. The rationale behind the favourable effect of reduced macronutrient concentration is that the concentration of nitrogen ions needed for root formation is much lower than for shoot formation and growth (Driver and Suttle, 1987).

In some plant systems, it is possible to treat the *in vitro* formed shoots as micro cuttings and induce the rooting under natural conditions. The small cuttings are transferred directly to a potting mix. The basal cut end is dipped in rooting powder and maintained under conditions of high humidity. This method is becoming increasingly popular as it saves a step in aseptic manipulation thus reducing the cost of plant production. Also, it is much simpler to pull out a shoot cluster from the media rather than a rooted plant (where roots are embedded in the media). In most plant species, *in vitro* formed roots die after sometime in the potting mix and new set of roots are formed. Thus it is not only economical but also the quality of roots is far superior.

Sucrose concentration also have an effect on rooting, as reported in apple that root initiation decreased proportionately with decreasing sucrose level (Lane, 1978) and shoots grown in the absence of sucrose did not survive during acclimatization (Zimmerman, 1993). Ray and Bhattacharya (2008) treated *Eclipta alba* microshoots with chlorocholine chloride and observed that it resulted in increased number of roots, increase in plant biomass and made the plants sturdier and more suitable for acclimatization.

Stage-4: Hardening and Transplantation

A micropropagation system becomes successful only when the *in vitro* raised plants are able to survive and grow vigorously when transferred to soil. The plantlets under *in vitro* are subjected to minimal stress in the controlled growth condition and so a large number of micropropagated plants do not survive the transfer from *in vitro* conditions to the variable and stressful field conditions. The process of gradually preparing the plants to survive under field conditions is called 'acclimatization'. The plants produced in tissue culture although green, are heterotrophic. Oliveira and Paiva (2008) studied the effect of cytokinins on *in vitro* development of autotrophism and acclimatization of *Annona glabra* L., a tropical fruit tree and found that chlorophyll and carotenoid content, photosynthesis and chloroplast differentiation were greater in plants cultivated in presence of either KN or BAP. They also reported that during *in vitro* culture of *A. glabra* L., the use of KN and BAP favoured the development and survival of plants during acclimatization. But, since the humidity inside the culture vessel is high, cuticle is lacking or insufficiently developed in tissue culture propagated plants. Therefore, immediately after transfer, plants are maintained under high humidity (above 90 per cent) which is gradually reduced over a period of 6 to 8 weeks. During this time, the plants attain the ability to synthesize food and develop the cuticular covering. Potting mixes also play an important role in determining the survival percentage of tissue-cultured plants at the hardening stage. Sahoo and Chand (1998) used four different substrates (vermiculite, vermi-compost, and soilrite mix and garden soil) separately for hardening of *Vitex negundo* plantlets inside the environmentally controlled growth chamber and observed the percentage survival to be highest (93 per cent) in vermi-compost. Plants are maintained under shade for 2-3 weeks and are then ready to be transferred to the field.

Applications and Limitations

Application

In vitro micropropagation techniques have wide applications and are now highly preferred to conventional practices of asexual propagation because of the following advantages:

1. A small amount of plant tissue is needed as the initial explant for regeneration of millions of clonal plants.
2. Many plant species are highly resistant to conventional bulk propagation practices, known as 'recalcitrant' plants are amenable to micopropagation techniques. Micropropagation helps in bulking up of new cultivars of important trees that would otherwise take many years to bulk up by conventional methods.
3. The *in vitro* techniques provide a method for speedy international exchange of disease or virus free plant materials and quarantine.
4. Micropropagated stocks provide a year round nursery for ornamental, fruit and tree species.
5. A major limiting factor in some of the crops is the sexual incompatibility, incomplete embryo development and so seed production leading to genetic erosion. In such cases micropropagation (axillary bud proliferation method) can be used. For example, production of F, hybrid seed lines in crops like cauliflower where individual parent clones can be bulked for the production of more uniform seed, the production of male sterile lines in crops like

onion to provide an alternative to difficult backcrossing methods, and the production of *Asparagus* for producing high quality supermale and female homozygous lines from which desirable all male hybrids can be produced.

6. The artificial or synthetic seeds (*i.e.* encapsulated embryos) produced by somatic embryogenesis is analog to true seeds, which is based on the similarity of SE with zygotic embryo with respect to gross morphology, physiology and biochemistry.
7. Plant breeding programmes rely heavily on the germplasm. Preservation or conservation of germplasm is a means to assure the availability of genetic materials as the need arises. Most seeds and vegetative organs have a limited storage life; research on germplasm preservation has concentrated on the development of procedures to extend usable life spans. Since meristem cells are highly cytoplasmic and non-vacuolated, a high percentage of cells can be expected to survive cryopreservation procedures.
8. Shoot multiplication can be achieved in a small space because small shoots are multiplied in test tubes or culture vessels.
9. Propagation is carried out under sterile conditions, therefore, no losses should occur from pests or diseases, and the plantlets finally produced should be free from bacteria, fungi and nematodes.
10. Reproducibly high rates of propagation can be achieved inder controlled culture conditions.
11. Labour economy: No attention is required between sub cultures. (*e.g.* watering, weeding, spraying). Mechanization may be possible to automate propagation procedures for some species and so reduce labor requirements.
12. Less expensive glass house space is required.
13. Stock plants can be stored *in vitro*.
14. Plants may acquire some transient characteristic through micropropagation which makes them more commercially desirable than the conventionally propagated plants as reported in case of *Fragaria sp.* that *in vitro* raised strawberry plants produced more runners, more leaves and inflorescences in comparison to vegetatively propagated plants.
15. Micropropagation from the female plant is highly desirable in case of dioecious plants to ensure fruit production as in in case of papaya.

Limitations

In spite of immense progress in the field of micropropagation, there lies limitations that hinders the utilization of the full potential of this technique. The process is still laborious, slow and its use is restricted in many crops due to its relatively high production costs.

1. The process is cost intensive because of expensive and sophisticated facilities *e.g.* culture rooms, laminar flow hoods etc., and also needs highly skilled trained manpower is required.

2. There is frequent chance of contamination by microbes and pathogens, leading to heavy losses.
3. The problem of genetic infedility of callus cultures due to chimeral breakdown, aberrant cell division in callus or cell suspension, epigenetic effects and pre-existing genetic variability.
4. Vitrification or hyperhydration of leaves in the regenerating shoots due to morphological, physiological and metabolic rearrangements occurring during intensive shoot multiplication in culture is one of the causes for high rate of plantlet mortality. Some preventive measures to reduce hyperhydration includes:
   (i) Increasing concentration of agar;
   (ii) Overlaying medium with paraffin;
   (iii) Using desiccant such as $CuSO_4$;
   (iv) Bottom cooling of the culture vial to improve aeration;
   (v) Lowering cytokinin level or replacing one kind of cytokinin by another;
   (vi) Manipulating NH_4^+ or salt concentrations in the culture medium.
5. Specific methods for efficient micropropagation may need to be developed for each species, including the conditions for rooting and plantlet establishment.
6. Browning of the medium is very common because of the leaching of phenols, more pronounced in case of tree explants, which became inhibitory for the grow of tissues. It can be neutralized with the application of antioxidants.
7. Environmental stresses causing death of a large number of plants during acclimatization or transferring *in vitro* grown plant (lets) to *ex vitro* conditions might lead to sever mortality rate.

Conclusion

Successful *in vitro* propagation of large numbers of plants, namely ornamentals, fruits, crops, vegetables, aromatic and medicinal plants etc. is achieved. Annually, millions of ornamental plants are routinely produced *in vitro*. Ornamental plants are produced mainly for their aesthetic value, thus the propagation and improvement of quality attributes such as leaf types, flower colour, longevity and form, plant shape and architecture, and the creation of novel variation are important economic goals for floriculturists. There are many commercial laboratories and National Institutes worldwide use *in vitro* culture system for rapid plant multiplication, germplasm conservation, elimination of pathogens, genetic manipulations, and for secondary metabolite production (O'Riordain, 1999). The great potential of micropropagation for large-scale plant multiplication can be tapped by cutting down the cost of production per plant by applying low-cost tissue culture methods, without compromising the quality.

References

Ammirato, P.V. (1983). Embryogenesis. In: D.A. Evans *et al.* (Eds.), *Handbook of Plant Cell Culture.* Vol. 1. Macmillan, New.York. pp. 82-123.

Bajaj YPS (1988). Regeneration of plants from apical meristem tips of some Legumes *Curr. Sci.,* **48**: 906- 907.

Bajaj, Y.P.S.(1986). Biotechnology of tree improvement for rapid propation and Biomass Energy production. In: Biotechnology in Agriculture and Forestry I. (Ed. Y.P.S. Bajaja), *Springer Verlag,* New York, pp. 1-23.

Balachandran S.M., Bhat S.R. and Chandel K.P.S (1990): *In Vitro* Clonal multiplication of turmeric (*Curcuma spp.*) and ginger (*Zingiber officinale*). *Plant Cell Rep.,* **8**: 521-524.

Baskaran, P. and Jayabalan, N. 2005. An efficient micropropagation system for *Eclipta alba*-A valuable medicinal herb. *In Vitro Cellular and Developmental Biology-Plant,* **41**(4): 532-539.

Borthakur, M., Hazarika J., Singh. R. S. (1998) A Protocol for Micropopagation of *Alpinia galangal. Plant Cell Tiss. Organ Cult.,* **55:** 231-233.

Bryan, C.P. and Dodds J.H. (1992)**:** *In Vitro* Micropropagation of potato plants, *J. Plant Cell,* **68:** 259-284.

Capellades, M.Q., Beruto, M., vanderschaeghe, A. and Deberg, P.C. (1991). Ornamentals. In P.C. Deberg and R.H. Zimmerman (Eds.), Micropropagation: Technology and application. Kluwar, Dordrecht, pp. 215-229.

Chand, S. Sahrawat, A. K. (2002). Somatic embryogenesis and plant regeneration from root segments of Psoralea corylifolia L. an endangered medicinally important plant. *In Vitro Cellular and Developmental Biology-Plant,* **38**(1): 33-38.

Chaudhuri, K.N. Ghosh, B. and Jha, S. (2004). The root: a potential; new source of competent cells for high-frequency regeneration in *Tylophora indica*. *Plant Cell Rep.* **22**: 731-740.

Cieslinska, M. (2002). Elimination of Apple chlorotic leaf spot virus (ACLSV) from pear by in vitro thermotherapy and chemotherapy. *Acta Hortic.,* **596**: 481–484.

Dave, A. Bilochi, G. and Purohit, S.D. (2003). Scaling-up production and field performance of micropropagated medicinal herb 'Safed Musli'(*Chlorophytum borivilianum*). *In Vitro* Cellular and Developmental Biology-Plant, **39(4)**: 419-424.

Debergh P.C. and Maene L.J. (1981). A scheme for commercial propagation of ornamental plants by tissue culture. *Sci. Hortic.,* **14**: 335-345.

Driver, J.A. Suttle, G.R.L. (1987). Nursery handling of propagules. Forestry Sciences, Martinus Nijhoff Publishers, (Netherlands), **25**: 320-335.

Gangopadhyay, G., Bandyopadhyay, T., Modak. B.K., Wongpornchai, S. and Mukherjee, K.K. (2004). Micropropagation of Indian Pandan (*Pandanus amaryllifolius* Roxb.), a rich source of 2-acetyle-l-pyrroline. *Curr.Sci.,* **87**: 1589-1592.

Gautheret, R.J. (1959). La culture des tissus vegetaux: Techniques et realisations. Masson, Paris.

Gella, R. and Errea, P. (1998). Application of *in vitro* therapy for ilarvirus elimination in three *Prunus* species. *J. Phytopathol,* **146**: 445–449.

George, M.M. and Morel, R.B. (1999). High frequency regeneration of *Vetiveria zizanioides* L. via mesocotyl culture. *Phytomorphology,* **49**: 309-313.

Guha, S. and Maheshwari, S.C. (1964). *In vitro* production of embryos from anthers of *Datura. Nature,* **204**: 497.

Gupta, P.P., Sharma, A.K. And Chaturvedi, H.C. (1978). Multiplication of *Lilium longiflorum* Thunb. by aseptic culture of bud scales and their segments. *Indian J. Exp. Biol.,* **16**: 940-942.

Hackett, W.P. (1969). Control of bulblet formation on bud scale of *Lilium longiflorum. Hort. Sci.,* **4**: 4-171.

Harikrishnan, K.N. and Hariharan, M. (1996). Direct shoot regeneration from nodal explants of *Plumbago rosea* Linn. – a medicinal plant. *Phytomorphology.,* **46**: 53-63.

Hasbullah, N.A. Taha, R. M. and Awal, A. (2008). Growth optimization and organogenesis of Gerbera jamesonii Bolus ex. Hook f. *In Vitro. Pakistan Journal of Biological Science,* **11** (11): 1449-1454.

Hatano, K. Shoyama, Y. Nishioka, I. (1987). Somatic embryogenesis and plant regeneration from the anther of *Aconitum carmichaeli* Debx. *Plant Cell Reports,* **6**(6): 446-448.

Ignacimuthu, S. Arockiasamy, S. and Antonysamy, M. (1999). Plant regeneration through somatic embryogenesis from mature leaf explants of *Eryngium foetidum,* a condiment. *Plant Cell, Tissue and Organ Culture,* **56**(2): 131-137.

Isogai, S. Touno, K. and Shimomura, K. (2008). Gibberellic acid improved shoot multiplication in Cephaelis ipecacuanha. *In Vitro Cell. Dev. Biol. Plant.,* **44**: 216-220.

Joshi, M. and Dhar, U. (2003). *In vitro* propagation of Saussurea obvallata (DC.) Edgew.- an endangered ethnoreligious medicinal herb of Himalaya endangered ethnoreligious. *Plant Cell Reports,* **21**(10): 933-939.

Karesova, R. Janeckova, M. and Paprstein, F. (2002). Elimination of Raspberry bushy dwarf virus from raspberry cv. 'Gatineau'. *Acta Hortic,* **585**: 359–362.

Karthikeyan, K. Chandran, C. Kulothungan, S. (2009). Rapid clonal multiplication through in vitro axillary shoot proliferation of *Centella asiatica* L. *Indian Journal of Biotechnology*. **8**: 232-235.

Kaur, K. Verma, B. Kant, U.(1998). Plants obtained from the Khair tree (*Acacia catechu* Willd.) using mature nodal segments. *Plant Cell Reports,* **17**(5): 427-429.

Knapp, E. Hanzer, V. and Weiss, H. da Ca^mara Machado, A. Weiss, B. Wang, Q. Katinger, H. Laimer da Ca^mara Machado, M. (1995). New aspects of virus elimination in fruit trees. *Acta Hortic,* **386**: 409–418.

Krishnan, P.N. Sudha, C.G. and Seeni, S. (1995). Rapid propagation through shoot tip culture of *Trichopus zeylanicus* Gaertn. a rare ethnomedicinal plant, a rare. *Plant Cell Reports,* **14**(11): 708-711.

Kumar, R. Sharma, K. and Agrawal, V. (2005). In: *In Vitro Cellular and Developmental Biology-Plant,* **41**(2): 137-144

Lane, W.D. (1978). Regeneration of apple plants from shoot meristem tips. *Plant Sci. Lett.,* **13**: 281-285.

Liu, C.Z. SJ Murch, S.J. and El-Demerdash, M. Saxena, P.K. (2003). Regeneration of the Egyptian medicinal plant *Artemisia judaica* L. *Plant Cell Reports,* **21**(6): 525-530.

Martin, K. P (2003): Clonal propagation, encapsulation and reintroduction of *Ipsea malabarica* (Reichb. f.) J. D. Hook., an endangered orchid. Biomedical and Life Sciences. *In Vitro Cellular and Developmental Biology - Plant.,* **39**(3): 322-326.

Misra, M. (1996). *In Vitro* Micropropagation of *Pogostemon cablin* Benth. through callus culture. *Plant Cell Reports,* **15**: 991-994.

Misra, M. and Misra, A. N. (1993). Genetic transformation of grass pea. In: DAE *Symposium on Photosynth. and Plant Molecular Biology*, BRNS/DAE, Govt. of India, May 1993, pp. 246-251.

Misra, M., Addis, G. and Narayan, R. K. J. (1994b). Method for callus induction and differentiation of *Lathyrus sativus* and embryo rescue in interspecific crosses by Tissue culture. *J.Univ. Wales Agric. Soc.,* **74**: 129-144.

Misra, M., Misra, A. N. and Patnaik, S. N. (1994a). *In Vitro* clonal propagation of chick pea (*Cicer arietinum* L.) through callus culture. In: *Advances in Plant Tissue Culture in India*, Ed. Tandon P. Pragati Prakashan, Meerut, India.

Morel, G. and Martin, C. (1952). Virus free Dahlia through meristem culture. C.R. *Acad. Sci.,* **235**: 1324-1325.

Murashige, T. (1974). Plant propagation through Tissue cultures. Ann. Rev. *Plant Physiol.* **25:** 135-166.

Normah, M.N., Hamidah S, Ghani, F.D. (1997). Micropropagation of Citrus halmii– an endangered species of S.E. Asia. *Plant Cell Tissue Org. Cult.,* **50**: 225-227.

Oliveira, L.M. Paiva, R. Santana, J.R.F. Alves, E. Noguera, R.C. and Pereira, F.D. (2008).Effect of cytokinins on *In Vitro* development of autotrophism and acclimatization of *Annona glabra* L. *In vitro Cellular and Developmental Biology-Plant,* **44**: 128-135.

O'Riordain, F. (1999). Directory of European plant tissue culture laboratories, 1996–97. COST Action, vol. 822. Brussels: Commission of the European Communities; 1999.

Pattnaik, S.K. and Chan, P.K. (1996). *In Vitro* propagation of the medicinal herbs *Ocimum americanum* L. and *Ocimum sanctum* L. *Plant Cell Rep.,* **15**: 846-850.

Pradhan, C. Kar, S. Pattnaik, S. Chan, P.K. (1998). Propagation of *Dalbergia sissoo* Roxb. through *In Vitro* shoot proliferation from cotyledonary nodes. *Plant Cell Reports.,* **18**: 122-126.

Reinert, J. (1958). Morphogeneses and in hre, kontrolle angewebekulturenous carotten. Natusuissen schaften, **45:** 344-345.

Reinert, J. (1959). Uber die kontrolle der morphogenese and die induction Von Advientive embryonen and Gewebekuluren aus kartten". *Planta,* **58**: 318-333.

Reuther, G. and Bhandari, N.N.(1981). Organogenesis ans histogenesis of adventitious organs induced on leaf blade segme t of Begonia-Elatior-Hybrids (*Begonia x hiemalis*) in Tissue culture. *Gartenbuwissenschft.,* **46:** 241-249.

Roy, A.T. and Bhattacharya. (2008). Cell suspension culture and plant regeneration in latex producing plant, *Calotropis gigantean* (Linn.) R.Br. *Plant Cell Tiss. Org. Cult.,* **63**: 15-22.

Sahoo, Y. Chand, P.K. (1998). Micropropagation of *Vitex negundo* L., a woody aromatic medicinal shrub, through high-frequency axillary shoot proliferation. *Plant Cell Reports,* **18**(3-4): 301-307.

Schenk, R.U. and Hildebrandt, A.C. (1972). Medium and techniques for induction and growth of monocotyledonous and dicotyledonous plant cell cultures. Can. *J. Bot.,* **50**: 199-204.

Sharp, W.R., Sondahl, M.R., Caldas, L. S. and Maraffa, S. B. (1980). In: Horticultural Reviews Vol. 2. (Janick, J. ed.), AVI Publishing Co, Westport, Conn, USA, p. 268.

Singh, A.K. Chand, S. Pattnaik, S. and Chand, P.K. 2002. Adventitious shoot organogenesis and plant regeneration from cotyledons of *Dalbergia sissoo* Roxb., a timber yielding legume. *Plant Cell, Tissue and Organ Culture,* **68**: 203-209.

Skoog, F. and Miller, C.O. (1957). Chemical regula';On of growth and organ formation in plant tissues cultivated in vitro. In: Biological Action of Growth Substances. Symp. *Soc. Exp. Biol.,* **11**: 118-131.

Steward, F.C., Mapes, M.O. and Mears, K. (1958). Growth and organized development of cultured cells. II. Organization in cultures grown from freely suspended cells. *Am. J. Bot.,* **45**: 705-708.

Sudha, G.G. and Seeni, S. (1996). *In Vitro* propagation of *Rauwolfia micrantha,* a rare medicinal plant. *Plant Cell, Tissue and Organ Culture,* **44**(3): 243-248.

Thakur, R. Rao, P.S.and Bapat, V.A.(1998). *In Vitro* plant regeneration in *Melia azedarach* L. *Plant Cell Reports,* **18**(1-2): 127-131.

Tiwari, V., B.D. Singh and K.N. Tiwari, (1998). Shoot regeneration and somatic embryogenesis from different explants of Brahmi [*Bacopa monniera* (L.)Wettst]. *Plant Cell Reports,* **17**: 538-543.

Watad, A.A.Kochba, M. Nissim, A. and Gaba, V. (1995). Improvement of Aconitum napellus micropropagation by liquid culture on floating membrane rafts. *Plant Cell Reports,* **14**(6): 345-348

Wawrosch, C. Malla, P.R. and Kopp, B. (2001). Clonal propagation of *Lilium nepalense* D.Don, a threatened medicinal plant of Nepal. *Plant Cell Reports,* **20**(4): 285-288.

White, P.R. (1963). The Cultivation of Animal and Plant Cells. Ronald Press Co., New York.

Zimmerman, J. L. (1993). Somatic Embryogenesis: A Model for Early Development in Higher Plants. *Plant Cell,* **5:** 1411-1423.

2015, Major Constrains and Verdict of Crop Productivity Pages 414–420
Editor: Dr. U.N. Bhale
Published by: ASTRAL INTERNATIONAL (P) LIMITED, NEW DELHI

Chapter 30

Efficacy of Acetone Neem Plant Parts Extract on Mortality of *Meloidogyne incognita*, Chitwod Juveniles

Naresh Nayak and J.L. Sharma*

P.G. Department of Botany,
Govt. College, Kota – 324 001, Rajasthan

ABSTRACT

The Root-knot nematode Meloidogyne incognita is a serious problem for the cultivation of agricultural crop throughout the world. Natural products are the alternative approaches to control root knot nematode. An experiment was carried out in vitro to test the Neem plant parts acetone extract. Different concentration of shade-dried bark leaves and seeds of Neem were tested for mortality of Meloidogyne incognita juveniles increase with the concentration and exposure period resulted in increased mortality rate of juveniles. Acetone Neem seed (5 ml concentration) proved to be the most effective among used plant parts viz. bark, leaves, and seed. Neem bark was the least effective among the three.

Keywords: *Meloidogyne incognita, Acetone, Plant extract, Mortality.*

Introduction

The root-knot nematode *Meloidogyne incognita* is widely distributed in India and cause severe damage to vegetable crops. Use of chemical nematicides to control

* *Corresponding Author.* E-mail: nayak.nema2009@gmail.com

nematodes always posses costly and human health hazards. As an alternative to chemical pesticides specially for the purpose of protecting crops against nematode and also for the conservation of biodiversity, botanicals may stand as the most promising source of bioactive products of plant origin. Extract of plant product contain nematicidal compounds. Thousand of plants possessing insecticidal properties are known today (Banerji *et al.*, 1985). Use of neem and neem product has been advocated for the management of root-knot nematode by many workers (Dekha and Rehman, 1998; Jain and Gupta, 1998). Plant extracts having the nematicidal properties and they not only reduce the *nematode* population but also enhance the plant growth (Hussain *et al.*, 2011). The present investigation to study the nematicidal effect of different plant parts acetone extracts of neem on mortality of *M. incognita* juveniles.

Materials and Methods

Preparation of Extracts

The Neem (*Azadirachta indica*) plant parts *viz.* neem bark, neem leaves, neem seeds were shade dried and then oven dry at 60°C for overnight, powder of plant parts were prepared by using a mixer or blander. Five gm powder of neem bark, neem leaves and neem seeds were soaked in 50 ml acetone for 7-9 days for evapouratation after that it was washed with distilled water and absolute organic solvents. Extract were filtered through four ply muslin cloth and then passed through Whatman filter paper no.1 filtered extract were make up 50 ml with required amount of distilled water and then centrifuged at 4000 rpm for 10 minutes then again filtered through Whatman filter paper no.1. The extract so obtained was put in conical flasks and steam sterilized in an autoclave. The plant part extract *viz.* neem bark, neem leaves and neem seeds designated as stock solution for preparing different dilution *viz.* (5 ml, 2.5ml, 2 ml, 1.5 ml, 1 ml,.50 ml,.25 ml) of each neem plant part extracts were prepared by adding the required amount of distilled water. The sterilized distilled water served as control.

Effect of Acetone Extracts on Juvenile Mortality of *M. incognita*

Seven dilution *viz.* (5 ml, 2.5ml, 2 ml, 1.5 ml, 1 ml, 0.50 ml, 0.25 ml) were prepared from stock solution. Five ml of each plant parts acetone extracts were prepared by adding the required amount of distilled water were taken in sterile beaker of 50 ml capacity. Two hundred freshly hatched second stage juveniles (J_2) of *M. incognita* were transferred in beaker. Equal number of J_2 was also transferred to separate beaker containing sterilized water to serve as control. Three replicated were taken completely randomized design (CRD). Observations of both live and dead nematodes were made at 24 hrs, 48 hrs, 72 hrs, 96 hrs and 120 hrs after inoculation.

Results and Discussion

Result present in Tables 30.1–30.3 revealed the maximum mortality was recorded in 5ml concentration which increased with increase in acetone extract con. More number or percentage of juveniles mortality with increase in exposure period. Minimum mortality per cent was observed in Neem bark acetone extract (Table 30.1

Table 30.1: Effect of Neem Bark Extract (Acetone) on Juveniles Mortality of *Meloidogyne incognita* at different Time Intervals

Extract Con.	*Per cent Mortality at different Exposure Period*				
	24 hrs.	*48 hrs.*	*72 hrs.*	*96 hrs.*	*120 hrs.*
0.25 ml	0.000	11.333	16.000	17.333	21.000
0.50 ml	4.333	15.667	19.667	25.667	31.333
1 ml	12.333	22.000	37.333	44.000	51.000
1.5 ml	20.667	38.333	54.000	57.667	70.000
2 ml	33.000	38.667	57.000	62.333	75.333
2.5 ml	37.333	54.000	59.000	78.000	83.667
5 ml	75.667	86.667	88.000	91.667	93.000
Control	0.000	0.000	0.000	0.000	0.000
SEM +	2.560	2.014	5.265	1.599	1.500
C.D. at 5%	5.904	4.644	12.142	3.686	3.459
C.D. at 1%	8.590	6.756	17.665	5.363	5.033

Observations are mean of three replicates.

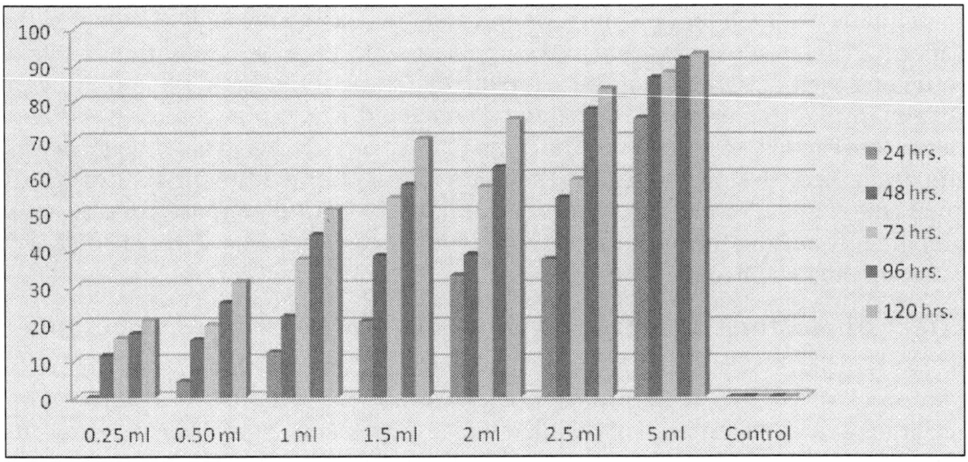

Figure 30.1: Effect of Neem Bark Extract (Acetone) on Juveniles Mortality of *Meloidogyne incognita* at different Time Intervals.

and Figure 30.1) there was 75.66 per cent mortality at its highest 5ml con. and 24 hrs exposure period. Only 21 per cent mortality recorded at its lowest conc (0.25ml) after 120 hrs of exposure period. Highest conc of Neem leaf extract (5ml) revealed that cent per cent mortality after 72 hrs of exposure period and 40.33 per cent mortality recorded at it lowest con. (0.25ml) after 120 hrs of exposure period (Table 30.2 and Figure 30.2). Neem seed acetone extract on the other hand was found to be the most effective in juvenile's mortality (Table 30.3 and Figure 30.3), the highest con. (5ml) were found to

Table 30.2: Effect of Neem Leaf Extract (Acetone) on Juveniles Mortality of *Meloidogyne incognita* at different Time Intervals

Extract Con.	*Per cent Mortality at different Exposure Period*				
	24 hrs.	*48 hrs.*	*72 hrs.*	*96 hrs.*	*120 hrs.*
0.25 ml	3.000	14.000	23.000	31.000	40.333
0.50 ml	7.000	19.000	27.333	36.667	49.000
1 ml	13.667	29.000	40.667	49.333	59.333
1.5 ml	25.667	43.333	53.667	59.000	71.667
2 ml	46.667	56.667	70.667	81.000	89.333
2.5 ml	73.667	85.000	92.000	95.333	100.000
5 ml	94.667	98.333	99.333	100.000	100.000
Control	0.000	0.000	0.000	0.000	0.000
SEM ±	1.462	1.848	1.247	1.155	1.130
C.D. at 5%	3.373	4.262	2.876	2.663	2.607
C.D. at 1%	4.907	6.201	4.184	3.874	3.792

Observations are mean of three replicates.

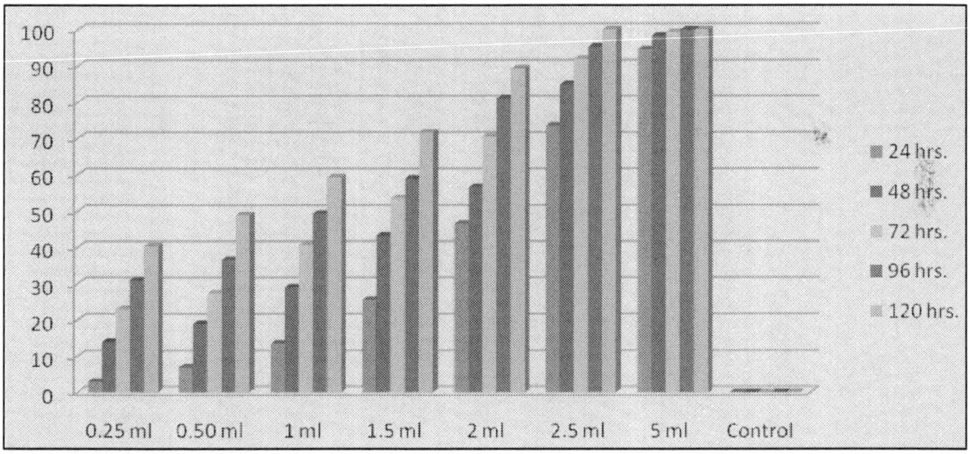

Figure 30.2: Effect of Neem Leaf Extract (Acetone) on Juveniles Mortality of *Meloidogyne incognita* at different Time Intervals.

be most effective caused cent per cent mortality from 24 hrs exposure and its lowest conc had only 52.66 per cent mortality after 120 hrs of exposure period.

Cumulative per cent mortality of *M. incognita* was maximum in neem seed extract and minimum in neem bark extract which could be due to the fact that Neem bark does not possess sufficient active ingredient which are nematicidal. Neem leaves and neem seed having more concentrated nematicidal/nematostatic properties. Acetone extract of leaf, flower, root and stem of five different varieties of tagetes were

Table 30.3: Effect of Neem Seed Extract (Acetone) on Juvenile Mortality of *Meloidogyne incognita* at different Time Intervals

Extract Con.	*Per cent Mortality at different Exposure Period*				
	24 hrs.	*48 hrs.*	*72 hrs.*	*96 hrs.*	*120 hrs.*
0.25 ml	8.667	19.000	37.000	46.667	52.667
0.50 ml	17.000	29.667	46.333	48.667	65.667
1 ml	29.667	41.667	47.333	65.333	75.333
1.5 ml	39.667	54.333	68.667	76.333	85.667
2 ml	71.333	74.667	84.333	90.667	95.000
2.5 ml	81.667	90.667	99.000	100.000	100.000
5 ml	100.000	100.000	100.000	100.000	100.000
Control	0.000	0.000	0.000	0.000	0.000
SEM ±	1.491	1.434	4.177	1.951	1.590
C.D. at 5%	3.438	3.306	9.631	4.499	3.666
C.D. at 1%	5.001	4.810	14.013	6.545	5.334

Observations are mean of three replicates.

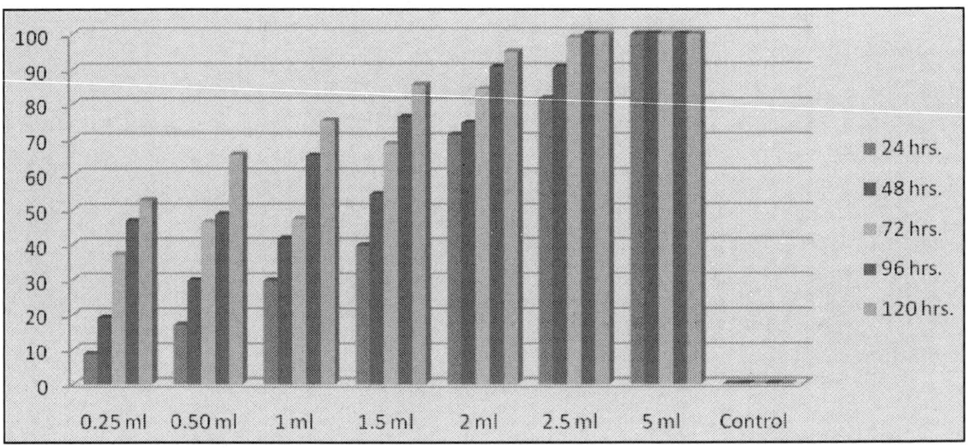

Figure 30.3: Effect of Neem Seed Extract (Acetone) on Juvenile Mortality of *Meloidogyne incognita* at different Time Intervals.

tested against egg hatching and mortality of second stage juveniles of *M.incognita* (Meena *et al.*, 2010). Leaves extract of *A. indica* Vijayalaxmi *et al.* (1979) significantly increased the *M. incognita* juvenile Mortality. Haseeb *et al.* (2007) reported that *A. indica* seeds powder were found effective against *M. incognita*. Neem producte (Neem Kernel, Neem leaf, Neem bark extract) are known to posses nematicidal activity against nematode population (Zaki and Bhatti, 1989, Derkar *et al.*, 1990). Different concentration of shade-dried leaves, bark, and seed of neem were tested for mortality of *M. incognita* juveniles increases in concentration and exposure period resulted in increased mortality rate of juveniles (Nayak and Sharma, 2013). The present study also revealed that the mortality rate was increased with increased expose period and concentration as reported by Siddiqui and Alam (1988). Azadirachtin is the principal active ingredient from neem seed kernel and a well known nematicidal action. The neem bitters (Teranotriterpinoid) such as epinimbin, sulanin, deacetylsalanin, azadiirachin, and many active principle like nimbidine, thionemone and limonoide from leaves have been reported the be highly nematicidal and nematostatics these neem chemicals are known to act against plant parasitic nematode (Mojumder, 2002).

In view of general awareness about hazardous effect of chemical pesticides, there in now an emphasis on the non chemical nematode management technologies.

Acknowledgement

One of the authors Naresh Nayak is highly grateful to Dr. J.L. Sharma, Head, PG Department of Botany, Government College, and Kota for Laboratory facilities, supervision and guidance also thankful to University Grants Commission for financial assistance.

References

Banerji, R., Mishra, G. Nigam, S. K. (1985). Role of indigenous plant material in pest control., *Pesticides*, **19**: 32-38.

Deka, B. C. and Rahman, M.F. (1998). Management of *Meloidogyne incognita* infecting okra by using neem part and products. *J. Agric. Sci., Sco.* North East India, **11**: 124-126.

Derker, K. S. Mhase, N. L. and Shelka, S. S. (1990) effect of placement of non edible oilseed cake on the control of root-knot nematode. *Indian Nematol.* Netw. *Newslwtter.*, **7**: 5-70.

Haseeb, A., Amen, A. and Sharma, A. (2007). Management of *Meloidogyne incognita*. *Fusariun oxysporum* Disease complex of *Pisum sativum*. *Indian Journal of Nematology*, . **37**(1): 19-22.

Hussain, M.A., Mukhtar, T. and Kayani, M.Z. (2011). *Pakistan Journal of Botany*, **43**: 197-204.

Jain, R. K. and Gupta, D.C. (1988). Efficacy of neem cake as nursery bed treatment in the management of root knot nematode infecting tomato. *Indian J. Nematod.*, **27**: 249-251.

Mojumder, V. (2002). Nematode, Nematodes. In "The Neem Tree: Azadirachta indica, A. Juss. And other meliacious plants: Source of Unique Natural product for integrated pest Management Medicine, Industry and other purpose." Revised Edition (Ed. Schmotterer. A.) Neem Foundation, Mumbai, India, pp. 168-196.

Meena, S.K., Sivakumar, M. and Prabhu, S. (2010). Efficacy of acetone extract of Tegetas species on egg hatching and mortality and larval mortality of meloidogyne incognita. *Indian Journal of Nematology.*, **40**(1): 88-89.

Nayak Naresh and Sharma, J.L. (2013).Effect of methanol neem extract on mortality of *Meloidogyne incognita* juvenile. *Indian Journal of Nematology*., **43**(1): .43-46.

Siddiqui, M.A. and Alam, M.M. (1988). Toxicity of different plant parts of *Tagetes lucida* to plant parasitic nematodes. *Indian J. Nemalo.*, **1**(18): 181-185.

Vijayalakshmi, K., Mishra, S.D. and Prasad, S.K. (1979). Nematicidal properties of some indigenous plant materials against second stage juveniles of *Meloidogyne incognita* (Kofoid and white, 1919) chitwood, 1949. *Indian J. Eng*., **41**: 326-333.

Zaki, F.A. and Bhatti, D. S. (1989). Phytotherapeutic effect of some plant leaves on *Meloidogyne javanica* infecting tomato plante. *Nematol Medit*., **17**: 71-72.

2015, Major Constrains and Verdict of Crop Productivity Pages 421–426
Editor: Dr. U.N. Bhale
Published by: ASTRAL INTERNATIONAL (P) LIMITED, NEW DELHI

Chapter 31

Pests and Diseases in Muga Silkworm: A Major Constrain in Conservation of Muga Silkworm Crop

Ranjana Das* and K .Das

Central Muga Eri Research and Training Institute, Lahdoigarh – 785 700, Assam, India

ABSTRACT

Non mulberry silk Muga – Golden silk produced only in Brahmaputra valley of Assam province in NE Region. The ranges of biological resources including their genetic resources are renewable in nature or in similar ex- situ conservation with their proper management can fulfill human needs in larger extent. Developing and strengthening of in situ mechanism for seri-biodiversity conservation in the protected areas is the need of the hour. Diseases and pests are the main constrains in management for conservation of muga silkworm. Flacherie, muscardine, uzi and yellow fly are the most common diseases and pests on muga silkworm which cause significant yield loss in the vanya silk industry. The aerial and ground spraying of bio-control insecticides like Beauveria bassiana and Bacillus thuringiensis to eradicate various kinds of moth caterpillars is an increasingly widespread practice throughout the world. These bio-control agents severely affect the vanya silkworms ie. the non target insects through different mode of transmission. Aspergilosis in muga silk moth during "Bhodia" (August) seed crop causes 35-40 per cent loss in seed production. Therefore, to understand the main constrains in management for conservation of vanya silk are discussed in the study.

Keywords: *Constrain, Disease, Pest, Vanya silkworm.*

* *Corresponding Author:* E-mail: drranjanadas2@gmail.com

Introduction

North East India is considered as hot spot of seri-biodiversity particularly in case of non-mulberry (*vanya*) silk sector which play a significant role in sustainable rural livelihood and poverty alleviation in the country. The vanya silk cultivation is an eco friendly and women friendly occupation that provides high employment, vibrancy to village economies and ideal programme for weaker section of the society. Muga, golden silk produced only in Brahmaputra valley of Assam province in NE Region. The ranges of biological resources including their genetic resources are renewable in nature or in similar *ex- situ* conservation with their proper management can fulfill human needs in larger extent. Anthropogenic climate change is an important issue which needs to be taken seriously in the context of Lepidoptera conservation and pest research (Porter, 1995; Woiwod, 1997; Watt and Woiwod, 1997). The term *conservation* refers to the preservation, restoration, or protection of the planet's natural resources and ecosystems. Muga silkworm is an insect which needs proper conservation. There are many constrains in conservation of muga silkworm. The abiotic and biotic factors of the environment during different seasons greatly influence the growth and development of muga silkworm in the form of cocoon weight, pupa weight, shell percentage, potential fecundity, reelability and denier of the silk (Chiang, 1985; Yadav, 2000). The aerial and ground spraying of bio-control insecticides like *Beauveria bassiana* and *Bacillus thuringiensis* to eradicate various kinds of moth caterpillars is an increasingly widespread practice throughout the world. These bio-control agents severely affect the muga silkworms *ie*. the non target insects through different mode of transmission. Therefore to understand the main constrains in management for conservation of vanya silk are discussed in the study.

Materials and Methods

Geography of the Study Area

The study was conducted in different muga growing districts of Assam, India. The state experiences a very hot–humid weather during summer with an average temperature of 30°C (max. 38.5°C. min. 7°C). The annual rainfall ranges between 1,500 and 2,600 mm with moderate humidity (75 per cent). The disease survey was conducted in the farmers filed of Goalpara, Kamrup, Jorhat, Golaghat, Sivsagar, Lakhimpur and Dibrugarh during the rearing period and the infected larvae were aseptically collected for isolation of the pathogen.

Collection of Data

The diseased larvae showing typical symptoms of muscardine and flacherie disease were collected from muga growing areas for isolation of the pathogens from all the districts. The typical diseased larvae were collected from the rearing sites in sterile polythene bags and brought to the laboratory immediately. The diseased samples were collected in the different crop seasons *viz.* Chatua Seed crop (March-Apri) Jethua commercial crop(May-June), Aherua Pre-seed crop (June-July), Bhodia Seed crop (August-September), Katia commercial(October-November) and Jarua Preseed crop (December- February. The diseased cadavers were treated with 70-90 per cent ethyl alcohol for 2 minutes, then 5.25 per cent Sodium Hypochloride for 3-5 minutes, 19 per cent Sodium thiosulphate for 3-5 minutes and later three successive washing with sterile water. The surface sterilized muga silkworms were homogenized aseptically. A loop- full homogenate were streaked on potato dextrose agar and nutrient agar plates and the plates were incubated for 48 hours at 25° ± 2°C and 30° ± 2°C separately. The growths of microbial colonies were observed and mycelium and conidia were studied. Pure culture was maintained for the isolates on potato dextrose agar and nutrient agar slant. At the same time the pests were also collected to know the constrains for conservation of muga silkworm.

Results and Discussion

The major constraints experienced by the respondents were lack of knowledge about diseases and pests which were expressed by the respondent in muga silkworm rearing. The main disease and pests of muga silkworm are discussed below.

Fungal Diseases

Three different species of entomopathogenic fungi *viz., Beauveria bassiana, Aspergillus niger* and *Fusarium moniliforme* are found to cause White muscardine, Aspergillosis and Fusarosis diseases of muga silkworm.

White Muscardine

White fluffy and frosty mycelial growth emerging out through inter-segments, lateral as well as ventral sides of whole body, ultimately covering the whole body and appendages. Initially the larvae loose appetite, become inactive and on death, gradually, they become harder and finally mummified (Das, *et al.*, 2007). The infected worms become harder, paler and completely inactive followed by bending of the body dorsally. It dies in about 72-96 hours of infection. Dead worms are compressed, reduced body fluid and spongy with fragile skin. A white encrustation appears round the body becomes laterally compressed, dry, hard, and brittle and mummified. Continuous cloudy and foggy weather for 3-4 days, medium temperature (22 ± 1°C) with high humidity (>80 per cent) are congenial for white muscardine disease.

Aspergillosis

This disease is found in almost all the growing season. White fluffy and frosty mycelial growth emerging out through inter-segments, lateral as well as ventral sides of whole body, ultimately covering the whole body and appendages. Initially the larvae loose appetite, become inactive and on death, gradually, they become harder and finally mummified (Das, *et al.*, 2007). The infected worms become harder, paler and completely inactive followed by bending of the body dorsally. It dies in about 72-96 hours of infection. Dead worms are compressed, reduced body fluid and spongy with fragile skin. A white encrustation appears round the body becomes laterally compressed, dry, hard, and brittle and, mummified. Aspergillosis generally infect early instar silkworm. The early instar silkworm becomes inactive and dies without clear morphological symptoms. Late instar silkworm develops black spot at the site of infection and dies due to Aflatoxin produced by the fungus.

Fusarosis

Fusarosis is found to occur severely in the month of Feb-March. Fusarosis mostly infect 3rd, 4th and early 5th stage silkworm. The infected worms become lethargic and dies suddenly. After 24 hrs of dyeing dirty white mycelial growth appear on the cadavars later the infected and died worms turns to black colour.

Though Aspergillosis and Fusarosis may occur in all type of weather but high temperature (30 ± 2°C) with high humidity (>85 per cent) are congenial for the disease development of infection. These fungus can over winter on soil and plant debris or as epiphytes by producing chlamydospore and on reaching favourable they germinate, multiply, spread and cause infection to considerable ranges. The continuous use of bio-pesticides *e.g.* Baba also responsible for fungal diseases in muga silkworm.

Table 31.1: Various Diseases and Pests of Muga Silkworm and their Field Occurrence.

Diseases	*Field Occurrence (per cent)*					
	Season 1 (Dec-Jan)	*Season 2 (Feb-Mar)*	*Season 3 (May-Jun)*	*Season 4 (Jun-Jul)*	*Season 5 (Jul-Aug)*	*Season 6 (Oct-Nov)*
White muscardine	38.98	24.00	0.00	0.00	0.00	13.13
Aspergillosis	28.47	0.00	0.00	2.56	11.11	17.19
Fusarosis	23.73	17.65	0.00	0.00	12.44	20.11
Bacillus thuringensi	84.74	81.17	68.18	0.00	72.22	0.00
Aeromonas salmonicida	22.03	11.76	10.60	43.75	16.66	57.81
Uzi fly (*E. sorbillans* and *B. zebina*)	13.7	23.8	1.2	0.4	0.3	0.2
Yellow fly	0	0	5.0	4.0	0	0

Flacherie or Bacterial Disease

Flacherie is a generic name for the syndrome represented by flaccidity of silk worm. This disease infects the alimentary tract of the worms and blood cells, hypodermis and fat bodies also infected. Severe infection can cause yield loss up to 80-90 per cent in summer and rainy seasons. It is a flaccid condition symptomatically showing softness, dullness and lethargicity in the larvae. Occurrence of the disease was in peak during last stage of the worms, *i.e.*, mid 4th to 5th instar and the worms died before spinning. Maximum mortality was reported in February-March, November-December followed by June-July and July-August (CMER and TI and AAU Project progress report, 2007).

Three bacteria associated with flacherie disease are identified as *Bacillus thuringensis, Aeromonas salmonicida* and *Streptococcus bombycis*. Bacteral septicemia (*Bacillus thuringensis*) and Bacterial toxicosis (*Aeromonas salmonicida*) are the two most devastating bacteria. *Streptococcus bombycis* is non pathogenic. (CMER and TI and AAU Project progress report, 2007).

Factors Affecting Disease Development

Sudden fluctuation of weather, *viz.*, temperature, relative humidity and rainfall is the main cause of flacherie disease. High temperature (36 °C and above), high humidity (above 75 per cent) is suitable for disease development. Bio-pesticides *e.g.* using of *Bt* in the adjacent area accelerate the bacterial diseases of muga silkworm.

Uzifly

Exorista sorbillans and *Blepharipa zebina* are major pests of muga silkworm. These pests complete their life cycle in muga silkworm and occurred in winter and post winter during pre-seed and seed crop of Jethua commercial brood. It caused the considerable lost of muga sill worm crop loss 20-90 per cent in pre-seed (Jarua, Dec-Jan) and seed (Chatua, Feb-Mar) crop

Yellow Fly

Generally wild muga cocoons are infested by *xanthopimpla* predator commonly as yellow fly during spinning time at the natural condition. This is one of the major endoparasites of wild muga silkworm (Das and Das, 2011) The parasite completely fed on head region of the pupae and emerged from the same by making a hole which is about 1cm diameter.

During the time of conservation of muga silkworm the adjacent area should be free from alternate host of the diseases causing agents and pests. People of the adjacent area should be aware about the knowledge of muga culture along with its enemies. From this study, it is understood that the white muscardine disease pathogen *B. bassiana* might have disseminated from the other crop fields as now a day's these bio pesticides are commonly used in tea gardens, rice fields vegetable fields etc. The native strains of *B. bassiana* are not so infective and infection percentage was lower than the commercial strains. It is reported that muscardine diseases in muga silkworm were not found in earlier days (Choudhury, 1981). The bio-pesticides were not available in those days although the native strains of *B. bassiana* were present in soils. Probably, due to change of Climate it may have also some effect on the infection capability of microbes as pathogens. This factor is recognized as a major threat to the survival of species and integrity of ecosystems worldwide. The commercial bio-pesticides agents prepared by *B. bassiana, B. thuriengiensis* etc. are infective and can easily disseminate by air, water or by other medium. Again, these are applied in the other crop filed very frequently which increase the chance of more dissemination in a wide area. In muga silkworm, effect of pests and diseases also are more intense, being reared outdoor and it affects the economy directly. Hence long term sustainability measures needs to be taken. Conservation measures both in-situ and ex-situ accounts a vital role in maintaining the muga silkworm sustenance.

References

Anonymous (2007). Identification, Characterization and Diagnosis of some important muga silkworm diseases. *Annual Report* of CMERTI, Lahdoigarh, Jorhat.

Choudhury S.N. (1981). Muga silk Industry, Directorate of Sericulture, Govt of Assam.

Chiang, H.C. (1985). Insects and their environment. Fundamentals of applied Entomology. Macmillan Publishing Company, New York: 128-161.

Das K and Das Ranjana (2011) *Xanthopimpla*, a new endoparasite of wild muga silk worm, *Antheraea assamensis* Helfer (Lapidoptera: Saturniidae) in Assam. *Indian Silk*, **49** (11): 18.

Das K, Ranjana Das P.Dutta, R.Chakrovorty, P.Devnath, SAS Rahman, N.Neog, I.Sarmah and M.Islam (2007). First report on some important endo-mycoflora of Antherea assamensis, Helfer (Lepideptera: Saturnidae) from North-East India. *Sericologia*. **47**(2): 225-228

Porter, J. (1995). The effects of climate change on the agricultural environment for crop insect pests with particular reference to the European corn borer and grain maize. Insects in a Changing Environment. 17th Symposium of the Royal Entomological Society (R. Harrington and N.E. Stork, Eds) *London: Academic Press*, pp. 93–123.

Watt, A. D. and Woiwod, I. P.(1997). The effect of phonological asynchrony on population dynamics: analysis of fluctuations of British macrolepidoptera. Oikos (in press).

Woiwod, I. P. (1997). Monitoring population change and diversity. In The Butterflies and Moths of Bedfordshire (V.W. Arnold, C.R.B. Baker, D.W. Manning and I.P. Woiwod, Eds) pp. 61–78. Bedford: The Bedfordshire Natural History Society.

Yadav, G. S. (2000). Studies on association between cocoon weight and silk of muga silkworm (*Antheraea assama* Westwood) during different broods. *Internat. J. of Wild Silkmoths and Silk.*, **5**: 221-225.

2015, Major Constrains and Verdict of Crop Productivity *Pages 427–440*
Editor: Dr. U.N. Bhale
Published by: ASTRAL INTERNATIONAL (P) LIMITED, NEW DELHI

Chapter 32

In vitro Antimicrobial Activity of Medicinal Plants in Bikaner, Rajasthan: An Overview

S. Ashwani and G. Dipali*

Tanveer Malawat College of Biosciences, Bikaner – 334 001, Rajasthan, India

ABSTRACT

Medicinal plants represent a rich source of antimicrobial agents. Plants are used medicinally in different countries and are a source of many potent and powerful drugs. Bikaner is semi arid region of Rajasthan having little rainfall and extreme temperatures. In semi arid region of Rajasthan, plants are the major source of local medicines and have ethnological values. In present review, we emphasized the antimicrobial properties of 10 selected different species of arid zone medicinal plants of Bikaner. It was found that antimicrobial activities of different extracts obtained from the fruit, bark, latex, flowers and leaf of plants differ considerably with different solvents. Various strains of pathogenic and nonpathogenic bacteria show different response against these extracts. The crude ethanol extracts of leaves of all selected plants showed considerable antimicrobial potential.

Keywords: *Medicinal plants, Antimicrobial agents, Ethanol extracts.*

Introduction

Rajasthan Desert regions are characterized as having extremely arid climate with low and erratic rainfall, dry atmosphere and high wind velocities (Purohit *et al.*, 2002). Bikaner is situated in the middle of the Thar Desert. In the summer season it is

* *Corresponding Author.* E-mail: dipaligupta71@yahoo.com

very hot when the temperatures lie in the range of 28–48.5 °C (82–119 °F). In the winter, it is fairly cold with temperatures lying in the range of 5–23.2 °C (41–74 °F). Annual rainfall is in the range of 260–440 millimeters (10–17 in). Multipurpose medicinal plants are found in Bikaner. Medicinal plants are part and parcel of human society to combat diseases; from the dawn of civilization (Bandhopadhyay *et al.*, 2002). Herbal medicines are in great demand in the developed as well as developing countries for primary healthcare because of their wide biological and medicinal activities, higher safety margins and lesser costs (Cragg *et al.*, 1997; Padma, 2005). Microorganisms have potential to cause human diseases, much infectious disease have been treated with herbal remedies through out the history of mankind. An antimicrobial is an agent that kills microorganisms or inhibits their growth. For example, antibacterials (commonly known as antibiotics) are used against bacteria and antifungals are used against fungi. A number of arid zone medicinal plants have been studied for their antimicrobial activities and their antimicrobial principles have also been evaluated (Singh *et al.*, 1988; Ahmed*et al.*, 1999; Ahmed and Beg-Arina, 2001; Kapoor *et al.*, 2003; Kapoor*et al.*, 2006; Kapoor*et al.*, 2007; Kapoor *et al.*, 2011). Researches have shown that all parts of the plants which include stem, root, flower, bark, leaves etc. possess antimicrobial property. When a large number of plants in different location around the world have been extracted and semi-purified to investigate individually their antimicrobial principal, among the different plant derivatives, secondary metabolites proved to be the most important group of compounds that showed wide range of antibacterial and antifungal activity. (Ahmad *et al.*, 2002). Different types of organic and inorganic solvents have been used for carrying out the antimicrobial studies against various gram positive and gram negative bacteria like; methanol, ethanol, methyl acetate, acetone, benzene etc.(Gupta, 2013; Gupta and Rathore, 2012). At the same time the harmful side effects and the continuous increase in the number of microorganisms that are resistant to the chemical antibiotics highlights the need for new strategies and new classes of antibiotics with low toxicity and high selectivity in their action. Thus, the study ascertains the value of plants used in Ayurveda, which could be of considerable interest to the development of new drugs. The present attempt is to review, collect and compile the updated information on *in vitro* antimicrobial potential of some selected medicinal plants of Bikaner (Raj). In the present study, 10 medicinal plants *viz., Calotropis procera, Cordia dichotama Forst. F, Tecomella undulata, Capparis decidua Edgew, Aloe barbedensis, Ocimum sanctum* Linn., *Lawsonia inermis* Linn., *Azardirachta indica, Prosopis cineraria, Zizyphus nummularia*) of the respective families Asclepiadaceae, Boraginaceae, Bignoniaceae, Capparidaceae, Liliaceae, Lythraceae, Labiatae, shrub, Meliaceae, Mimosaceae, Rhamnaceae, were selected to analyze their medicinal nature.

Calotropis procera (AiT)-Ait

Calotropis procera (Ait) belongs to family Asclepiadaceae, used medicinally, to treat boils, infected wounds and other skin problems (Himalaya, 2002). It also yield ash for making gun powder, the latex is processed and used in treating vertigo, baldness, hair fall, tooth aches, intermittent fevers, rheumatoid/joints swellings, paralysis, bronchitis, asthma, leprosy, eczema and elephantiasis, hepatic and splenic enlargement and as eye tonic. (Vohra, 2004). The antimicrobial effect of ethanol, aqueous and chloroform extracts of leaf and latex of *Calotropis procera* on six bacteria namely, *Escherichia coli, Staphylococcus aureus, Staphylococcus albus, Streptococcus pyogenes, Streptococcus pneumoniae and three fungi: Aspergillus niger, Aspergillus flavus, Microsporium boulardii* and one yeast *Candida albicans were studied.* The ethanolic extracts of *C. procera* latex gave the widest zone of inhibition (14.1mm) against *E-coli* using agar well diffusion. The best antifungal activity was recorded in ethanol extract of *C.procera* latex against *Candida albicans*. This study revealed that the *C.procera* latex possessed stronger inhibitory effect on the test organisms than *C.procera* leaf. (Kareem *et al.*, 2008). For antibacterial activity testing, the methanol and aqueous extract of leaves of *Calotropis procera* were subjected to test its effectiveness against both Gram-positive and Gram-negative bacteria by agar diffusion method. The zones of inhibition produced by the crude methanol and aqueous extract against few sensitive strains were measured and compared with those of standard antibiotic Gentamycin. It was evident that both extracts were active against the bacteria at low concentrations. (Yesmin *et al.*, 2008). Bioassays for antimicrobial activities were carried out using stem, leaves and flowers of *Calotropis procera*(Ait.) also. The potential of antimicrobial property of plant is related to their ability to synthesize compounds by the secondary metabolism (Remya *et al.*, 2010). The antimicrobial activities of the organic solvent extracts were tested on the various fungi; *Alternaria alternata, Aspergillus flavus, Asperigellus niger, Bipolaris bicolor, Curvularia lunata, Pencillium expansum, Pseudomaonas marginales, Rhizoctonia solani, Ustilago maydis* by using agar well diffusion technique (Vadlapudi and Naidu, 2010). Aqueous and organic solvent extracts of the leaves, flowers and latex of *Calotropis procera* (Ait.) were tested by using the disc diffusion bioassay and the Minimal Inhibitory Concentration (MIC). The leaf and latex methanolic extracts showed the strongest activities against *Escherichia coli, Staphylococcus epidermides,* and *Bacillus* spp. All extracts showed biological activities against all of the tested fungal strains. The latex methanolic extract was the most effective extract against *Candida albicans, C. tropicalis, Penicillium chrysogenum* and *Saccharomyces cerevisiae*. To test any synergistic effect between the latex methanolic extract and Ciprofloxacin and Clotrimazole, the extract was added to the tested antibiotics at concentrations equal ½, ¼, $1/_8$ and $1/_{32}$ and 0 of the original MIC values. Results revealed that the MIC's of the two antimicrobial standards, were lowered indicating a synergistic interaction between the botanical and the conventional drugs. (Gomah and Essam, 2011).The phytochemical analysis and antibacterial activities of water, methanol and ethanol extracts obtained from the fruit and bark of *Calotropis procera* was also performed. Anti-bacterial activity was determined against *Pseudomonas aeruginosa, Salmonella typhi, Escherichia coli* and *Streptococcus pyrogenes*. Water extracts showed inhibition against the tested organisms at 30 mg/ml (Mainasaral *et al.*, 2011).

Cordia dichotama Forst. F

Cordiadichotoma (Lasora) from Boraginaceae is small to moderate-sized deciduous tree. The fruit contains about 70 per cent pulp, seed contains per 100 g; water 32 g, fat 46 g and the principal fatty acids (Srivastava and Srivastava, 1979). Chemical screening of both the leaves and the fruits showed the presence of pyrrolizidine alkaloids, coumarins, flavonoids,saponins, terpenes and sterols. Plant is diuretic.

The bark decoction is used to treat dyspepsia, mouth ulcers, fever, abscesses and tumors, dysentery, coughs and chest complaints, uterus and urethra disorders(Ficarra *et al.*, 1995). Ethanol extracts from fruits showed significant antioxidant activities due to the carotenoids but no antimicrobial activity against gram-positive or gram-negative bacteria. The crude ethanol extract of leaves of *Cordiadichotoma* showed zone of inhibition against both Gram negative and Gram positive bacteria including *Streptococcus aureus, Streptococcus pyogenes, Vibrio cholerae, Streptococcus epidermis, Hafnia and Escherichia coli* (Shazid *et al.*, 2009). Methanol and butanol extracts of the bark were tested against *Escherichia coli, Pseudomonas aeruginosa, S. pyogenes* and *Staphylococcus aureus*. The antifungal activity of the extracts was carried out against three common pathogenic fungi (*Aspergillu sniger, A.clavatus,* and *Candida albicans*) (Nariya *et al.*, 2011).

Tecomella undulata (Sm.) Seem

*Tecomella undulata*Seem (Bignoniaceae) commonly known as "Rohira has been used in the indigenous system of medicine for spleen, liver and abdominal complaints. The plant is also useful in urinary discharges due to "kapha" and "pitta", enlargement of spleen, leucorrhoea and lucoderma. The bark of the young branches is often employed in Sind as a remedy for syphilis. The antibacterial activity of aqueous and methanol extracts was determined by agar disc diffusion and agar well diffusion method. The methanol extracts were more active than the aqueous extract (Giridhar *et al.*, 1980; Rao *et al.*, 1989).

Capparis decidua Edgew

Capparis decidua or kareel (ker) is an indigenous medicinal plant of India (Rajasthan) belongs to the family of Capparidaceae. The young flower buds and fruits are used to make pickles. Plant is used as ailments to relieve variety of pains or aches such as toothache, cough and asthma heal. Plant contains quercetin which act as melanogenesis stimulator and also increase estyrosinase and protein expression. *Capparis* seeds contain lectins that exhibit potent anti HIV-1 reverse transcriptase inhibition activity. It shows anti-rheumatic, anti-diabetic, anti-arthritis and anti-gout agent. Different extracts of the plant have been demonstrated to possess pharmacological properties.Petroleum ether, chloroform and methanolic extracts of *Capparis decidua* showed inhibitory activity against *Bacillus subtilis* and *Aspergillus niger* (Iqbal and Khan, 2008).The plant has been reported for its antimicrobial properties (Upadhyay, 2011).

Aloe barbedensis Miller

*Aloe vera*Linne or *Aloe barbadensis* Miller is a succulent from the Aloe family (400 different species).The gel moisturizes the skin because it has a water holding capacity. *Aloe vera*juice is considered helpful for relieving many types of gastro intestinal irritation. Extracts of dried *Aloe leaves* are used as laxatives preceding rectal surgery and as a hemorrhoid treatment, sunburn and promote wound healing, stimulate the body's immune system.The comparative antimicrobial activities of the gel and leaf of *Aloe vera* were tested against *Staphylococcus aureus, Pseudomonas aeruginosa, Trichophyton mentagraphytes, T. schoeleinii,Microsporium canis* and *Candida albicans*. Ethanol was used for the extraction of the leaf after obtaining the gel from it. Antimicrobial susceptibility test showed that both the gel and the leaf inhibited the growth of *S. aureus* (18.0 and 4.0 mm, respectively). Only gel inhibited the growth of *T. mentagrophytes* (20.0 mm), while the leaf possesses inhibitory effects on both *P. aeruginosa* and *C. albicans* (Agarry *et al.*, 2005). Antibacterial activities of leaf and gel extracts of *Aloe vera* were tested against gram positive and gram negative skin infections isolates. For this purpose, one hundred and fifteen bacterial strains were isolated from skin wounds, and the strains were identified by conventional methods. Among the total isolates, 90 per cent of the organisms were gram positive while the remaining 10 per cent were gram negative. The gel extracts of *Aloe vera* showed antibacterial activity against both gram positive and gram negative isolates while the leaf extracts showed no such activity (Bashir*et al.*, 2011).Three different solvents such as aqueous, ethanol and acetone were used to extract the bioactive compounds from the leaves of *Aloe vera* to screen the antimicrobial activity against selected human clinical pathogens by agar diffusion method. The maximum antibacterial activities were observed in acetone extracts (12±0.45nm, 20±0.35nm, 20±0.57nm and 15±0.38nm). Antifungal activity of *Aloe vera* was analyzed against *Aspergillus flavus* and *Aspergillus niger*. The maximum antifungal activity was observed in acetone extracts (15±0.73nm and 8±0.37nm).*Aloe vera* plant extract with acetone can be used as antimicrobial agents. The antibacterial activity of *Aloe barbadensis* was studied by using agar diffusion assay and gel filtration chromatography. The bacterial strains used in this research work were *Escherichia coli, Bacillus subtilis, Salmonella typhi, Pseudomonas, Klebsiella pneumoniae, Staphylococcus epidermidis. Aloe vera* plant leaves and gel were macerated in different organic solvents including ethanol, methanol and distilled water. The *Aloe Vera* extract of methanol showed the maximum antibacterial activity as compared to other solvent extracts (Saba*et al.*,2011).

Ocimum sanctum Linn.

Tulsi (*Ocimum sanctum* L.) is a traditional plant considered sacred by the Hindus belonging to the family Lamiaceae are used for medicinal purpose. Different parts of plant are used in ill nesses and cold, headache, cough, flu, earache, fever, colic pain, sore throat, bronchitis, asthma, hepatic diseases, malaria fever, as an antidote for snake bite and scorpion sting, flatulence, migraine, fatigue, skin diseases, wound, insomnia, arthritis, digestive disorders, night blindness, diarrhea and influenza. Chewing of tulsi leaves also cures ulcers and infections of mouth and to sharpen memory (Pandey and Madhuri, 2010).*Ocimum sanctum* has been extensively studied for therapeutic potentials in various areas like immuno-stimulation, anticancer antioxidant, as adjuvant to radiotherapy, antiulcer, analgesic and antidiabetic (Hammer *et al.*, 1999). Oil extracted from leaves of this plant possesses significant insecticidal properties (Nanasombat and Lohasupthawee, 2005).*Ocimum sanctum'*sessential oil has significant antifungal activity against Candida, includingazole-resistant strains.Antibacterial activity of crude extracts was found to be reliant on the nature of extract and the bacterial strains evaluated. Methanol extract was found to have comparatively higher activity than other organic and aqueous extracts. Gram positive bacteria showed variable susceptibilities while gram-negative *Salmonella typhi*has shown to be completely resistance to all the tested extracts.

Minimum inhibitory concentration data showed hopeful results as some of the extracts exhibited significant inhibitions of bacteria even at low concentrations. This study indicated that leaves of *Ocimum sanctum* L. have significant antibacterial activity (Pankaj and Purshotam, 2011). *E. coli* was highly susceptible to ethanol extract of Tulsi leaves at 250 µg concentration (Gajendrasinh *et al.*, 2012).

Lawsonia inermis Linn.

Heena (*Lawsonia inermis* syn. *L. alba*) is a flowering plant belonging to the family Lytharaceae. The main uses of henna are as a cooling agent, astringent, antifungal and anti-bacterial herb for the skin and hair. The anti-mycobacterial activity of quinonoid compounds, particularly those isolated from natural sources, has remained unexplained. Inhibitory action of henna was shown against both gram negative and gram positive microbes. In one report the inhibitory action was greatest against *B. anthracis* as it stood out from other tested bacteria (Malekzadeh, 1968).Lawsone has shown significant antifungal antibiotic effect (Dixit *et al.*, 1980). In nosocomial infection, *Staphylococcus aureus* is one of the most prevalent microorganisms worldwide. In general they are active against gram positive bacteria such, as *Staphylococcus aureus Enterococcus faecium* and *Bacillus subtilis,* as *Staphylococcus aureus, Enterococcus faecium* and *Bacillus subtilis,* but are inactive against gram negative bacteria (Papageorgiou *et al.*, 1999).They exerted inhibitory effects upon common nosocomial urinary tract pathogens such as *Escherichia coli, Proteus mirabilis, Klebsiella pneumoniae, Pseudomonas aeroginosa* and *Staphylococcus aureus* at certain concentrations. The antibacterial activity of the natural naphthoquinone products alkannin and shikonin and their derivatives has been investigated (Riffel *et al.*, 2002). Essential oil obtained by hydro-distillation from leaves of *L. inermis* growing in Iran were showed an antifungal activity (Aghel *et al.*, 2005) Aqueous, methanol and chloroform crude extracts of leaf showed the *in-vitro* antimicrobial activity to inhibit the growth of 6 human pathogenic fungi and 4 types of bacteria in dose dependent manner (Habbal *et al.*, 2005; Saadabi, 2007). Ethanol extract could be used as alternative source of antifungal agents for protection of plants or crops against fungal infection. Alcoholic extracts had the highest antibacterial activity with a MIC of 0.125-0.150 µg/ml against ß-hemolytic *streptococci* and against CONS was 0.125-175 µg/ml. Both alcoholic and oily extracts had the same MIC (0.5 µg/ml) on *Staphylococcus aureus*. However, alcoholic extracts were more effective on *Pseudomonas aeruginosa* with a MIC of 0.5-0.57 µg/ml than oily extract (MIC of 0.20-0.28 µg/ml) (Kathem *et al.*, 2008). Henna samples demonstrated antibacterial activity against all isolates but the highest susceptibility was against *P. aeruginosa* with henna samples obtained from Al-sharqyia region (Habbal *et al.*, 2011). Antibacterial activity of aqueous, ethanol, methanol, ethyl acetate and chloroform extracts of *Lawsonia inermis* Linn. leaves exhibited profound antibacterial activity against the bacterial pathogens tested (Vinoth*et al.*,2013).

Azardirachta indica

Neem (*Azadirachta indica*) belonging to family Meliaceae. Twig of its stem is widely used for chewing (brushing) and proved to have anti plaque and related antibacterial properties. The antibacterial activity of aqueous and acetone extract of neem bark was examined on agar plate by using *Streptococcus sobrinus*, it was found that acetonic extract had an effect on growth of *Streptococcus sobrinus* with MIC value of 0.05 percent (Mohashine *et al.*, 1997). Few authors have reported antimicrobial activity of neem oil on *Escherichia coli* and *Klebsiella pneumonia* with methanol but not with chloroform and hexane extracts (Almas, 1999). To study the antifungal activity of 10 different extracts of seed kernels of *Azadirachta indica* A. on *Candida* sps. isolated from immune compromised patients, The ethanol extract of commercial neem seed oil, ethanol extract of neem seed kernels and the hexane extract showed best results. All strains were resistant to methanol: chloroform: water extracts and chloroform extracts of the successive extraction procedure (Charmaine *et al.*, 2005). Antibacterial activity of the extracts of *Azadirachta indica* was effective on *Escherichia coli, Klebsiella pneumonia* and *Bacillus subtilis*etc. and needs no further confirmation (Jagannadh and Radhika, 2006). Antimicrobial activity in leaf extracts of neem *(Azadirachta indica)* against *E.coli, Staphylococcus aureus, Pseudomonas aeruginosa, Salmonella typhimurium, Bacillus pumilus* was assayed in alcoholic extracts. Varying concentration of each extracts 200mg/ml, 150 mg/ml, 100mg/ml, 50mg/ml, 25mg/ml prepared, tested by using disc diffusion method. An *in vitro* experimental study was performed using neem leaf, properly identified and verified, was subjected to ethanol extraction of its active ingredients then diluted to produce (25 per cent, 50 per cent, 75 per cent, and 100 per cent) concentrations. Standard strains of *Staphylococcus aureus* and clinical isolates of MRSA were inoculated on blood agar plates and subjected to the standardized disc susceptibility testing method. The ethanol extract from neem leaves exhibited *in vitro* antibacterial activity against both *Staphylococcus aureus* and MRSA with greatest zones of inhibition noted at 100 percent concentration.Leaf and bark methanol and ethyl acetate extracts of *A. indica* can be considered as good sources of natural antioxidants and antimicrobial compounds and can be incorporated into the drug formulations (Joji Reddy *et al.*, 2012). When compared with gentamycin 200mg and gentamycin10mg, the methanol and ethanol extract showed maximum inhibition on *Bacillus pumillus, Pseudomonas aeruginosa* and *Staphylococcus aureus* in an ascending order(Maragathavalli *et al.*,2012). Dichloromethane and methanol extracts of *Azadirachtaindica* showed very little activity against *S. aureus* and *S. epidermidis,* but did not show any antimicrobial activity against *P. acnes*.

Prosopis cineraria

Prosopis cineraria (Mimosaceae) is a small to moderate sized tree found in the regions of Rajasthan. The common name of the plant is Khejari (Sangri) in Rajasthan. The smoke of the leaves is good for eye troubles. The bark is used as a remedy for rheumatism, cough, common cold asthma and scorpion stings. Prosogerin A and B were isolated from flowers. Acetone extract of aerial part exhibited moderate antibacterial activity against 3 bacterial strains *Staphylococcus aureus, Staphylococcus epidermidis* and *Escherichia coli* while no antifungal activity against *Aspergillus niger* and *Candida albicans* at concentration of 175 mg/ml using Agar well diffusion assay (Salar and Dhall,2010).Methanolic and aqueous extracts of the stem bark exhibited moderate antibacterial activity against 8 bacterial strains(*Salmonella typhi, Bacillus Lintus, Klebsiella pneumonia, Streptomyces griseus, Bacillus subtilis, Staphylococcus aureus, Staphylococcus albus, Escherichia coli*) at concentration of 250 mg/ml in comparison with standard ciprofloxacin (100 mg/ml) using agar well diffusion assay

(Velmurugan*et al.*,2010). The methanolic leaf extracts were tested against four strains of bacteria (*Bacillus subtilis, Escherichia coli, Vibrio cholera and Enterobacter aerogenes*) and two strains of fungi (*Aspergillus niger* and *Aspergillus fumigatus*). Maximum radical scavenging activity (per cent RSA) was observed in *P. cineraria* (Abdul *et al.*, 2012). Methanolic extract of leaf was evaluated for antibacterial activity against *Bacillus subtilis, Escherichia coli, Vibrio cholera* and *Enterobacter aerogenes* at concentration of 15 mg/ml using agar disc diffusion assay showed moderate antibacterial activity (Napar *et al.*, 2012).

Zizyphus nummularia

Ziziphus nummularia(Rhamnaceae) also called Jharber, is a species of *Ziziphus* native to the Thar Desert of western India.Dried fruit used medicinally as astringent in bilious affliction in India. The leaves are used to treat scabies and other skin diseases. Anti-inflammatory and analgesic activity of ethanolic extract of *Zizyphus nummularia* have been analyzed by Goyal *et al.* (2012). Callus of *Ziziphus nummularia* were examined for potential antimicrobial activity by preparing their crude extracts in water and ethanol and testing them against gram- positive, gram-negative bacteria and fungal pathogens. The results obtained indicated that the ethanol extracts were potent with a broad spectrum of activity. The results of this study suggest that a number potentially active antimicrobialagents may be obtained from this arid zone plant (Amit *et al.*, 2011).*Zizyphus nummularia* found to have potential antibacterial and antifungal activity against four medically important bacterial and fungal strains respectively. *Staphylococcus aureus, Streptococcus pyogenes, Bacillus subtilis, Pseudomonas aeruginosa, Aspergillus niger, Aspergillus flavus, Candida albicans* and *Trichophyton rubrum*. The antimicrobial activity of aqueous and ethanol extracts was determined by agar disk diffusion method and dry weight method. The ethanol extract was more active than the aqueous extract (Gautam *et al.*, 2011). In case of extracts of ripe *Zizyphus mauritiana* against *Escherichia coli* and *Staphylococcus aureus* widest inhibition zones (14-15 mm) were obtained (Suman,2012).

Table 32.1: Plant Parts and Solvents Showing Antimicrobial Potential in some Arid Zone Medicinal Plants

Sl.No.	*Plants*	*Plant Parts*	*Solvents*	*Reference*
1.	*Calotropis procera*	Latex, leaves	Ethanol	Kareem *et al.*, 2008
		leaves	Aquous, Methanol	Yesmin *et al.*, 2008
		Leaves, Stem, Flowers	Organic	Remya *et al.*, 2010
		Leaves, Latex, Flowers	Water, Methanol	Gomah and Essam, 2011
		Fruit, Bark	Water, Methanol, Ethanol	Mainasaral *et al.*, 2011
2.	*Cordia dichotoma*	Leaves	Ethanol	Shazid *et al.*, 2009
		Bark	Methanol, Butanol	Nariya *et al.*, 2011
3.	*Tecomella undulata*	Bark, Leaves	Water, Methanol	Giridhar *et al.*, 1980; Rao *et al.*, 1989

Contd...

Table 32.1–*Contd...*

4.	*Capparis decidua*	Leaves, Fruits	Petroleum ether, Chloroform, Methanol	Iqbal and Khan, 2008; Upadhyay, 2011
5.	*Aloe barbadensis*	Leaves, Gel	Ethanol	Agarry *et al.*, 2005
			Water, Ethanol, Acetone	Bashir *et al.*, 2011
			Ethanol, Methanol, Water	Saba *et al.*, 2011
6.	*Ocimum sanctum*	Leaves	Methanol Ethanol	Pankaj and Purshotam, 2011; Gajendrasinh *et al.*, 2012
7.	*Lawsonia inermis*	Leaves	Water, Methanol, Chloroform	Habbal *et al.*, 2005; Saadabi, 2007
			Alcoholic, Oil	Kathem *et al.*, 2008
			Aqueous, ethanol, methanol, ethyl acetate and chloroform	Vinoth *et al.*, 2013
8.	*Azadirachta indica*	Bark	Aquous, Acetone	Mohashine *et al.*, 1997
		Oil	Methanol	Almas, 1999
		Seed kernels	Ethanol, Hexane	Charmaine *et al.*, 2005
		Leaves, Bark	Ethanol	Joji Reddy *et al.*, 2012
		Leaves	Methanol, Ethanol	Maragathavalli *et al.*, 2012
9.	*Prosopis cineraria*	Leaves, Stem	Acetone	Salar and Dhall, 2010
		Stem bark	Water, Methanol	Velmurugan *et al.*, 2010
		Leaves	Methanol	Abdul *et al.*, 2012; Napar *et al.*, 2012
10.	*Ziziphus nummularia*	Callus	Water, Methanol	Amit *et al.*, 2011
		Leaves	Water, Ethanol	Gautam *et al.*
		Fruits	Ethanol	Suman, 2012

Conclusion

TharDesertof Bikaner (Rajasthan)is a potential source of medicinal plants. In present review,ten selected arid zone medicinal plants have been reviewed for their antimicrobial activities.It was found that plants and plant parts including stem bark, gel, leaves and oil possessed strong antimicrobial property with varying potential. Difference in antimicrobial efficacy of various plant parts is due to unequal distribution of secondary metabolites inside plant.Various Solvents like; Water, Methanol, Ethanol, ethyl acetate, butanol, acetone, hexane, chloroform etc. were used for extract preparation. Among different organic solvents,ethanolic and methanolicextracts were found more effective than other solvents (Table 32.1). Agar disc diffusion technique and Agar well diffusion technique were applied for analysis of antibacterial activity against different bacterial and fungal strains. All selected desert zone plant's extracts showed antibacterial activity against both gram positive and gram negative bacteria and fungi.

References

Abdul, A. N., Hadi, B., Muhammad, A. Z., Muhammad, Z. A., Arshad, I., Sohaib, R.Izhar, M. and Sabir, H. S. (2012). Antimicrobial and antioxidant activities of Mimosaceae plants; *Acacia modesta* Wall (Phulai), *Prosopis cineraria* (*Linn.*) and *Prosopis juliflora* (*Swartz*). *Journal of Medicinal Plants Research,* **6** (15): 2962-2970.

Agarry, O. O., Olaleye, M. T. and Bello-Michael, C. O. (2005). Comparative antimicrobial activities of *Aloe vera* gel and leaf. *African Journal of Biotechnology*, 4 **(12)**: 1413-1414.

Aghel, N., Ameri, A. and Ebrahimi, P. (2005). Essential oil of *Lawsoniainermis*growing in iran: chemical composition and antifungal activity. First Seminar of Medicinal and Natural Products Chemistry Shiraz, Iran, pp. 10-11.

Ahmed, A. M. A., Rahman, M. S. and Anwar, M. N. (2002). Antimicrobial activity of extracts and crude alkaloids isolated from the leaf of *Adhatoda vasicanees. Bangladesh Journal of Life Science,* **15**(2): 125-128.

Almas, K. (1999). The antimicrobial effects of extracts *Azadirachta indica* (Neem) and *Salvadora persica* (Arak) chewing sticks. *Indian Journal Dental Research.,* **10:** 23-26.

Ahmed and Beg-Arina, Z. (2001). Antimicribial and phytochemical studies on 45 Indian Medicinal Plants Against Multi drug Resistant Human Pathogens. *Journal of Ethnopharmacology*, **74** (2): 113-123.

Ahmed-El-Sawi, S., Abd-El-Megeed, H. F. and Ali, A. M. (1999). Flavonoid and antimicrobial volatiles from *Adhatoda vasica* Nees. *Pharmaceutical and Pharmacological Letters*, **9** (2): 52-56.

Almas, K. (1999). The antimicrobial effects of extracts *Azadirachta indica* (Neem) and *Salvadora persica* (Arak) chewing sticks. *Indian J Dent Res,* **10**: 23-26.

Amit, S., Sapna, Manju, B. and Nag, T. (2011). Antimicrobial screening of plant parts and tissue culture of *Ziziphus nummularia* (Burm. f.) Wight and Arn. *Asian Journal of Traditional Medicines,* **6** (2): 83-88.

Bandyopadhyay, U., Biswas, K., Chattopadhyay, I. and Banerjee, R. K. (2002). Biological activities and medicinal properties of neem (*Azadirachta indica*). *Current Science,* **82** (11): 1336-1345.

Bashir, A., Bushra, S. Talat, Y., Mujahid, and Nayar, J. (2011). Comparative study of antimicrobial activities of *Aloe vera* extracts and antibiotics against isolates from skin infections. *African Journal of Biotechnology,* **10**(19): 3835-3840.

Charmaine Lloyd, AC., Menon, T., Umamaheshwari, K. (2005). Anticandidal activity of *Azadirachta indica*. *Indian journal of Pharmacology*, **37**(6): 386-389.

Cragg, G. M., Newman, D. J. and Sander, K.M. (1997). Natural products in drug discovery and development. *Journal Natural Product,* **60:** 52-60.

Dixit, S.N., Srivastava, H.S., Tripathi, R.D. (1980). Lawsone, the antifungal antibiotic from the leaves of *Lawsoniai nermis* and some aspects of its mode of action. *Indian Phytopathology*, **31:** 131-133.

Ficarra, R., Ficarra, P., Tommasini, S., Calabro, M.L., Ragusa, S., Barbera, R. and Rapisarda, A. (1995). Leaf extracts of some *Cordia* species: analgesic and anti inflammatory activities as well as their chromatographic analysis. *Farmaco,* **50**(4): 245-256.

Gajendrasinh, P., Rathod, B., Kotecha, M., Sharma, R., Hetal, A. and Prajapati, P. K. (2012). *In vitro* antibacterial study of two commonly used medicinal plants in Ayurveda: Neem (*Azadirachta indica* L.) and Tulsi (*Ocimum sanctum* L.). *International Journal of Pharmaceutical and Biological Archives,* **3**(3): 582-586.

Gautam, S. J., Ashok, K. and Ajay, K. (2011). Potential antimicrobial activity of *Zizyphus nummularia* against medically important pathogenic microorganisms. *Asian Journal of Traditional Medicines,* **6**(6): 267.

Giridhar, R., Rao, K.K., Banerjee, S.K. (1980). Antibacterial activity of the extract from *Tecomella undulata* (G.Don) Seem. *Indian Drugs,* **17:** 176.

Gomah, N. E. andEssam, A.M. (2011). Antimicrobial activity of extracts and latex of *Calotropis procera* (Ait.) and synergistic effect with reference antimicrobials. *Research Journal of Medicinal Plant,* **5:** 706-716.

Goyal, M. D., Sasmal and Nagori, B. P. (2012). Analgesic and anti-inflammatory activity of ethanolic extract of *Zizyphus nummularia*. *Research Journal of Medicinal Plant,* **6**(7): 521-528.

Gupta, D. (2013). Antibacterial activity of leaf extracts of *Nyctanthes arbor-tristis* against different strains of bacteria. *Biologix,* **2**(I): 16-18.

Gupta, D. and Rathore, V. (2012). Contradictory Effects of *Chrysanthemum cinerariae folium* Leaf Extracts against Different Bacteria. *Biologix,* **1**(1): 85-86.

Habbal, O. A., Al-Jabri, A. A., El-Hag, A.H., Al-Mahrooqi, Z. H., Al-Hashmi, N. A. (2005). *In-vitro* antimicrobial activity of *Lawsonia inermis* Linn (henna) - A pilot study on the Omani henna. *Saudi Medical Journal,* **26:** 69-72.

Habbal, O., Hasson, S.S., El-Hag, A. H., Al-Mahrooqi, Z. H., Al-Hashmi, N. A., Bimani, Z., Al-Balushi, M.S. and Jabri, A.A. (2011). Antibacterial activity of *Lawsonia inermis* Linn (Henna) against *Pseudomonas aeruginosa*. *Asian Pacific Journal Tropical Biomedical,* **1**(3): 173–176.

Hammer, K. A., Carson, C. F. and Riley, T.V. (1999). Antimicrobial activity of essential oils and other plant extracts. *Journal Applied Microbiology,* **86:** 985-990.

Himalaya, (2002). Herbal monograph, Himalaya herbal health care. URL: www. Himalayahealthcare/researchpaper/index.hlm.

Iqbal, H. and Khan, Z. U. (2008).Antimicrobial activity of *Capparis decidua* (Forssk.) Edgew. *Pakistan Journal of Plant Sciences,* **14**(1): 29-33.

Jagannadh, J. H. and Radhika, M. (2006). Antimicrobial emulsion (coating) based on biolymer containing neem (*Meliaaz adirachata*) and Turmeric (*Curcuma longa*) extract for wound healing. *Biomedical Mater Engineering*, **16**(5): 329-336.

Joji Reddy, L., Spandana, G., Beena, J. and Reshma D. J. (2012). Evaluation of antibacterial and antioxidant activities of the leaf and bark extracts of *Azardirachta indica* A.Juss. *World Journal of Pharmaceutical Research*, **1**(3): 661-672.

Kapoor, B. B. S., Bhumika, and Khatri, J. S. (2007). Antimicrobial activities of some medicinal tree species of Hanumangarh District of Rajasthan. *Journal Phytology*, **20**(2): 325-326.

Kapoor, B. B. S., Acharya, S., Mishra, R. and Swami, S. (2011). Antimicrobial screening of some arid tiliaceous plants. *Journal Phytology*, **24**(2): 255 – 256.

Kapoor, B. B. S. and Kumar, S. (2006). Herbal plants of Rajasthan desert: A potential source of antimicrobial principles. *Journal of Arid Land Studies*, **15**(4): 423-426.

Kapoor, B. B. S. and Ranga, P. (2003). Flavonoids from asteraceous medicinal plants of Rajasthan desert, *J. Phytol. Res.*, **16** (1): 101-102.

Kareem, S. O., Akpan, I. and Ojo, O. P. (2008). Antimicrobial activities of *Calotropis procera* on selected pathogenic microorganisms. *African Journal of Biomedical Research*, **11:** 105 – 110.

Kathem, K., Al-Rubiay Nawres, N., Jaber Al-Mhaawe, B.H. and Laith, K. (2008). Antimicrobial efficacy of Heena extracts. *Oman Medical Journal*, **23**(4): 253-256.

Mainasara, M. M., Aliero, B. L., Aliero, A. A. and Dahiru, S. S. (2011). Phytochemical and antibacterial properties of *Calotropis procera* (Ait) R. Br. (Sodom Apple) fruit and bark Extracts.*International Journal of Modern Botany*, **1**(1): 8-11.

Malekzadeh, F. (1968). Antimicrobial activity of *Lawsonia inermis* L. *Applied Microbiology*, **16**(4): 663-4.

Maragathavalli, S., Brindha, S., Kaviyarasi, N.S., Annadurai, B. and Gangwar, S. K. (2012). Antimicrobial activity of leaf extract of Neem (*Azardirachta indica* Linn). *International Journal of Science and Nature*, **3**(1): 110-113.

Mohashine, M.D., Bhuiyan, M. N., Seishi, M. and Tsutomu, S. (1997). Antimicrobial effect of crude *Azadirachta indica* neem bark extract on *Streptococcus sobrinus*. *Pediatric Dental Journal*, **7**(1): 71-74.

Nanasombat, S. and Lohasupthawee, P. (2005). Antibacterial activity of crude ethanolic extracts and essential oils of spices against Salmonellae and other enterobacteria. *KMITL Science Technique Journal*, **5**(3): 527-538.

Napar, A.A., Bux, H., Zia, M.A., Ahmad, M.Z., Iqbal, A. and Roomi, S. (2012). Antimicrobial and antioxidant activities of Mimosaceae plants; *Acacia modesta* Wall (Phulai), *Prosopis cineraria* (Linn.) and*Prosopis juliflora* (Swartz). *Journal Medicinal Plants Research*, **6**(15): 2962-2970.

Nariya, P. B., Bhalodia, N. R., Shukla, V. J. and Acharya, R. N. (2011). Antimicrobial and antifungal activities of *Cordia dichotoma* (Forster F.) bark extracts. *Ayuvedic*, **32**(4): 585–589.

Padma, T.V. (2005). India Ayurveda. *Nature*, pp. 436-486.

Pandey, G. and Madhuri, S. (2010). Pharmacological activities of *Ocimum sanctum* (tulsi): a review. *International Journal of Pharmaceutical Sciences Review and Research*, **5**(1): 61-66.

Pankaj, G. and Purshotam, K. (2011). *In vitro* Evaluation of Antibacterial activity of various crude leaf extracts of Indian sacred plant, *Ocimum sanctum* L. *British Microbiology Research Journal.*, **1**(3): 70-78.

Papageorgiou, V. P., Assimopoulou, A. N., Couladouros, E. A., Hepworth, D. and Nicolaou, K.C. (1999). The chemistry and biology of alkannin, shikonin and related naphthazarin natural products. *Angewandte Chemie International,* **38:** 270-300.

Purohit, U., Mehar, S.K. and Sundaramoorthy, S. (2002). Role of *Prosopis cineraria* on the ecology of soil fungi in Indian desert. *Journal of Arid Environments*, **52:** 17-27.

Rao, A. V., Bala, K., Lahiri, A.N. and Bala, K. (1989). Influence of trees on microorganisms of arid soil and its fertility. *Indian Forest,* **115:** 680-683.

Remya, M., Jyotirmoy, R. and Malcolm, N.(2010). Anti HIV-1 and antimicrobial activity of the leaf extracts of *Calotropis procera. International Journal of Scientific and Research Publications* (*IJSRP*), **4**(4): 242-246.

Riffel, A., Medina, L. F., Stefani, V., Santos, R.C., Bizani, D. and Brandelli, A. (2002). *In vitro* antimicrobial activity of a new series of 1, 4-naphthoquinones. *Braz. Journal Medicinal Biological Research,* **35:** 811-18.

Saadabi, M.A.A. (2007). Evaluation of *Lawsonia inermis* L. (Sudanese Henna) Leaf extracts as an antimicrobial agent. *Research Journal Biological Science,* **2**(4): 419-423.

Saba, I., Muneeba, B. and Hira, Y. (2011). *In-vitro* antibacterial activity of *Aloe barbadensis Miller* (Aloe Vera). *International Research Journal of Pharmaceuticals,* **1**(2): 59-64.

Salar, R.J. and Dhall, A. (2010). Antimicrobial and free radical scavenging activity of extracts of some Indian medicinal plants. *Journal Medicinal Plants,* **4**(22): 2313-2320.

Shazid, Md., Sharker, K. P. and Shahid, I.Z. (2009). Analgesic antibacterial and cytotoxic activity of *Cordia dichotoma*. *Pharmacologyonline,* **2:** 195-202.

Singh, V., Sethia, M., Mathur, K. and Nag T.N. (1988). Flavonoids of some arid zone plants of Rajasthan. *Ind. J. Pharm. Sciences*, **88**: 133.

Srivastava, S.K. and Srivastava, S.D. (1979). Taxifollin 3, 5-dirhamnoside from the seeds of *Cordia dichotoma*. *Phytochemistry.* 205-208.

Suman, D. (2012). Antimicrobial and antioxidant activities of green and ripe fruits of *Averrhoa carambola* Linn. and *Zizyphus mauritiana* Lam. *Asian Journal of Pharmaceutical and Clinical Research,* **5**(3): 102-105.

Upadhyay, R.K. (2011). Kareel plant: A natural source of medicines and nutrients. *International Journal Green Pharm,* **5:** 255-65.

Vadlapudi, V. and Naidu, C. K.(2010). *In vitro* bioactivity of Indian medicinal plant *Calotropis procera. Journal of Global Pharma Technology,* **2**(2): 43-45.

Velmurugan, V., Arunachalam, G. and Ravichandran, V. (2010). Antibacterial activity of stem barks of *Prosopis cineraria* (Linn.) Druce. *Archives of Applied Science,* **2**(4): 147-150.

Vinoth, J., Samiraj, R. andSathish, K. R. (2013). Preliminary phytochemical screening and antibacterial activity of *Lawsonia inermis* leaf extract against reference bacterial strains and clinically important Ampc beta-lactamases producing proteus mirabilis. *International Journal of Pharmacy and Pharmaceutical Sciences*, **5**(1).

Vohra, R. (2004). *Calotropis* the medicinal weed. Online medicinal book store, India.

Yesmin, M. N., Uddin, S.N., Mubassara, S. andAkond, M. A. (2008). Antioxidant and antibacterial activities of *Calotropis procera* Linn. *American-Eurasian Journal Agriculture and Environmental Science,* **4** (5): 550-553.

2015, Major Constrains and Verdict of Crop Productivity *Pages 441–444*
Editor: **Dr. U.N. Bhale**
Published by: **ASTRAL INTERNATIONAL (P) LIMITED, NEW DELHI**

Chapter 33

Effect of some Insecticides on Haemocytes of Bean Beetle, *Callosobruchus chinensis* L. (Coleoptera : Chrysomelidae : Bruchinae)

C.S. Disale[1]*, S.D. Disale[1] and R.S. Dubal[2]

[1]*Department of Zoology, D.B.J. College, Chiplun, Dist. Ratnagiri, Maharashtra, India*
[2]*Department of Zoology, Y.C. Institute of Science, Satara, Maharashtra, India*

ABSTRACT

The bean-beetle, Callosobruchus chinensis L. (Syn. Bruchus chinensis) is serious pest of Moong seeds (Vigna radiata (L.) R. Wilczek) and some other pulses in storage conditions. Effect of sub-lethal concentrations of synthetic insecticides such as Monodhan 36 and Sevin on haemogram was studied after determining the LC_{50} values of these insecticides. After 24 hours treatment, the haemocytes were studied by employing standard methods. The present study revealed that total haemocyte count (THC) was increased in response to these insecticides, proving their role in detoxification. Out of insecticides employed, Monodhan 36 caused highest increase in total cell count, followed by Sevin. Relative cell counts or differential haemocyte counts (DHC) were also calculated and found that the cells involved in defense mechanisms, such as granulocytes and plasmatocytes increased significantly as compare to others.

Keywords: *Haemocytes, THC, DHC, Detoxification, Defense mechanism.*

---

* *Corresponding Author.* E-mail: drsddisale@rediffmail.com

Introduction

The moong bean (also known as mung bean or green gram or golden gram) is the seed of *Vigna radiata* (L.) R. Wilczek). It is native to the Indian subcontinent and mainly cultivated in India, China, Pakistan, Thailand, Philippines, Indonesia, Burma, Bangladesh, Vietnam, Laos and Cambodia, but also in hot and dry regions of Southern Europe and the Southern United States. It is protein rich and one of the commonest Indian cuisines, used as an ingredient in both savory, sweet dishes, etc.

The bean-beetle, *Callosobruchus chinensis* L. (Syn. *Bruchus chinensis*) (Coleoptera: Chrysomelidae: Bruchinae) is one of the serious pests of Moong seeds and some other pulses in storage conditions (Roomi, *et al.*, 1997). The infestation begins when the crop is ripening in farm. The grubs and adults are responsible for considerable damage to the whole beans. There are three aspects of bruchid damage, which are of particular importance: (i) the overall weight loss; (ii) changes in nutritional quality and presence of off-smelling by-products of insect infestation; and (iii) loss in seed viability (Talekar, 1912). Inspite of various pest control strategies used to protect moong (Talekar, 1912), *C. chinensis* causes considerable economic losses.

The insects have open circulatory system that circulates haemolymph, containing various types of cells, called haemocytes, such as prohaemocytes, granulocytes, plasmatocytes, spherulocytes, oenocytoids and coagulocytes. These cells confer major immunity to the insects and are responsible for defense mechanisms like phagocytosis, encapsulation, nodule formation, coagulation, formation of other tissues, intermediary metabolism, etc.

Recently some information has gathered on effect of insecticides on the haemocyte morphology and their populations duè to notable contributions by Pilat (1935); Fisher (1936); Arvey *et al.* (1950); Arnold (1952); Jones and Tauber (1954); Trehan and Pajni (1961); Gupta and Sutherland (1968); Disale (1997); Bharti Prakash *et al.* (2007); Pugazhvendan and Soundararajan (2009), etc. However, there is no such information available on the present beetle, which is serious pest of Moong and some other beans.

Looking at above facts, in present investigation, attempts are made to find out the effect of commonly used two different insecticides on the haemocytes populations of this insect.

Materials and Methods

The adult bean beetles were collected from the infested storages of local shop keeper in the second week of September 2013. They were acclimatized to laboratory conditions by rearing in tin containers using their natural food. After three days, the beetles were taken out from rearing containers and divided in to two groups as 'A' and 'B'. Group 'A' was used to determine LC_{50} (for 24 hour) values of Monodhan 36 and Sevin by using probit analysis (Fenny, 1971). The 'B' group of beetles was further divided into three sub-groups as B1, B2 and B3. Group B1 and B2 were further treated with sub-lethal concentrations of Monodhan 36 (10 ppm) and Sevin (10 ppm) respectively, and group B3 was given untreated food and used as control. After 24 hours treatment, the beetles were used to determine haemogram.

The beetles were bleed in a drop of 2 per cent versene ringer fixative taken on slide (Rosenberger and Jones, 1960). The haemocytes were identified by using wet smears and haemocyte classification of Gupta (1979). Giemsa-stained preparations were used to determine DHC's and THC's were determined by Neubaeur's haemocytometry (Jones 1962).

Results and Discussion

Microscopic observations of fresh and permanent, Giemsa stained preparations revealed six types of haemocytes in present species, as per scheme of haemocyte classification (Gupta, 1979). These haemocytes are: prohaemocytes (Pro), granulocytes (Gra), spherulocytes (Sph), plasmatocytes (Pls), oenocytoids (Oes) and coagulocytes (Cos). The results of THC and DHC are tabulated in Table 33.1.

Table 33.1: Effect of Insecticides on Haemogram of *C. chinensis* L.

Category	*Sex*	*DHC's (Percentage)*						*THC's*
		Pro	*Gra*	*Sph*	*Pls*	*Oes*	*Cos*	*(Mean and SD)*
Monodhan 36	M	11.2	24.5	24.7	22.6	12.6	4.4	5100 ± 130
	F	10.4	23.4	24.3	28.3	10.1	3.5	4950 ± 97
Sevin	M	11.3	23.2	23.1	26.4	13.5	2.5	4850 ± 55
	F	10.6	22.6	23.5	25.9	14.3	3.1	4500 ± 60
Control	M	14.3	20.1	25.3	21.3	14.3	4.7	3640 ± 77
	F	13.8	18.5	24.3	23.8	16.4	3.2	3270 ± 62

Table 33.1 shows considerable increase in THC's in both sexes due to effect of 10ppm concentrations of insecticides. Comparatively, Monodhan 36 was found to be more effective than Sevin in increasing the total haemocyte counts. In general, the insecticides are responsible for increase in THC's. Similar kind of observations were also reported by several authors including Arvy, *et,al*, (1950) in *C. decimlineata* and Gupta and Sutherland (1968) in *P. Americana*. However, regarding the cause of increase in cell number, Pilat (1935), Fisher (1936) and Trehan and Pajni (1961) have expressed different views. They attributed the increase in number to the fact that the cells are released from adhered sites and brought in general circulation. This happens in response to insecticide, as these cells are directly involved in defense mechanisms (Gupta, 1979).

The DHC's were also changed in both sexes and in both cases of treatment too. The data given in table reveal that comparatively, Grs and Pls populations are increased, proving their role in detoxification of insecticides as mentioned by Gupta (1979).

Such changes in haemogram may be attributed to the haemocyte transformations and release of adhered haemocytes in to the general circulation due to the effect of insecticides for defense purpose.

Acknowledgement

The authors are thankful to the Principal, Y. C. Institute of Science, Satara, (Maharashtra) for his kind co-operation and encouragement while carrying out the research work leading to this article.

References

Arnold, J. W. (1952). The haemocytes of the Mediterranean flour moth, *E. kuhniella*. Zell. (Lepidoptera: pyralididae) Can. *J. Zool.*, **30:** 352 - 365.

Arvey, L. M., Gabe, M. and Lhoste, J. (1950). Action de quelques insecticides sur le sang du Doryophore (*C. decimlineata* Say). *Bull. Soc. Entomol.* Fr., **55**: 122 – 127.

Bharti Prakash, Bhargava, S. and Rawat, K. (2007). Effect of Penfluron on Total Haemocyte Count of *Dysdercus koenigii*. *Asian J. Exp. Sci.*, **21**(1): 151 - 154.

Disale, S. D. (1997). Coleopteran Haemocytes: A Comparative Study. *Ph.D. Thesis*, Shivaji University, Kolhapur, India.

Fenny, D. J. (1971). Probit Analysis. 2nd Ed, London and New York Cambridge University Press.

Fisher, R. A. (1936). The effect of a few toxic substances upon the Total blood cell count in the cockroach, *B.* orientalis L. Ann. *Entomol. Soc. Amer.*, **29**: 334 – 340.

Gupta, A. P. and Sutherland, D. J. (1968). Effect of sublethal doses of chlordane on the haemocytes and midgut epithelium of *P. Americana*. *Ann. Entomol. Soc. Amer.*, **61**(4): 910 – 918.

Gupta, A. P. (1979). *In* Insect Haemocytes, Cambridge University Press, Cambridge.

Jones, J. C. and Tauber, O. E. (1954). Abnormal haemocytes in mealworm (*T. molitor* L.) Ann. *Entomol. Soc. Amer.*, **47**(3): 428 – 444.

Jones, J. C. (1962).Current concepts concerning insect haemocytes. *Amer. Zool.*, **2**: 209 – 246.

Pilat, M. (1935). The effects of intestinal poisoning on the blood of locust (*L. migratoria*). *Bull. Entomol. Res.*, **26**: 283.

Pugazhvendan, S. R. and Soundararajan, M. (2009), Effect of Penfluron on Total Haemocyte Count of *Chrysocoris purpureus. Middle-East Journal of Scientific Research.*, **4** (4): 338-340.

Roomi, M. W., Shah, A. H. and S. A. Qureshi. (1997). Studies on biological control of the Bean-beetle (*Bruchus chinensis* L.) by *Apanteles flavipes* (Cam.) (Hymen., Braconidae). *Anzeiger für Schädlingskunde, Pflanzenschutz, Umweltschutz*. **70** (2): 37- 38.

Rosenberger, C. and Jones, J. C. (1960). Studies on total blood cell counts of the southern army worm larva *Prodenia irridania* (Lepidoptera). *Ann. Entomol. Soc. Amer.*, **53**: 351 – 355.

Talekar, N. S. (1912). Insect Pests of Mubgbean and Their Control. http: / /www.peipfi-komdasulsel.org/wp-content/uploads/2012/03/eam0121.pdf.

Trehan K. N. and Pajni, H. R. (1961). Mode of action of insecticides. *Beitr. Entomol.*, **11**: 386.

2015, Major Constrains and Verdict of Crop Productivity *Pages 445–454*
Editor: Dr. U.N. Bhale
Published by: ASTRAL INTERNATIONAL (P) LIMITED, NEW DELHI

Chapter 34

Morphology and GCMS Analysis of *Stachybotrys* spp.

S.R. Kadu[1], D.V. Hande[1]* and K.P. Suradkar[2]

[1]*P.G. Department of Botany,*
Shri Shivaji Science College, Amravati, M.S., India
[2]*Sant Gadgebaba*
Amravati University, Amravati – 444 604, M.S., India

ABSTRACT

The fungal species Stachybotrys was isolated from decaying wood from Melghat forest of Amravati. The fungus Stachybotrys is a cellulolytic saprophyte. The present fungi were grown in vitro and the extract of the culture medium was fractionated by different solvent for GCMS analysis. Morphological characters such as size, length and width of conidia and phialides of species were observed using light microscope. In this study author compare two species of Stachybotrys on basis of morphology and also study secondary metabolite content of species.

***Keywords**: Stachybotrys spp., Morphology, Secondary metabolite, GCMS analysis.*

Introduction

The genus *Stachybotrys* was erected with *S. atra* as type species (Corda,1837). *S. atra* (*S.chartarum*) is a greenish-black mold assigned to the fungal division Deuteromycota (Nelson, 2001). The genus is worldwide in distribution; some species are restricted to the tropical and subtropical regions (Pinruae *et al.*, 2004) and studied 11 species of *Stachybotrys* (Jong and Davis, 1976). *Stachybotrys* species are saprobes

* *Corresponding Author.* E-mail: dvhande@gmail.com

common in soil, decaying plant material, submerge wood and water damage buildings etc. *Stachybotrys* produces different types of metabolites. *Stachybotrys* is capable of producing at least six chemical classes of toxins including thichothecenes, stachybocins, cyclosporins, atranone and stachylysin (NIST Webbook). The objective of the present study deals with morphological identification of two species of *Stachybotrys* as well as isolation of secondary metabolites by using different solvent (nonpolar, medium polar and highly polar).

Materials and Methods

Morphology of *Stachybotrys*

Fungal samples were collected from different sites preferably during rainy season of monsoon to winter *i.e.* from the month of July to February end from different locations of Melghat forest. Decaying branches and twigs of plants were collected, wrapped in butter paper and brought to laboratory. Samples were placed in incubator for spore development at about 25-30°C. Samples were regularly examined under stereomicroscope for spore development. Slides were prepared by mounting in Lactophenol cotton blue. Measurements were made with the help of ocular micrometer. The taxonomic systems of fungi were followed for identification of samples Subramanian, 1957). Pure cultures were obtained by transferring spores on potato dextrose agar medium.

GCMS Analysis

Preparation of Extract

After 8 days, colonies from PDA were transfer to Potato Dextrose Broth for about 8-10 days. Mycelium mat was observed after10 days. Mycelium was removed from the conical flask, the media used for extraction of different metabolites. For extraction of secondary metabolites three extraction solvents were used ethylacetate, dichloromethane and chloroform (non polar, middle polar and polar). 25 ml of medium added for separating funnel to which 50 ml of ethylacetate was added and after shaking two layers of solvent observed.Transperant layer was separate out to which sodium sulphate was added to whole moisture get removed. Extract was kept in incubator at about 37°C until solvents get evaporated. Compounds were then dissolved in pure methanol for GCMS analysis. The procedure was repeated for dichloromethane and chloroform.

Gas Chromatography – Mass Spectroscopy Analysis

The analysis was carried out using gas chromatography – high resolution mass spectrophotometer. All samples were worked out at SAIF, IIT Powai, Mumbai, Maharashtra, India. 2 µl of previously concentrated extract solution was employed for GC – MS analysis. The GC-MS analysis was carried out using Alegent Hp 7880 with column of 30 meter length, 0.25 mm ID, and 0.32 thicknesses. Helium gas was used as carrier gas at constant flow rate of 1ml/minute. Injector temperature was set at 100°C. The oven temperature was programmed from 50°C to 280° C at 10° C/ minute to 200°C then 10°C/3 minutes to 250°C ending with a 5 minutes isothermal at 280°C. The sample was injected in split mode as 50:1. Three solvents used for preparing the sample for GC-MS analyses namely ethylacetate, dichloromethane and chloroform. Three samples of crude extract were analyzed by Gas Chromatography – Mass Spectroscopy.

Result and Discussion

Morphology of *Stachybotrys*

Stachybotrys nilgirica Subram. (Figures 34.1 and 34.2)

During present study, mycelium was creeping, hyaline or pale colored, branched hyphae.Conidiophores (phialophores) observed on substratum, erect straight, hyaline, slightly narrow above than below,2-3 septate (septa 21.28-39.9μ apart),swollen at base,110.7-289.1 μ long,smooth, terminating in a cluster of about 6-7 phialides,apical cell of philophore 13.3-17 μ long and 2-6 μ wide and subhyaline. Conidia when young appeared pale olive-green, conidia borne singly, acrogenously at the tip of phialides.1-celled,globose,tuberculate,dark,greeninsh black at maturity 15.32-23.98 μ in diameter.

Habitat: Saprobic on *Eucalyptus* sp.

Distribution: India, Japan, NewZealand, Peru, Cuba

Material examined: Maharashtra,Melghat Forest, India, on dead twig of *Eucalyptus* sp.,15/07/12, D.V.Hande, S.R.Kadu and K.P.Suradkar.

Among the species circumscribed in the genus, *S.microspora* (Jong and Davis, 1976) and *S. nilagirica* Subram. are close to *S. globosa* having spherical or globose conidia (Misra and Srivastava, 1982). *Stachybotrys sphaerospora* (Morgan-Jones and Sinclair, 1980) has spherical conidia similar to *S. nilagirica,* however, it differs by smaller conidia and branched conidiophores.

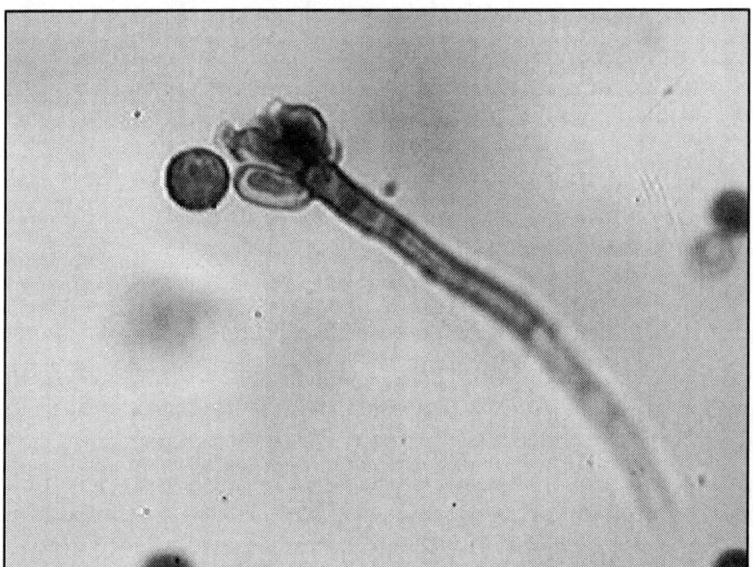

Figure 34.1

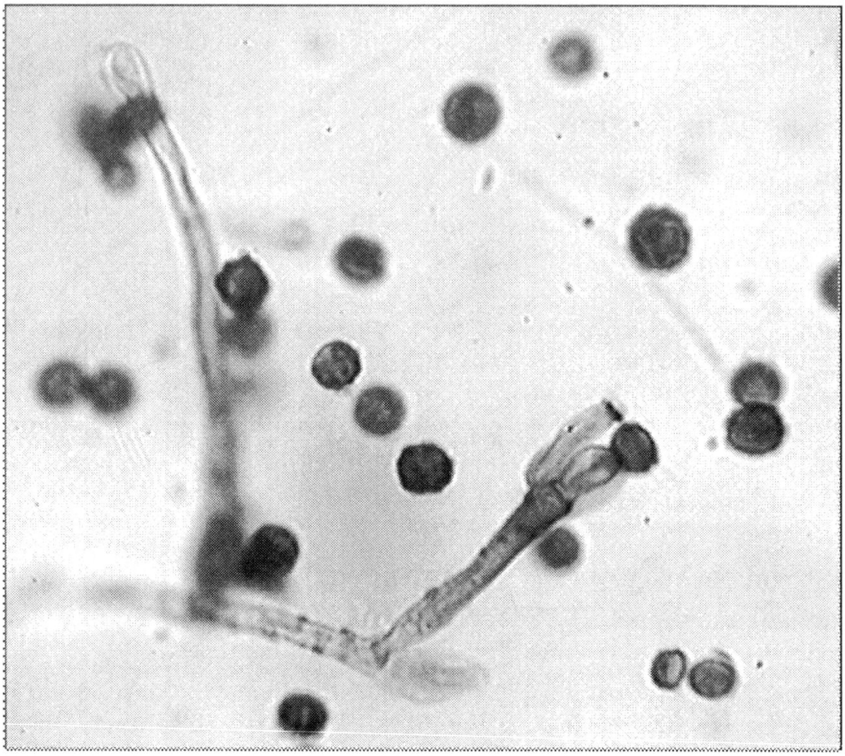

Figure 34.2

Stachybotrys chlorohalonata **Andersen and Thrane (Figures 34.3 and 34.4)**

During morphotaxonomical investigation, colonies appeared to thick, dark black coloured.Mycelium composed of hyaline, septate, branched, hyphae.Conidiophore arising erect, straight on the hyphae,almost hyaline towards base,septate,fuliginous near tip,73-91 μ long,13-16 μ wide.Conidiophore terminating in 3 to5 cluster of phialides.Phialides 11-16 μ long ×4-6 μin diameter,. The phialides were hyaline at base and dark toward apex. Conidia single, aseptate, smoth, ellipsoidal to broadly ellipsoidal, dark coloured,8-16 ×3-9 μ diameter.

Habitat: Saprobic on *Zea mays*

Distribution: India, Belgium, Denmark, Finland, Iraq, New Guinea, Spain and U.S.A.

Material examined: India, Maharashtra,Melghat Forest, on dead stem of *Zea mays*,20/07/12, D.V.Hande, S.R.Kadu and K.P.Suradkar.

Morphologically on PDA, *S. chlorohalonata* most closely resembles with *S. chartarum.* The conidiophores apparatus of *S. chlorohalonata* are similar in appearance to that of *S. chartarum*, but individual conidiophores are shorter (up to 70 mm and 90 mm, respectively). The phialides of *S. chlorohalonata* also resemble those of *S. chartarum*

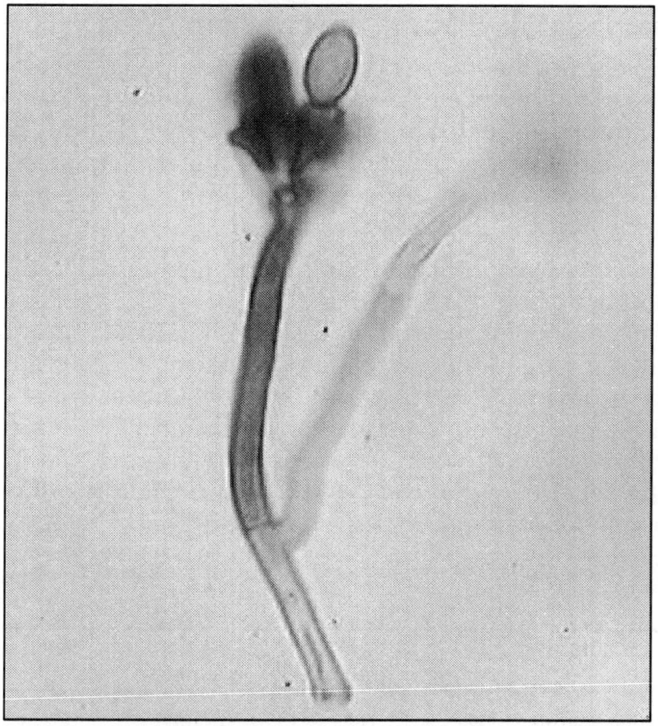

Figure 34.3

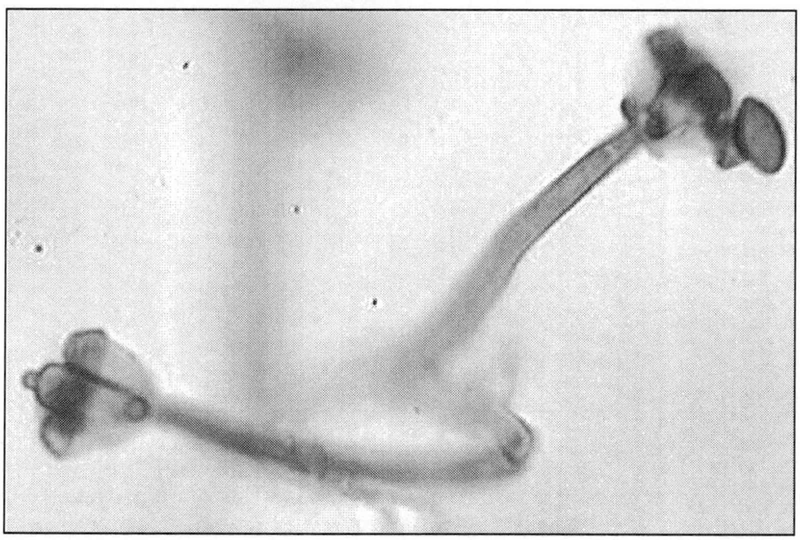

Figure 34.4

but are shorter. The conidia of *S. chlorohalonata* are ellipsoidal to broadly ellipsoidal with a smooth surface, in contrast to those of *S.chartarum*, which were slightly longer, ellipsoidal and have a rough surface.

Gas Chromatography – Mass Spectroscopy Analysis

The principle aim of the study was to develop a simple and reliable method for investigation of secondary metabolites excreted by the cultured *S.chlorohalonata*. The fungus was culture and incubated to avoid oxygen limitation during fungus growth and then allow samples for extraction, chromatographic separation and spectra detection. The GCMS data were interpreted using NIST library, results were matched with entries in the library of chemical compounds.

S.chlorohalonata grown on Potato Dextrose Broth media produced a wide variety of volatile compounds. More than 35 chromatographic signals (as shown in chromatograms) could be detected, and among these more than 25 volatile compounds were identified. After analysis of sample signals, the most significant signals with high abundance are listed in Tables 34.1–34.3. Numerous minor peaks were observed and could not be identified. Thus saturated and unsaturated hydrocarbons, alcohols, aromatic hydrocarbons, esters, aldehyde ketons,etc. could be detected. The compound classes of aldehyde, alcohols, ketones, esters, and sequiterpenes from the fungal origin has been previously reviewed (Meruva *et al.*, 2004).

Table 34.1: Secondary Metabolites Identified from *S. chlorohalonata* using Ethylacetate Solvent

Sl.No.	*Name of Compound*	*Chemical Structure*	*Retention Time (min)*	*MW*	*Abundance (Per cent)*
1.	Cyclopentanol, 2,4,4-trimethyl-	$C_8H_{16}O$	11.05	128	11.52
2.	2-Hexanone,3-methyl	$C_7H_{14}O$	11.13	114	2.03
3.	Hexyl octyl ester	$C_{14}H_{30}O$	11.57	214	8.48
4.	Cyclohexanol,1-ethyl	$C_8H_{16}O$	12.46	128	4.60
5.	2-Undecene,4,5-dimethyl	$C_{13}H_{26}$	14.29	182	9.09
6.	Phenol,2,4-bis(1,1-dimethyl)	$C_{14}H_{22}O$	16.55	206	4.45
7.	Heptadecane	$C_{17}H_{36}$	17.50	240	3.24
8.	1-Nanodecane	$C19H_{38}$	19.71	266	8.19
9.	Octadecane	$C_{18}H_{38}$	19.78	254	10.21
10.	Tetradecyltrichloro silane	$C_{14}H_{29}Cl_3Si$	21.358	330	2.47
11.	3-Methylnonadecane	$C_{20}H_{42}$	21.56	282	1.88
12.	1-Eicosene	$C_{20}H_{42}$	21.83	282	6.71
13.	1-Docosanol	$C_{22}H_{46}O$	23.77	326	3.95

Separation of compounds from *S. chlorohalonatum* was obtained by using non polar, medium polar and high polar solvents. Over 30 secondary metabolites were identified during the experiment. None of these compounds has been reported earlier by any researches from present fungus. The method was applied on a micro-scale method ("LC-UV-MS for detection) for metabolites screening of more than 400 fungal metabolites (Nelsen, 2002). *S. chartarum* is capable of producing six classes of chemical compounds, which are highly toxic (Scott, 2004).

Table 34.2: Secondary Metabolites Identified from *S. chlorohalonata* using Dichloromethane

Sl.No.	*Name of Compound*	*Chemical Structure*	*Retention Time (min)*	*MW*	*Abundance (Per cent)*
1.	Pentanoic acid,5-hydroxy-2,4-di-t-butylphenyl ester	$C_{19}H_{30}O_3$	11.8	306	20.29
2.	3-Octadecene	$C_{18}H_{38}$	13.0	254	10.91
3.	5-Octadecene	$C_{18}H_{36}$	15.8	254	29.57
4.	1-Nonadecene	$C_{19}H_{38}$	19.0	266	22.52
5.	3-Eicosene	$C_{20}H_{42}$	22.5	282	6.23
6.	Nonadecene	$C_{19}H_{40}$	22.6	268	0.65
7.	3-Eicosene	$C_{20}H_{42}$	25.2	282	3.49
8.	1,2-Benzene dicarboxylic acid, mono(2-ethylhexyl)ester	$C_{16}H_{22}O4$	27.3	278	5.16

Table 34.3: Secondary Metabolites Identified from *S. chlorohalonata* using Chloroform

Sl.No.	*Name of Compound*	*Chemical Structure*	*Retention Time (min)*	*MW*	*Abundance (Per cent)*
1.	Pentanoic acid,5-hydroxy-2,4-di-t-butylphenyl ester	$C_{19}H_{30}O3$	11.8	306	5.59
2.	5-Octadecene	$C_{18}H_{36}$	15.8	254	6.09
3.	1-Nonadecene	$C_{19}H_{38}$	19.0	266	25.67
4.	Hexadecene	$C_{16}H_{34}$	19.1	226	1.63
5.	3-Eicosene	$C_{20}H_{42}$	22.5	282	27.38
6.	Nonadecene	$C_{19}H_{40}$	22.6	268	1.32
7.	1,2-Benzene dicarboxylic acid, mono(2-etjulhexyl) ester	$C_{16}H_{22}O4$	27.3	278	26.84

Identification of volatile metabolites of *S.chloroholonata* includes commonly obtained hydrocarbons with predominance of carbon. According to the results, C12-C20 alkanes were mostly identified. Most interesting was the appearance of series of straight chain alkane acetyl ester including hexyl, heptyl, decyl alcohols. 2-nonadecene and 2-undecene were abundant metabolites. These metabolites have been described to produce by *T. aureoviride* (Bruce *et al.,* 2000). Eicosene and octadecene were also found in higher percentage in *S. chlorohalonata*. Octadecene and Eicosene are said to be found in marine fungi *A.ustus* (Oleinikova *et al.,* 2011). Octadecene is said to posses various activities such as anticancer, antioxidant and anti microbial activities (Lee *et al.,* 2007;. Kumar *et al.,* 2011).

Phenol,2,4-bis(1,1dimetylethyl)was found in ethylacetate extract. Phenol, 2, 4-Bis (1, 1-Dimethyl ethyl) has good antioxidant activity (Ajayi *et al.,* 2011). 2-Hexanone is known as methyl-n-butyl ketone or propyl acetone used as an important solvent for varnishes, lacquers, ink thinner, nitrocellulose, resin, oils, waxes and fats.2-Hexanone is also used as an intermediate in the manufacture of organic chemicals.

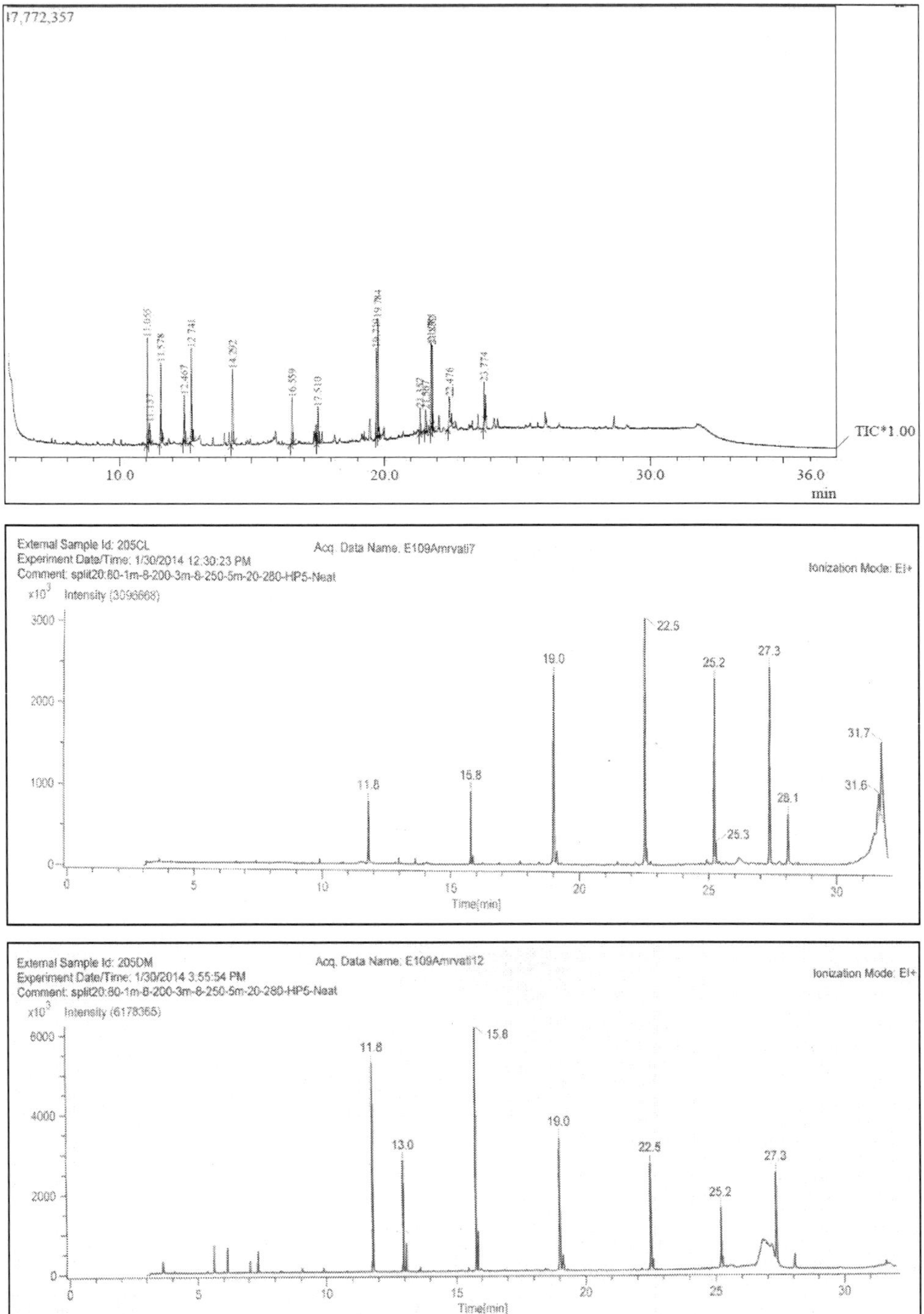

Figure 34.5: Chromatograms of *S. chlorohalonata* Showing Peaks of Secondary Metabolites in Three different Solvent.

Cyclohexanol is used in the production of nylon, lacquers, paints, varnishes, degreasers,plastics, soaps, detergents, and insecticides. It may be used also in the production of a highly deadly chemical warfare agent cyclosarin (Proctor *et al.*, 2006).

Many hydrocarbons are the major substances in diesel fuel, including octane; 1-octene; heptane, 2-methyl; hexadecane; undecane, 4-methyl; nonane,3-methyl; and benzene, 1,3-dimethyl (Strobel *et al.*, 2008).

It is found that it is first application using three different solvent (non polar, medium polar and high polar) to indentify massive number of volatile compounds from *S.chlorohalonata*. The determination of secondary metabolites by GCMS is very useful because of its highly sensitive detection and separation capability.

Conclusion

A morphological observation reveals the two species of *Stachybotrys* from Amravati reported. Volatile compounds are involved in interaction between fungi and their environment. Thus GCMS methods for the identification of volatile compounds are very important in investigation of their formation and functions in biological interactions. The present study deals with successful separation of secondary metabolites using different extraction solvent. GCMS offers powerful tool for study of dynamic range of metabolites. The present method resulted in identifications of more than 25 compounds (Tables 34.1–34.3) from *S.chlorohalonata* which is not reported.

Acknowledgement

The authors are grateful to Sophisticated Analytical Instrumentation Facilities, IIT Powai, Mumbai; Maharashtra, India for analysis of fungal metabolites using gas chromatography – high resolution mass spectrophotometer.

References

Ajayi, G.O., Olagunju J.A., Ademuyiwa O. and Martins, O.C. (2011). Gas chromatography-mass spectrometry analysis and phytochemical screening of ethanolic root extract of Plumbago zeylanica Linn. *Journal of Medicinal Plants Research,* **5**(9): 1756-1761.

Bruce, A., Wheatley R.E., Humphris S.N., Hackett C.A., Florence, M.E.J. (2000); Production of volatile organic compounds by Trichoderma in media containing different amino acids and their effect on selected wood decay fungi; *Holzforschung,* **54:** 481–486.

Corda A.C.J. (1837) Icones fungorum hucusque cognitorum, vol. 1. Prague. 32 p.

Jong, S.C. and Davis, E.E.(1976). Contribution to the knowledge of *Stachybotrys* and *Memnoniella* in culture. *Mycotaxon,* **3:** 409-486.

Kumar, Vinay A.K., Bhatnagar J and Srivastava, J.N. (2011). Antibacterial activity of crude extracts of *Spirulina platensis* and its structural elucidation of bioactive compound. *Journal of Medicinal Plants Research,* **5**(32): 7043-7048.

Lee, Y.S., M.H. Kang, Y.S. Y.S. Cho and Jeong, C.S. (2007). Effects of constituents of *amomum xanthioides* on gastritis in rats and on growth of gastric cancer cells. *Archives of Pharmaceutical Research*, **30**: 436-443.

Meruva, N.K., Penn J.M., Farthing, D.E.(2004). Rapid identification of microbial VOCs from tobacco molds using closed-loop stripping and gas chromatography/time-of flight mass spectrometry; *Journal of Industrial Microbiology and Biotechnology*, **31**: 482–488.

Misra, P.C. and Srivastava, S.K. (1982). Two undescribed *Stachybotrys* species from India. *Transactions of the British Mycological Society*, **78:** 556-559.

Morgan-Jones G. and Sinclair, R.C. (1980). Notes on hyphomycetes. XXXIII. *Stachybotrys sphaerospora* sp.nov. from South Africa. *Mycotaxon*, **10**: 372-374.

Nelsen, K.F. (2002). Mould growth on building materials. Secondary metabolites, mycotoxins and biomarkers. Internets add. http: //www.biocentrum.dtu.dk/mycology/publication/phd-kfn.pdf

Nelson, B.D. (2001). *Stachybotrys chartarum*: The toxic indoor mold. APBSnet feature story. Internet address: http: //www.apsnet.org/online/feature/stachybotrys/

NIST (2005). Webbook. Internet adds. http: //webbook.nist.gov/chemistry/

Oleinikova, G.K., N.N. Slinkina and Sh. Sh.Afiyatullov (2011) Nonpolar compounds and free fatty acids from marine fungi *Aspergillus ustus. Chemistry of Natural Compounds*, **47**(5): 805- 806.

Pinruae, U., Mckenzie E.H.C., Jones E.B.G. and Hyde, K.D. (2004). Two new species of *Stachybotrys*, and a key to the genus. *Fungal Diversity*, **17:** 145-157.

Proctor, S.P., Newton K.J., Heren T., White, R.F. (2006). Effects of sarin and cyclosarin exposure dutiny the 1991 Gukf War on neurobehavioral functioning in US army veterans. *NeuroToxicology*., **27**(6): 921-939.

Scott, A. Masten (2004) Stachybotrys chartarum Review of toxicological literature. Research Triangle Park, North Carolina.

Strobel, G. A., Berk Knighton, Katreena Kluck, Yuhao Ren (2008) The production of myco-diesel hydrocarbons and their derivatives by the endophytic fungus *Gliocladium roseum* (NRRL 50072). *Mycologia*, **154**: 3319–3328

Subramanian, C.V. (1957). Hyphomycetes IV. *Proceedings of the Indian National Sciences Academy*, **46**: 324-335.

2015, Major Constrains and Verdict of Crop Productivity *Pages 455–461*
Editor: **Dr. U.N. Bhale**
Published by: **ASTRAL INTERNATIONAL (P) LIMITED, NEW DELHI**

Chapter 35

Looming of *Trichoderma* spp. against *Alternaria alternata* Provoking Leaf Spot of *Trigonella foneum-graecum* L.

M.G. Ambuse[1] and U.N. Bhale[2*]

[1]*Department of Botany, Shrikrishna Mahavidyalaya, Gunjoti, Tq. Omerga, Dist Osmanabad – 413 613, M.S., India*
[2]*Research Laboratory, Department of Botany, Arts, Science and Commerce College, Naldurg, Tq. Tuljapur, Dist Osmanabad – 413 602, M.S., India*

ABSTRACT

Leaf spot of fenugreek (Trigonella foneum-graecum L.) is caused by Alternaria alternata. This paper describes the efficacy of Trichoderma spp against sensitive and resistant isolates of A. alternata by dual culture method under in vitro conditions. Trichoderma viride T. harzianum, T. virens, T. koningii and T. pseudokoningii species were used for antagonistic study. Results indicated that all Trichoderma species showed antagonistic activity. But among them, T.viride, T. harzianum and T.pseudokoningii showed 87 per cent antagonistic activity than others in case of sensitive isolate of test fungus. Resistant isolate of pathogen was restricting the antagonism in some extent.

Keywords: *Trigonella foneum-graecum, Alternaria alternata, Trichoderma species dual culture.*

---

* *Corresponding Author.* E-mail: unbhale2007@rediffmail.com, mallama.a10@gmail.com

Introduction

Fenugreek (*Trigonella foenum-graecum* L.) of family Fabaceae is an important leafy vegetable and its seeds are used as condiments in India. Fenugreeks are herb that is commonly found growing in the Mediterranean region of the world. The seeds and leaves are primarily used as a culinary spice; it is also used to treat a variety of health problems in Egypt, Greece, Italy and South Asia. (Morcos *et al.*, 1981).

In India and China it has also been used to treat arthritis asthma, bronchitis, improve digestion; maintain a healthy metabolism, cure skin problems, treat sore throat and cure acid reflux. Recent studies have shown that fenugreek helps lower blood glucose and cholesterol levels and may be an effective treatment for both type 1 and 2 diabetes. It is also being studies for its cardiovascular benefits (Patil *et al.*, 1997; Abdel-Barry *et al.*, 2000; Rao, 1996).

India is the world's second largest producer of vegetables next to China. Vegetables are more susceptible to insect's pests and diseases due to their softness as compared to other crops and virtual absence of resistance characters because of intensive hybrid cultivation (Chiranjeevi *et al.*, 2002). The native value of healthy vegetables are altered because of fungal attack and sometimes fungi produce certain mycotoxin in them and make them unsuitable for human consumption

During Oct 2009 to Jan 2011 surveys was conducted in Fenugreek growing areas and markets of Marathwada region of Maharashtra. It suffers severely by leaf spot disease incited by *Alternaria alternata* (Fr) Keissler). The *in vitro* screening with our arbitrary system of bio-antagonists effective against soil borne pathogens is a simplistic approach to understand a small sector of biological system in disease control. Therefore, biological control of plant pathogens has been considered as a potential control strategy in recent years and search for these biological agents is increasing. *Trichoderma* is the most commonly used fungal biological control agent and have long been known as effective antagonists against plant pathogenic fungi (Chet *et al.*, 1981; Kumar and Mukerji, 1996; Bhale *et al.*, 2012). During the last decade, species of *Trichoderma* have emerged as the most powerful bioprotectants for the management of a wide variety of plant diseases by virtue of their broad spectrum action against a number of plant diseases caused by fungi, bacteria, viruses and even nematodes (Mukhopadhyay, 2005). Thus the present study was aimed to evaluate the antagonistic activity of *Trichoderma* spp. in laboratory conditions.

Materials and Methods

Isolation and Identification of Test Pathogen

Leaf spot infected materials were collected and cut into small pieces (2mm) by sterilized blade then surface sterilized with mercuric chloride (1 per cent $HgCl_2$) for 1 min. The pieces were then washed with sterilized distilled water thrice and dried by sterilized blotting paper. In each case, surface disinfested tissue plated on potato dextrose agar (PDA) medium and produced an *Alternaria* species (Simmons 2007; Subramaniam1971: Ellis, 1971).

Isolation of *Trichoderma* spp.

Rhizosphere soils of irrigated and non irrigated plants were collected from different parts of Marathwada region of Maharashtra. From the rhizosphere soil samples, desired *Trichoderma* species were isolated by using potato dextrose agar (PDA) and *Trichoderma* selective medium (TSM) by dilution plate technique (Johnson, 1957). The isolated species were identified by reculturing on another petriplates containing sterilized TSM. The isolated species were identified up to species level based on colony characters, growth of fungus and structure of mycelium, conidiophores and conidia (Kubicek and Harman, 2002). All *Trichoderma* spp. were purified by hyphal tip technique. The isolated strains of *Trichoderma* spp. were maintained throughout the study by periodical transfers on PDA and TSM slants under aseptic conditions to keep the culture fresh and viable.

Dual Culture Experiment

The Antagonistic efficacy of different species of *Trichoderma. viride, T. harzianum, T. virens,* T. *koningii* and *T. pseudokoningii* were tested against the isolated sensitive and resistant pathogenic fungus by dual culture experiment (Morton and Stroube, 1955). *Trichoderma* spp. and test fungus was inoculated at 6 cm apart. Three replicates were maintained for each treatment and incubated at 28 ± 2° C for 7 days. Monoculture plates of both served as control. Seven days after incubation (DAI), radial growth of test fungus and *Trichoderma* isolates were measured. Colony diameter of test fungus in dual culture plate was observed and compared with control. The growth inhibition was calculated by using the formula: 100 X C - T/C, Where C = growth in control and T = growth in treatment (Vincent, 1947).

Statistical Analysis

Statistical analyses of the experiments were performed by using the book of Mungikar (1997).

Results and Discussion

Isolation and Identification of Test Pathogen

Diseased leaf spots were found as a typical blackish spot on leaves, leaf spot vary in size from pin point up to one to two cm in diameter and dark ash colored spot with concentric rings appeared. *Alternaria alternata* (Fr.) Keissler, Colonies usually black conidia formed in long, often branched chain, overall length 20-63(37) µm, 9-18(13) µm thick in the broadest part; beak pale, 2-5µm thick. Such symptoms were collected from different locations of Marathwada region of Maharashtra and ten isolates of *Alternaria alternata* were isolated. The culture was deposited at Department of Botany, Arts, Science and Commerce College, Naldurg.

Isolation of *Trichoderma* spp.

Isolates of five species of *Trichoderma*, T. *viride* Pers. *T. harzianum* Rifai, *T. virens* J. Miller, Giddens and Foster, *T. koningii* Oud. and *T. pseudokoningii* Rifai. were isolated from irrigated and non-irrigated rhizosphere soil of Marathwada region of

Maharashtra. Isolates are deposited at Dept. of Botany, Arts, Science and Commerce College, Naldurg.

Dual Culture Experiment

Results indicated that all *Trichoderma* species showed antagonistic activity. But among them, *T.viride, T. harzianum* and *T. pseudokoningii* showed 87 per cent antagonistic activity than others in case of sensitive isolate of test fungus. Resistant isolate of pathogen was restricting the antagonism in some extent. Overall, all *Trichoderma* species were found more than 50 per cent antagonistic nature (Table 35.1 and Figure 35.1).

Table 35.1: Effect of *Trichoderma* Species against *A. alternata*

Trichoderma Species	*Isolates*	*Radial Growth of A. alternata*	*Radial Growth of Trichoderma Species*	*Per cent Inhibition*
T.viride	SR	1116	7974	87.7782.22
T.harzianum	SR	2428	6662	73.3368.88
T.koningii	SR	3040	6050	66.6655.55
T.pseudokoningii	SR	2531	6559	72.2265.55
T.virens	SR	3236	5854	64.4460.00
CD (p=0.05)				6.95

S: Sensitive; R: Resistant.

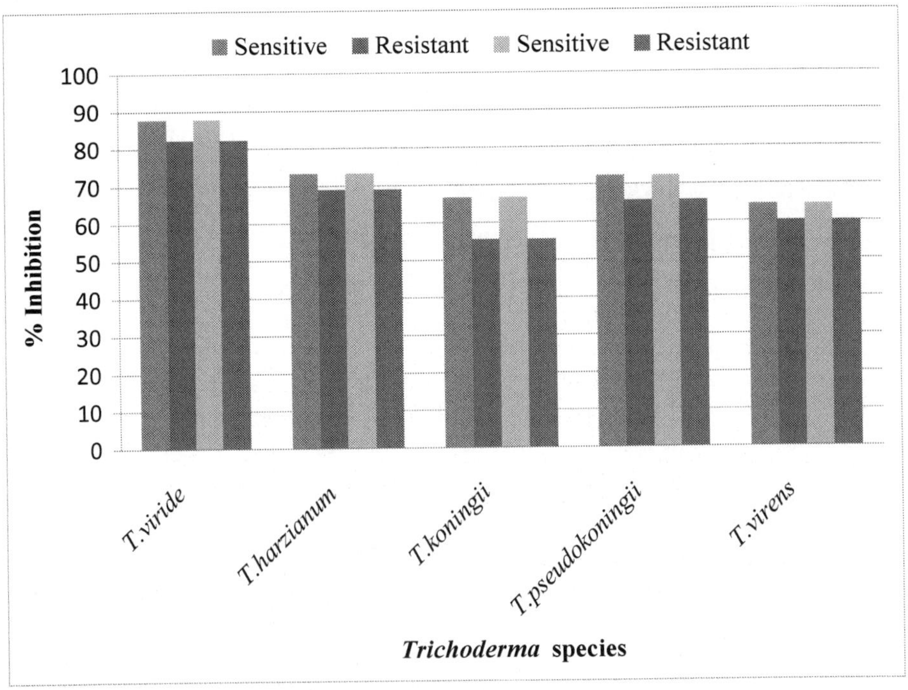

Figure 35.1: Influence of *Trichoderma* Species against *Alternaria alternata.*

Several workers have been reported that the use of *Trichoderma* species against number of plant pathogenic fungi (Brisa *et al.*, 2007; El-Mougy *et al.*, 2007; Harman, 2006).Akbari and Parakhi (2007) reported *T.viride*-I and *T.hamatum*-IV and V isolates showed strong antagonism against *Alternaria alternata* causing blight of sesame. High inhibitory effect of volatile toxic substances emitted by *Trichoderma* spp. on the radial growth of *Fusarium* spp. has also been reported by Dubey *et al.* (2007).The inhibition was high with the direct use of *Trichoderma* spp. in dual culture against *Fusarium oxysporum f sp psidii* (61-69 per cent) and *F.solani* (58-68 per cent) (Gupta and Mishra, 2009). Kumar *et al.* (2007) tested three species of *Trichoderma i.e. T.virens, T. viride* and *T. harzianum* against *F.moniliforum var subglutinas* and found them effective. Among the *Trichoderma* species *T. viride* showed the best performance *in vitro* biological control of *Fusarium oxysporum* followed by others (Irfan and Khalid, 2007). *Trichoderma viride* reached the confluence of the Petri dish four days after sowing, so that different fungal isolates occupy a surface of 29 per cent to *Fusarium roseum* (Bouziane *et al.*, 2011). Waghmare and Kurundkar (2011) reported efficacy of *Trichoderma* species against *Fusarium oxysporum* f. sp. *carthami* causing wilt of safflower and isolates no. 29 and 33 were found to minimum growth of the pathogen as compared to others. The species of *Trichoderma* significantly inhibited the mycelial growth of plant pathogenic fungi (Rajkonda *et al.*, 2011).

Conclusion

Trichoderma species play an important role in controlling fungal plant pathogens. The use of *Trichoderma*-based products is not only safe for the farmers and consumers but it is also good for the environment. However, much more work needs to be done to develop stable, cost effective, easy to produce and easy to apply formulations. Our results concluded that the tested *Trichoderma* spp reduced the growth of pathogen.

Acknowledgement

Authors are thankful to the Director, Agharkar Research Institute (ARI) Pune for the identification of *Trichderma* spp.And also thankful to UGC, New Delhi for financial assistance for Major Research Project.

References

Abdel-Barry, J. A., Abdel-Hassan, I. A., Jawad, A. M. and Hakiem, M. H. (2000). Hypoglycaemic effect of aqueous extract of the leaves of *Trigonella -foenum-graecum* in healthy volunteers. *EastMediterr Health J.*, **6**: 83-88.

Akbari, L.F. and Parakhi, A.M. (2007). Eco- friendly approaches to manage blight of seasame. *J.Mycol. Pl.Pathol*. **37**(3): 389-400.

Bhale, U. N., M. G. Ambuse, V. S. Chatage and J. N. Rajkonda (2012**)**. Bioefficacy of Trichoderma isolates against pathogenic fungi inciting spinach (*Spinacea oleracea* L.). *JBiopest*. **5**(2): 222-227.

Bouziane, Z., Dehimat, L., Abdel Aziz, W., Benabdelkader, M., Kacem Chaouche N. (2011). The antagonism between *Trichoderma viride* and other pathogenic fungal strains in *Zea mays. Agric. Biol. J. N. Am*. **2**(4): 584-590.

Brisa, R., Fernando, M.A., Asunción, G.S., Noemí, M.R., Arturo, P.E. and José, M.D. (2007). The gene coding for a new transcription factor (*ftf1*) of *Fusarium oxysporum* is only expressed during infection of common bean. *Fungal Genetics and Biology*. **44**: 864-876.

Chet, I., Harman, G.E. and Baker, R. (1981). *Trichoderma hamatum*: its hyphal interaction with *Rhizoctonia solani* and *Pythium* spp. *Microbial Biology*. **7**: 29-38.

Chiranjeevi, C., Reddy, I. P., Naryanamma, M. and Neerja. (2002).Effect of shoot clipping and insecticides on the incidence of fruit borer in brinjal. *International conferece on Vegetables. Nov.11-14 Bangalore, India*. p.271.

Dubey, S.C., Suresh, M. and Singh, B. (2007). Evaluation of *Trichoderma* spp against *Fusarium oxysporum f sp ciceri* for integrated management of chickpea wilt. *Biol.Control*.**40**: 118-127.

Ellis, M.B. (1971). Dematiaceous Hypomycetes. *Commonwealth Mycological. Institute, Kew, Surrey, England.*

El-Mougy, S.N., Nadia, G.E.and Abdel-Kader, M.M. (2007). Control of wilt and root rot incidence in *Phaseolus vulgaris* L. By some plant volatile compounds.*Journal of Plant Protection. Reearchs*, **47:** 255-265.

Gupta, V.K. and Mishra, A.K. (2009).Efficacy of biogents against *Fusarium* wilt of guava. *J.Mycol Pl. Pathol.* **39**(1)**:** 101-106.

Harman, G.E. (2006).Overview of mechanisms and uses of *Trichoderma* spp. *Phytopathol*., **96:** 190–194.

Irfan Yousaf Sahi, Khalid A.N. (2007). *In vitro* biological control of *Fusarium oxysporum* causing wilt in *Capsicum annuum*. *Mycopath.* **5**(2): 85-88

Johnson, LA.(1957). Effect of antibiotics on the number of bacteria and fungi isolated and fungi isolated from soil by dilution plate method. *Phytopathology*, **47**: 21-22.

Kubicek, C. P. and Harman, G. E. (2002). *Trichoderma* and *Gliocladium* (vol. I). *Basic biology, taxonomy and genetics.* pp.14-24.

Kumar, P., Mishra, A. K. and Panday, B.K. (2007). *In vitro* evaluation of *Trichoderma* spp against vegetative mango malformation pathogen *Fusarium moniliformae var subglutinas. J. Eco-Friend Agri.* **2**: 187-189.

Kumar, R.N. and Mukerji, K.G. (1996). Integrated disease management futureperspectives, pp. 335-347. *In:* K.G.Mukerji, B. Mathur, B.P. Chamala and C.Chitralekha (Eds.), *Advances in Botany*. APH Publishing Corporation, New Delhi.

Morton, D.J.and Stroube, W.H. (1955).Antagonistic and stimulating effects of soil microorganisms upon sclerotium. *Phytopathol*. **45**: 417-420.

Mukhopadhyay, A.N. (2005). Comparative antagonistic properties of *Gliocladium* virens and *Trichoderma harzianum* on *Sclerotium rolfsii* and *Rhizoctonia solani*. *Indian Phytopath*. **40**(2): 276.

Mungikar, A.M. (1997). An Introduction to Biometry. *Saraswati Printing Press, Aurangabad.*, pp.57-63.

Morcos, S.R., Elhawary, Z. and Gabrial, G. N. (1981). Protein rich food mixtures for feeding the young in Egypt. 1. Formulation. *Z Ernahrungswiss,* **20**: 275-282.

Patil, S.P., Niphadkar, P.V. and Bapat, M.M. (1997). Allergy to fenugreek *Trigonella foenum graecum*). *Ann Allergy Asthma Immunol,* **78**: 297-300.

Rajkonda, J. N., Sawant, V.S., Ambuse, M. G. and Bhale, U N. (2011). Inimical potential of *Trichoderma* species against pathogenic fungi. *Plant Sciences Feed.,* **1**(1): 10-13.

Rao, P.U., Sesikeran, B., Rao, P. S. (1996). Short term nutritional and safety evaluation of fenugreek. *Not Res,* **16**: 1495-1505.

Simmons, *E.G.*(2007). *Alternaria*: *An identification Manual*. Utrecht, Netherlands: CAB fungal Biodiversity Center.

Subramanian, C.V. (1971)."Hypomycetes an account of Indian species except *Cercospora*". Indian Council of Agricultural Research, New Delhi, p.810.

Vincent, J.M. (1947). Distortion of fungal hyphae in the presence of certain inhibitors. *Nature.* **150** : 850.

Waghmare, S. J. and Kurundkar, B. P. (2011). Efficacy of local isolates of *Trichoderma* sppagainst *Fusarium oxysporum* f. sp. *carthami* causing wilt of safflower. *Ad. Plant Sci.* **24**(1): 37-38.

2015, Major Constrains and Verdict of Crop Productivity *Pages 462–475*
Editor: Dr. U.N. Bhale
Published by: ASTRAL INTERNATIONAL (P) LIMITED, NEW DELHI

Chapter 36

Arbuscular Mycorrhizal Fungi: An Environo-friendly Bioinoculant for Sustainable Agriculture

D.D. Namdas*

P.G. Research Laboratory,
Department of Botany, Dahiwadi College, Dahiwadi – 415 508
Tal. Man, Dist. Satara, Maharashtra, India

Introduction

The term mycorrhizae refer to a plant root associated with symbiotic fungus, which is functionally distinct organ involved in mineral nutrient uptake from soil. Microbial communities of soil include bacteria actinomycetes, fungi, algae, protozoans and others which play important role in cycling of elements, biological conversions, humas formation, ecosystem sustains, geo-chemical cycling and productivity. Among the microbial communities, the most important and widespread symbiotic microorganisms are the vesicular arbuscular mycorrhizal (VAM) fungi. Occurrence of VAM fungi is ubiquitous. Distribution of VAM is markedly influenced by climatic and edaphic factors and also land use. These includes soil fertility, soil moisture, pH, temperature, altitude, soil organic matter, plant susceptibility, light intensity and depth in addition to disturbance (Mohankumar and Sivaswamy, 1992).

Research efforts during past few years throughout the world made their attention towards commercial exploitation of microbes for sustainable agriculture. Mycorrhizae forms mutualistic symbiotic association between the fine roots of most plant species

* E-mail: ddnamdas@gmail.com

and fungi. Plant provides nutrients to the fungus and in return fungus provides inorganic nutrients to the plant through hyphal colonization in cortical tissue of the root during the growth. Mycorrhizae plays significant role in phosphate solubilization, mobilization and nutrient uptake in growth and development of plants hence, called 'Phosphorous gathering fungi'.

Now days, modern agricultural practices such as excessive chemical fertilizers, pesticides, fungicides and other sprays adopted by the farmers to achieve short term increase in yield and productivity. It makes soil polluted and infertile. In this context use of soil inoculants can represent an attractive eco-friendly alternative to maintain good health of soil and sustainable cropping system. The research work is undertaken to know the status of soil inoculants (AMF) its reaction on health of soil and plant productivity and to know more about Mycorrhizal status in soil with respect to, its occurrence, distribution, abundance, associations and effect on growth response in agricultural crops *viz.*, tomato (*Lycopersicon esculentum* Mill.) and Onion (*Allium cepa* L.). It could help to understand the potential role and benefits of AMF as soil inoculant. Hence it will be possible to express role of arbudcular mycorrhizae for sustainable agriculture.

1. **Tomato** (*Lycopersicon esculentum* Mill.) is one of the most important vegetable crops commercially grown all over India. It is also known as "Queen of the vegetables". It is important and essential item in our diet. It is rich in carbohydrate, protein, fat, vitamin-C minerals, fibers, sodium, calcium, phosphorus and magnesium. It increases the taste and flavor of our food. Tomato is grown on a large scale in our country for local consumption as vegetable and export purpose.

2. **Onion** (*Allium cepa* L.) is one of the most important vegetable crops commercially grown all over India. It is widely used in salads, chutneys and as a spice in cooking. It is premier among bulb crops used by both rich and poor people. It is an indispensable item in every kitchen as a condiment, spice and vegetable. It adds flavor to various vegetable preparations. Onion is believed to possess certain antibiotics and medicinal properties. It contains 87.5 per cent water and food value for 100g of edible portion gives energy 49 calories, proteins 1.4g, calcium 32g, vitamins, ascorbic acid 12g, thiamine 0.03mg, riboflavin 0.12mg, albuminoides 1.2g and ash 0.4mg.

Review

The literature on mycorrhizal status in the soil is reviewed to view the role of mycorrhizal diversity in soil from Ahmednagar for sustainable agriculture. The microbial forms in soil are known as soil inoculants or bioinoculants. They are believed to be an eco-friendly alternative over application chemicals which are considered to be 'safe' for agricultural application as they are derived from rhizosphere or roots (Rashi Gupta, 2013). AM fungi forms mycorrhizal association, it determines soil health and productivity of plant. (Louis and Lim 1987). Distribution of arbuscular mycorrhizal fungi (AMF) shows variation over the range of soil depth which determines AMF diversity (Shukla *et al.*, 2009). The distribution and abundance of AMF can be altered by several factors such as soil pH, moisture content etc. (Yang *et al.*, 2010, Sturmer and Siqueira, 2011). Intraradical development of AM fungi is highly influenced by plant species, soil pH, phosphorous content etc. (Carrenho *et al.*, 2007).

A report presented by Ghazi *et al.*, in 2003 states that AM fungi plays an important role in nutrient uptake, growth and yield in wheat plant. AM fungal colonization influences increase in water absorbing area in the soil, plant resistance and in leaf elasticity changes (Auge *et al.*, 1978a).

The isolation of extramatrical spores of Endogone in common cultivated soils from Scotland was carried out by Gerdemann and Nicolson (1963) with the help of wet sieving and decanting method. They reported six types of spores which represent distinct species. Extramatrical spores were much more abundant than spores formed in sporocarps. Mosse and Bowen (1968) worked for occurrence of Endogone spores in 250 soil samples from Australian, Newzealand and some Rothamsted soils and described diagnostic features of nine Endogone spore types and their developmental stages. Seven of nine spore types formed VAM with inoculated host plants. Khan (1971) from West Pakistan reported Endogone spores in 200 soil samples. All spore types were formed vesicular arbuscular mycorrhiza with red clover seedlings. Introduction of mycorrhiza, occurrence, ectomycorrhiza, ectomycorrhizal structure (mantle-heart), characterization and identification, systematics, inoculum and inoculum technique, inoculation, selection of symbiont, growth and development seedlings, ectomycorrhiza, nutrient uptake and research in India was documented by Lakhanpal and Anandsagar (1994).

Phillips and Hayman (1970) successfully assessed mycorrhizal infection in the root cortex. The technique of root staining and clearing by simmering in 0.1 per cent trypan blue in lacto-phenol greatly increases the accuracy and speed of estimating fungal infection in roots. Aldwell and Hall (1986) studied the lateral spread of VAMF hyphae in white cover (*Trifolium repens*) infected with *Glomus mosseae* and *Acaulospora laevis*, found that, *Glomus mosseae* was competent against *Acaulospora laevis* and spread throughout soil. Kunwar and Manoharachary (1998) reported extra-radical and intra-radical colonization in roots and scale leaves of garlic.

Biermann and Linderman (1983) supported role of intraradical vesicles in increment inoculum potential. Intraradical vesicles of *Glomus fasciculatum, Glomus mossae* and *Acaulospora spinosa* cause more infectivity while *Gigaspora marginata* and *G. gigantea* were less effective in root colonization. According to Shivkumar *et al.* (2006) use of microbial inoculants in agriculture such as biofertilizer (mycorrhizae), phytostimulators and biopesticides represent an attractive eco-friendly alternative to mineral fertilizers and chemical pesticides. A continued exploration of the natural biodiversity of soil micro-organisms and the manipulation of microbial interactions in the rhizosphere of crops represents a pre-requisite step to develop more efficient microbial activities in soil is considered essential to improve farming system.

Boerner (1990) studied variation in seedlings of *Geranium robertianum* inoculated with *Glomus occulatum* and reported intraspecific difference among populations of *Glomus occulatum* differentially affecting growth and nutrient uptake. Reports obtained indicate that local adaptation and intraspecific variations in induced growth and nutrient uptake in host.

plants. Aher (2004) interpreted role of *Glomus fasciculatum, G. mossease. G. geosporum,* and *G. halon* for biomass production with seeding growth and vigour of

Pennisetum typhoides (Brns.) Stapf. and Habbard. He reported that *Glomus fasciculatum* and *G. mosseae* as most efficient in increasing the shoot growth, dry weight and per cent infectivity. AM inoculated seedlings were grown faster and healtheir than non-mycorrhizal.

Materials and Methods

Climatic Conditions

Study area is situated between 19°47 and 19°57 North latitude and between 74° 32′ and 70° 47′ East longitudes. The altitude varies from 495m to 569m above the mean sea level. This tract is in the eastern side of western ghat and falls under rain shadow area. Climatically, the area falls in semi-arid tropics with average annual rainfall 55cm, which is erratic, and unevenly distributed in 15 to 45 rainy days in different years. Agro-climatically, the area falls in scarcity zone of Maharashtra. The soil sample was collected in clean polythene bags from the different localities of study area growing tomato and onion crops. The composite mixture of 50g soil sample was used for experimental study.

Isolation of AM Spores (Gerdemann and Nicolson, 1963)

50 gm of soil was mixed in 200ml (for quantitative studies) by hand in luke-warm water in a 500 ml beaker until all soil aggregates disperse to form uniform suspension. Heavier particles were allowed to settle down. Most of the suspensions decanted through a 710 mm sieve to remove large organic matter and roots. 200 ml water was added, filtered through 710 mm sieve and again decanted through 250 mm sieve. The obtained suspension then filtered and decanted through 75 mm and 45 mm consecutively. Residues were collected from the 75mm and 45 mm sieve separately, washed them well with water and spores were collected on filter paper. Higher frequency of spores was observed from 75 and 100 mm sieves. The spores were observed under a bino head dissecting stereoscopic zoom microscope. The observations were recorded in terms of spore number and enumeration.

Quantification of AM Spores (Adholeya and Gaur, 1994)

A circular filter paper (Whatman No.1, size 11 cm diameter) was taken. Horizontal fold was made, followed by second vertical fold. Filter paper was reopened and two lines were drawn to divide into four equal quadrants. Vertical lines were drawn on one-half of the filter paper to divide it into approximately twenty columns about 0.5 cm apart. Soil suspension was filtered and the spores were collected only on the marked surface of the filter paper. Then filter paper was spread on a big petri-plate. Spores were counted from each column under bino-head dissecting stereoscopic zoom microscope.

Microscopic Observations

Spores were mounted in lacto-phenol on a clean cover slide with cover slip, sealed with DPX and observed under compound microscope. To preserve the spores in original form, permanent slides were prepared by mounting the spores in Polyvinyl Lacto Glycerol (Polyvinyl alcohol 16.6g, Lactic acid 100.0ml, Glycerol 10.0ml, distilled water 100.0ml).

Identification

Identification of different genera of mycorrhizal fungal species was done with help of manual of mycorrhizae identification of Schenck and Perez (1987, 1990), revised synoptic key of Berch and Trappe (1991) and standard transparencies of Hall and Abbott (1981). Microphotographs were taken with phase contrast microscope (Leitz orthophan, Germany) using Kodak film.

VA-Mycorrhizal Root Colonization

Root Clearing and Staining Technique (Phillips and Hayman, 1970 and Nicolson, 1960)

The roots of both the crops were analyzed for VAM colonization by staining with 0.05 per cent trypan blue in lacto-phenol (Phillips and Hayman, 1970) and counted VAM root colonization percentage by slide technique (Nicolson, 1960). Root pieces were washed thoroughly under tap water, boiled at 90°C for 1-2 hours in 10 per cent KOH, and bleached with alkaline H_2O_2 for 10-20 minutes. Root segments were washed in distilled water and acidified with 5N HCl for a few minutes. Root segments were stained in 0.05 per cent trypan blue or cotton blue in lacto-phenol and removed the excess stain in lacto-phenol. Root segments were mounted temporally on the slides containing acetic acid Glycerol (1:1v/v), the edges of the cover slip sealed with DPX mountant and observed under microscope.

Assessment of VAM Root Colonization

i. Slide Method (Giovannetti and Mosse, 1980)

For the slide method, 1 cm root pieces from a stained sample were randomly selected and mounted on cover slide in groups of ten and examined under compound microscope. The presence or and absence of colonization in each root segment was recorded. Averages of ten pieces were expressed as percentage. It was calculated as follows:

$$\text{Mycorrhizal colonization} = \frac{\text{Number of roots with mycorrhiza}}{\text{Number of root segments taken for observation}} \times 100$$

ii. Gridline Intersect Method (Giovannetti and Mosse, 1980)

Root samples were cut down to 1cm. pieces and stained in acid fuchsine-lactic acid solution then spread evenly in Petri- plate. The grids of lines were marked on the bottom of the dish to form 0.5 cm size squares. Vertical and horizontal gridlines were scanned and the presence or absence of infection in the root was recorded at each point where roots intersected a line. Percentage infectivity was calculated by the following formula.

$$\text{Infectivity (per cent)} = \frac{\text{No. of infected roots}}{\text{Total no. of roots}} \times 100$$

Production of AMF Inoculum: Soil Funnels Technique (Subbarao, 1988)

AM spores were collected by wet sieving and decanting method and dominant spore types were selected for AM culture by picking with for the capillary tubes or pasture pipette. Spores were surface sterilized with 200ppm streptomycin sulphate solution for 3-5 minutes, then washed with sterile distilled water and again resterilized with 2.0 per cent chloramines-T solution for 3-5 minutes. After surface sterilization, spores were serially washed with sterile water for 4-5 times. The funnels were prepared by plugging the tube at bottom with cotton. Sand and soil mixture (1:1) was added into the funnel. Mouth of the funnel was wrapped with aluminum foil and were sterilized at 15 lbs (121°C) pressure for 45 minutes in an autoclave and cooled. Surface sterilized spores were added in funnel by making small slit at the center of the aluminum foil. Few drops of sterile water were added into the soil. A hole of 1 cm deep was made with sterile glass rod. Spores were added in hole and covered with soil. Above the spores, 2-3 surface sterilized seeds were sown. The funnels were then kept in test tubes containing water to wet soil by capillary action. These funnels were kept for 45-50 days in glass house. After desired growth, the plants with soil medium were planted in pot containing sand-soil (1:1) mixture and watered regularly. After 55-60 days, the roots were examined for AM colonization. After colonization, root system was chopped into small bits. These small bits were mixed in potting mixture and used as an AM culture (Mycofertilizer).

Pot Culture Experiment

Pot culture experiment was carried out at Botanic garden of the college with was conducted in sterile soil comprising of following treatments which were replicated three times.

a. **Treatment details:** *Glomus fasiculatum* (T1), *Glomus mosseae* (T2) *Acaulospora laevis* (T3), Recommended (N.P.K.) and (T4) Uninoculated control (T5)
b. **Pot filling**: The earthen pots (30cm dia.) surface disinfected with 5 per cent copper sulphate were filled with sterile soil (3 parts): sand (1 part)@3 kg/ pots per requirement of treatment. The soil was mixed well with sand and sterilized in autoclaved at 121°C (15 lbs pressure) for three hours for three successive days.
c. **Method of inoculation**: Earthen pots were filled with soil 2 cm less than regular filling for the treatment of mycorrhizal inoculation. Mycorrhizal inoculums of VAM fungus *Glomus fasiculatum, Glomus mosseae, Acaulospora laevis* (150g/pot) was spread over the soil surface by hand in a thin layer and over this, last 2 cm layer of soil was added. Then 45 days old five seedlings per pot were transplanted as per the treatment details. The pots transplanted with seedlings devoid of VAM inoculation served as control.
d. **Irrigation**: The pots were lightly irrigated with tap water before and after transplanting and subsequently at every week up to harvesting stage.
e. **Plant protection measures:** For the management of diseases on tomato and onion crop plants, a mixture of 0.05 per cent malathion 50 EC + 0.2 per cent Blitox 50 WP was sprayed immediately after the onset of disease incidence.

The second spraying was followed fortnight. In all four sprays were given at interval of fortnight.

f. **Observations**: One plant per pot with three replicates was maintained for recording the observations. The following observations were recorded.
   1. Plant height (cm) was recorded 90 days after transplanting.
   2. Fresh weight of shoot (g/plant) was recorded immediately after harvesting.
   3. Dry weight of shoot (g/plant): The base of the shoot was cut at ground level and whole shoot was cut into small pieces with the help of stainless knife and dried in the sunlight. Similarly, the fruits were cut into small pieces with the help of stainless steel knife and dried in sunlight. The sun dried pieces of shoot and fruits were kept in hot air oven at 60°C temperature till constant weight was obtained.
   4. Mycorrhizal dependency (per cent) was calculated according to Menge *et al.* (1978) as the ratio of dry weight of mycorrhizal plant to that of nonmycorrhizal plant and has been expressed as percentage, *i.e.*

$$\text{Mycorrhizal dependency (per cent)} = \frac{\text{Dry weight of VAM inoculated plant}}{\text{Dry weight of non-inoculated plant}} \times 100$$

Bio-chemical Analysis of Plant

1. Nitrogen content of shoot and fruit (per cent) was determined with Micro-kjeldal's digestion and distillation method (Jackson, 1973)
2. Total phosphorous content of shoot and fruit (mg/plant) was estimated by Vanadomolybdate yellow colour method using colour intensity at 470 mμ wavelength on Spectronic-20 (Jackson, 1973)

Physico-chemical Analysis of Soil

1. Soil reaction (pH) : (Page *et al.*, 1985).
2. Available phosphorus : (Olsen *et al.*, 1954).
3. Available Nitrogen : (Jackson, 1973).
4. Organic carbon : (Nelson and Sommer, 1982)

Results and Discussion

During investigation, mycorrhizal diversity in soil of tomato and onion fields was studied at the growth stage of 90 days at selected localities from Ahmednagar. Observations of both the crops at each locality were recorded separately

The soil samples were analyzed for physico-chemical properties. The soil of the tomato and onion fields showed pH range 8.09-8.26, organic carbon 0.40-66 per cent, P_2O_5 26.09-113.0 kgha^{-1}, K_2O 890-962 kgha^{-1} and N 248-260 kgha^{-1} (Table 36.1.)

Table 36.1: Physico-chemical Properties of Soils of Tomato and Onion Fields

Sl.No.	Soil Propertities	Tomato Fields			Onion Fields		
		Min.	Max.	Avg.	Min.	Max.	Avg.
1.	pH	8.00	8.15	8.09	8.16	8.42	8.28
2.	Electrical conductivity (mmhos/cm)	0.45	0.72	0.540	0.32	0.45	0.38
3.	Moisture per cent	5.30	6.46	5.90	5.33	8.61	6.71
4.	Organic carbon per cent	0.24	0.66	0.45	0.25	0.67	0.40
5.	Ca (Meql)	8.82	20.58	14.50	11.17	17.64	13.71
6.	Mg (Meql)	4.96	13.37	10.50	10.69	14.43	12.60
7.	P_2O_5 (kg ha^{-1})	52.19	154.9	113.0	15.39	32.6	26.09
8.	K_2O(kg ha^{-1})	825	1101	962	872	923	890
9.	N(Kg ha^{-1})	216.3	258.7	260.0	240.6	255.5	248.0

Table 36.2: Estimation of AM Spores of Tomato and Onion Fields

F1	F2	F3	Mean ± S.D	SEM	C.D. at 5 per cent
		No. of AM spores/50g soil (Tomato fields)			
49	44	57	50.0 ± 7.0	5.65	16.28
77	68	73	72.6 ± 8.5	3.68	11.20
85	79	89	4.3 ± 9.1	4.11	12.50
		No.of AM spores/50g soil (Onion fields)			
41.0	44.0	32.0	39.0 ± 6.2	5.09	15.51
75.0	68.0	29.0	61.6 ± 7.8	14.20	43.19
90.0	89.0	62.0	80.3± 8.9	12.97	39.46

Mycorrhizal Status of Tomato and Onion

The exhaustive data generated from analysis of soil samples and their biological screening suggests that, the study area has very wide generic and species mycorrhizal diversity. In the present piece of work, soil samples were screened for isolation of AM fungi (Gerdemann and Nicolson, 1963) from both the crops. 16 AMF species assignable to 4 genera were isolated. *Glomus fasciculatum. G. constrictum, G. mosseae* and *Scutellospora fasciculatum* were found dominant. However, *G. microcrapum, S. gilmorei, S. heterogama* and *Acaulospora laevis* showed their subdominace. *Gigaspora gigantea, G. halon, S. corolloidea, S. dipapillosa* and *S. dipupurens* showed their normal appearance (Table 36.3). From the above results, it is concluded that *Glomus fasciculatum* and *Scutellospora fasciculatum* proved their dominance over *S. gilmorei, S. heterogama* as subdominant. However, *G. halon, S. corolloidea, S. dipapillosa* proved their normal existence. Dominance of *Glomus* spp. In was supported by Gupta *et al.* (2002), Nehra (2004), Mohan *et al.* (2005). Subdominace of *Scutellospora* spp. was supported by Miller *et al.* (1985). Uneven distribution of AMF supported the variation in physico-chemical properties of soils at the selected localities (Tables 36.1 and 36.2). It is evident from the present data that spore population of *Glomus* spp. was high at pH above 8.0. However, *Gigaspora* spp. favours neutral alkaline soils. The uneven distribution of AM fungal population in different growth stages of the crop plants may be attributed to different physico-chemical properties of soils from selected sites. The soil contains high potassium which was inhibitory to the AM fungal population. The result corroborated with earlier finding of Mathur and Vyas, 1997.

Table 36.3: Occurrence of AM Fungi in Tomato and Onion Fields

Sl.No.	*AM Fungal Species*	*Tomato*			*Onion*		
		F1	*F2*	*F3*	*F1*	*F2*	*F3*
1.	*Acaulospora laevis*	+	+	+	+	+	+
2.	*Gigaspora gigantean*	–	–	+	+	+	–
3.	*Glomus constrictum*	+	+	+	–	+	–
4.	*G. fasciculatum*	+	+	+	+	+	–
5.	*G. geosporum*	+	–	+	–	–	+
6.	*G. halon*	–	+	–	+	–	+
7.	*G. microcarpum*	–	+	+	+	+	–
8.	*G. mosseae*	+	+	+	–	–	+
9.	*G. versiformae*	+	+	–	+	+	+
10.	*Scutellospora auriglobosa*	+	+	+	+	–	+
11.	*S. coralloidea*	–	–	+	–	+	–
12.	*S. dipapillosa*	+	–	–	+	+	+
13.	*S. dipurpurascens*	–	–	+	–	–	+
14.	*S. fasciculatum*	+	+	+	+	+	–
15.	*S. gilmorei*	–	+	+	–	–	+
16.	*S. heterogama*	+	+	–	+	+	+

F: Field; +: Present; –: Absent.

Effect of AM Bioinoculant on Tomato and Onion Crops

The results in respect of (Table 36.4 and 36.5) supported to the significant action of *Glomus mosseae* AM bioinoculant. It was resulted in increase in plant height (44-72; 58.10-60.25cm), fresh weight (194-259.2;195.9-257g/plant), dry weight (14.0-25.8;14.63-25.80g/plant), N and P- uptake (254.6-500.5 and 10.97-22.31mg/plant;166-226 and 41.59-49.45mg/plant). It was followed by *Glomus fasciculatum* and *Acaulospoa laevis*, All the inoculation treatments recorded significant increase in plant height over RDF and uninoculated control. The investigation supported the remarkable soil habitat diversity that counterparts in set up of the soil microflora at selected sites. Soil microclimate at both the localities influences and bring out notable changes in the mycorrhizal population.

Table 36.4: Effect of AM Bioinoculant on Tomato Crop Plants

Sl.No.	Growth Parameters	T1	T2	T3	T4	T5	Mean±S.D
I	Plant height (cm)	64.0 (34.3)	72.0 (41.6)	49.0 (14.2)	44.0 (4.5)	42.0	54.2 ±7.3
Ii	Fresh weight (g/plant)	219 (17.3)	259.2 (30.1)	195 (7.1)	194 (6.7)	181	209.6±14.4
Iii	Dry weight (g/plant)	22.0 (37.4)	25.8 (46.6)	14.0 (1.6)	14.55 (5.3)	13.7	18.0±4.2
Iv	N-uptake (mg/plant)	419.2 (43.6)	500.5 (52.8)	268.9 (12.1)	254.6 (7.2)	236.1	335.8±18.3
V	P-uptake (mg/plant)	18.52 (48.9)	22.31 (57.5)	11.56 (18.1)	10.97 (13.7)	9.46	14.5±3.8
Vi	Mycorrhizal dependency (per cent)	45.1 (88.0)	46.6 (88.4)	37.94 (85.8)	–	–	43.2±6.5
Vii	VAM colonization (per cent)	77.1 (44.6)	80.9 (47.2)	66.18 (35.4)	–	–	85.7±9.2
viii	Chlamydospores (/50g)	162.1 (69.1)	180. (72.2)	135 (62.9)	–	–	159±12.6

T1: *Glomus fasciculatum*; T2: *Glomus mosseae*; T3; *Acaulospora laevis*; T4: NPK; T5: As control/uninoculated.

Table 36.5: Effect of AM Bioinoculant on Onion Crop Plants

Sl.No.	Growth Parameters	T1 (G.F.)	T2 (G.M.)	T3 (A.L.)	T4 (NPK)	T5 Control	Mean±S.D
i	Plant height (cm)	59.05 (54.63)	60.25 (57.52)	58.10 (51.89)	56.92 (48.81)	38.25	48.5±6.9
ii	Fresh weight (g/plant)	219.0 (20.99)	257.0 (42.09)	195.9 (8.23)	190.0 (4.97)	181	208.5±14.4
iii	Dry weight (g/plant)	14.63 (6.28)	25.80 (87.41)	22.19 (61.14)	14.55 (5.69)	13.77	18.1±4.2
iv	N-uptake (mg/plant)	207.35 (116.0)	226.0 (135.5)	166.0 (73.62)	168.2 (75.27)	95.99	172.7±13.1
v	P-uptake (mg/plant)	43.38 (55.32)	49.45 (191.4)	41.59 (144.7)	42.71 (151.3)	16.18	38.6±6.2
vi	Mycorrhizal dependency	158.40	169.44	140.70	–	–	156.1±12.4
vii	VAM colonization	85.18 (67.65)	91.20 (73.76)	80.75 (64.10)	–	–	85.7±9.2
viii	No.of chlamydospores/50g	175	184	162	–	–	173.6±13.1

T1: *Glomus fasciculatum*; T2: *Glomus mosseae*; T3; *Acaulospora laevis*; T4: NPK; T5: As control/uninoculated.

The results of plant height mentioned in Table 36.4 and 36.5 after transplanting in both the crops supported that significant increase in plant height from 4.5 per cent to 41.6 per cent in tomato and 51.89 per cent to 57.52 per cent in onion over the RDF and uninoculated control. The finding is in close agreement with the findings of Reddy *et al.* (1996) in *Citrus aurantifolia,* Selvraj and Malik (2004) in tobacco, Aher (2004) in bajra. The significant increase in fresh weight of shoot was from 6.7 to 30.1 per cent in tomato and 8.23 per cent to 42.09 per cent in onion. It supports to the effect of AM fungi on the growth of crop plants. The increase in fresh weight with AM inoculum was supported by Bagyaraj and Manjunath (1980) in cotton, cowpea and ladies finger, Meenu and Saralabai (2001) in *Sorghum vulgare.* Similarly, increase in dry weight of both the crops was from 1.6 per cent to 46.6 per cent and 6.28 to 87.41 per cent . Thus as stated by Garampalli *et al.* (2005) in *Cajanus cajan,* the growth responses in dry weight of shoot and total dry weight due to *Glomus aggregatum.*

Results in respect of mycorrhizal dependency value in tomato and onion showed significant increase from 85.8 to 88.4. It clears that mycorrhizae are beneficial to agricultural crops. The results from Tables 36.4 and 36.5 indicate significant increase in nitrogen uptake from 12.1 per cent to 52.8 per cent and 73.62 per cent to 135.5 per cent. It was supported by, Ames *et al.* (1990) in *Geranium robertianum,* Suri *et al.* (2006) in onion, Stribely *et al.* (1980) in *Trifolium subterranean;* Mitchell and Read (1981) in ericaceous plants, Bhadoria (2002) in *Cajanus cajan.*

Conclusion

It could be concluded from the above results and discussion, AM inoculum exists enormous scope for achieving increased yield in tomato and onion crops by application of *Glomus mosseae* as the crop exhibits satisfactory growth responses with respect to plant height, fresh and dry weight, mycorrhizal dependency, nitrogen and phosphorous content and uptake when compared to normal practice.

References

Adholeya, A and Gaur, A. C. (1994). Establishment of VAM spores in soil. A modified method, *Mycorrhiza News,* **6** (1): 10-11.

Aher, R. K.(2004). Role of four VA-mycorrhizal fungi on the growth, biomass production and root colonization in *Pennisetum typhoides. Asian Jr. of Microbiol. Biotech. Env. Sci.,* **6**(3): 435-436.

Aldwell, F. E. B. and Hall, I. R. (1986). Monitoring spread of *Glomus mosseae* through soil infested with *Acaulospora laevis* using serological and morphological technique. *Trans. Br. mycol. Soc.,* **87**(1): 131-134.

Ames, R. N., Reid, C. P. P., Porter, L. K. and Cambardella, C. (1983). Hyphal uptake and transport of nitrogen from two ^{15}N-labelled sources by *Glomus mosseae,* a Vesicular- Arbuscular Mycorrhizal fungus. *New Phytol.,* **95**: 381-396.

Auge, R. M., Schekel, K.A. and Wample, R. L. (1987a). Rose leaf elasticity changes in response to mycorrhizal colonization and drought acclimation. *Physiol Plant* **70**: 175–182.

Bagayraj, D. J. and. Manjunath, A. (1980). Response of crop plants to VA-mycorrhizal inoculation in an unsterile Indian soil. *New phytol.*, **85**: 33-36.

Berch, S. M. and Trappe, J. M. (1991). Revision of Trappe's (1982). Synoptic Keys to Genera and Species of Endogenaceae: 1-30.

Bhadoria, P. B. S. (2002). Effect of lime and mycorrhiza on phosphate uptake. *Madras Agric. J.*, **89**(1-3): 126-128.

Biermann, B. and Linderman, R. G. (1983). Use of vesicular arbuscular mycorrhiazal roots, intraradical vesicles and extraradical vesicles as inoculum. *New Phytol.*, **95**: 97-105.

Boerner, E. J. (1990). Role of mycorrhizal fungus origin in growth and nutrient uptake by *Geranium robertianum*. *Amer. J. Bot.*, **77**(4): 483-489.

Carrenho, R., Trufem, S.F.B., Bononi, V.L.R. and Silva, E.S. (2007). The effect of different soil properties on arbuscular mycorrhizal colonization of peanuts, sorghum and maize. *Acta Bot Bras.*, **21**: 723-730.

Garampalli, R. H., Deene, S. and Reddy, N. C. (2005). Infectivity and efficacy of *Glomus aggregatum* and growth response of *Cajanas cajan* (L.) Millsp. in flyash amended sterile soil. *J. Enviorn. Biol.*, **20**(4): 705-708.

Gerdemann, J. W. and Nicolson, T. H. (1963). Spores of Mycorrhizal Endogone species extracted from soil by wet sieving and decanting method. *Trans. Brit. mycol. Soc.*, **46**(2): 235-244.

Ghazi A.L., Karaki, B., McMichael and John Zak, (2003). Field response of wheat to arbuscular mycorrhizal fungi and drought stress. *Mycorrhiza.*, **14**: 263-269.

Gupta, N., Basak, U. C. and Das, P. (2002). Arbuscular mycorrhizal association of mangrooves in saline and nonsaline soils. *Mycorrhiza News.*, **13**(4): 14-18.

Giovannetti, M. and Mosse, B. (1980). An evaluation technique for measuring vesicular-arbuscular mycorrhizal infection in roots. *New Phytol.*, **84**: 489-500.

Hall, I. R. and. Abbott, L. K. (1981). Photographic slide collection illustrating features of the Endogenaceae. Ed. 3, 1-27pp. Plus 400 colour transparencies. Invermay Agric. Res. Centre and Soil Science Department, University of W. Australia.

Hayman, D. S. (1970). Endogone spore number in soil and vesiculsr arbuscular mycorrhiza in wheat as influenced by season and soil treatment. *Trans. Br. Mycol. Soc.*, **54**(1): 53-63.

Jackson, M. L. (1973). Soil chemical analysis. Prentice Hall of India, New Delhi.

Khan, A. G. (1971). Occurrence of Endogone spores in West Pakistan soils. *Trans. Br. Mycol. Soc.*, **56**(2): 217-224.

Kunwar. I. K. and Manoharachary, C. (1998). Colonization by arbuscular mycorrhizal and mycophyllous fungi in roots and scale leaves of garlic. *J. Indian. Bot. Soc.*, **77**: 201-203.

Lakhanpal, T. N. and Anandsagar, (1994). The ectomycorrhiza - current status. *Indian Journal of Microbiology.*, **34** (1): 1-26.

Louis, I. M. and Lim, G. (1987). Spore density and root colonization of vesicular-arbuscular mycorrhizas in tropical soil. *Transactions of British Mycological Society.*, **88**: 207-212.

Mathur, N. and Vyas, A. (1987). Influence of arbuscular mycorrhizae on biomass production, nutrient uptake and physiological changes in *Ziziphus mauritiana* Lam. Under water stress. *J. Arid Environ.*, **45**: 191-195.

Meenu, R. and Saralabai, V. C. (2001). Effect of Vesicular Arbusular Mycorrhiza and Nitrogen fixing biofertilizers on plant growth in *Sorghum vulgare*. *Ad. Plant. Sci.*, **14**(1): 615-617.

Menge, J. A., Johnson, E. L. V. and Platt, R. G. (1978). Mycorrhizal dependency of several citrus cultivars under three nutrient treatments. *New Phytol.*, **81**: 553-559.

Miller, Diad Doud Domoto., Paul, A. and Walker, Christopher. (1985). Mycorrhizal fungi of eighteen apple root stock plantings in the United states. *New Phytol.*, **100**: 379-391.

Mitchell, D. T. and Read, D. J. (1981). Utilization of inorganic and organic phosphates by the mycorrhizal endophytes of *Vaccinium macrocsrpon* and *Rhododendron ponticum*. *Trans. Br. Mycol. Soc.*, **76**(2): 255-260.

Mohan, V., Bappammal, M., Malathy, N. and Manokaran, P. (2005). Distribution of arbuscular-mycorrhizal (AM) fungi in association with some important medicinal plants of Tamil Nadu. *Indian Forester*, June, pp.797-804.

Mohankumar, V., Narender, S. and Sivaswamy. (1992). Vesicular Arbuscular Mycorrhizae. *Ad. Plant Sci.*, **5**(1): 17-28

Mosse, B. and Bowen, G. D. (1968). A key to the recognition of some Endogone spore types. *Trans. Br. mycol. Soc.*, **51**(3-4): 469-483.

Nehra, Sampat. (2004). VAM fungi and organic amendments in the managements of *Meloidogyne incognita*. *J. Ind. Bot. Soc.*, **83**: 90-97.

Nelson, D. W. and Sommer. L. E. (1982). Total carbon, organic carbon and organic matter. In A. L. Page (ed.). Method of Soil analysis Part 2. Agron. 9: 539-579. *Am. Soc. of Agron. Inc., Soil. Sci. Soc. of Am*. Madison. Wisconsin, USA.

Nicolson, T. H. (1960). Mycorrhiza in the Gramineae. II. Development in different habitats, particularly sand dunes. *Trans. Br. Mycol. Soc.*, **43**(1): 132-145.

Olsen, S. R., Cole, C. V., Watanabe, F. S. and Dean, L. A. (1954). Estimation of available phosphorous in soil by extraction with sodium bicarbonate. *Circ. U. S. Dept. Agri* p.939.

Page, A. L., Milla, R. H. and Keeney, D. R. (1985). Method of soil analysis. Part 2, Agronomy monograph No. 9, American Society of Agronomy Inc. Soil Science Society of America, Inc. Publisher Madison, Wisconsin, USA.

Phillips, J. M. and Hayman, D. S. (1970). Improved procedures for clearing roots and staining parasitic and vesicular arbuscular mycorrhizal fungi for rapid assessment of infection. *Trans. Br. Mycol. Soc.*, **55**(1): 158-161.

Rashi Gupta, Virendra Swarup., Bisaria and Shilpi Sharma (2013), Bioinoculants: More than just plant growth promoting agents. *Journal of Endocytobiosis and Cell Research,* pp. 8-11.

Reddy, B., Bhagyaraj, D. J. and Mallesha, B. C. (1996). Selection of efficient VA-Mycorrhizal fungi for Acid Lime. *Indian Journal of Microbiology.,* **36**: 13-16.

Schenck, N. C. and Perenz Y. (eds) (1987). Manual for Identification of VA mycorrhizal fungi in VAM, University of Florida Gaines Ville, FL, pp.1-245.

Schenck, N. C., and Perez. Y. (1990). Manual for the identification of VA-Mycorrhizal fungi, University of Florida, Gainesville, FL, pp.1-286.

Selvraj, T. and Malik, A. M. (2004). Genotypical response of tobacco to inoculation with arbuscular mycorrhizal fungus, *Glomus aggregatum*. *Mycorrhiza News.,* **16**(3): 20-23.

Shivkumar, B. S., Jayashakara, M., Umesh, M.R. and Senthil Kumar, M.(2006). Microbial inoculants for sustainable agriculture. *Agrobios Newsletter.,* **5**(7): 15-16.

Shukla, A., Kumar, A., Chaturvedi, O. P., Prasad, R., and Gupta, A. (2009). Effects of shade on arbuscular mycorrhizal colonization and growth of crops and tree seedlings in Central India. *Agroforestry Systems*. **76**: 95-109.

Stribley, D. P., Tinker, P. B. and Rayner, J. H. (1980). Relation of internal phosphorous concentration and plant weight in plant infected by vesiculsr arbuscular mycorrhiza. *New Phytol,* **86** : 261-266.

Sturmer, S.L., and Siqueira, J.O. (2011). Species richness and spore abundance of arbuscular mycorrhizal fungi across distinct land uses in Western Brazilian Amazon. *Mycorrhiza*. **21**: 255-267.

Subbarao, N.S. (1988). Biofertilizers in agriculture (2nd edition), Oxford and IBH publication, New Delhi, p.207.

Suri, V. K., Choudhary, A. K., Chander, G. and Verma, T. S. (2006). Studies on VA-mycorrhizal fungi used as a potential biofertilizers in acid alfisol of Northestern Himalayas. 18th World congress of soil science (9-15 july), Philadelphia, Pargisylvania, USA.

Yang, F.Y., Li, G.Z., Zhang, D.E., Christie, P., Li, X.L., and Gai, J.P. (2010). Geographical and plant genotype on the formation of arbuscular mycorrhiza in *Avena sativa* and *Avena nuda* at different soil depths. *Biology and Fertility of Soils.,* **46**: 435-443.

2015, Major Constrains and Verdict of Crop Productivity *Pages 476–484*
Editor: Dr. U.N. Bhale
Published by: ASTRAL INTERNATIONAL (P) LIMITED, NEW DELHI

Chapter 37

Synergistic Effects of Agrochemicals on Carbendazim Resistance *Geotrichum candidus* Incitant of Fruit Rot of Ivy Gourd

V.S. Chatage[1] and U.N. Bhale[2]

[1]Department of Botany,
Kai. Rasika Mahavidyalaya, Rasika Campus,
Borol Road, Deoni, Tal. Deoni, Dist. Latur – 413 519, M.S., India
[2]Research Laboratory.Department of Botany,
Arts, Science and Commerce College, Nadurg.
Tal – Tuljapur, Dist. Osmanabad – 413 602, M.S., India

ABSTRACT

There was variation in the MIC of carbendazim among the Geotrichum candidus incitant of fruit rot of Ivy gourd (Cucurbitaceae) fruits and isolate on (CZA) Czapek Dox agar medium plates. Minimum inhibitory concentration (MIC) on CZA plates ranged from 4000 µg/ml while it was more resistant with resistant factor 2. Use of carbendazim with–other agrochemicals like Fungicides (mancozeb, copper oxychloride, chlorothalonil), Insecticides (phorate, chlorpyriphos and endosulfan), Herbicides (metribuzin, sodium salt and atrazine), Antibiotics (bacipen, ambistryn–s and penicillin), Salts (mercuric, chloride, stannous chloride and barium chloride), and Fertilizers (MOP, urea, 10:26:26) inhibited the growth of the pathogen in in vitro studies.

Keywords: *Coccinia indica, Geotrichum candidus, Agrochemicals, Radial growth.*

---

* *Corresponding Author.* E-mail: unbhale2007@rediffmail.com, vaishalichatage3@gmail.com

Introduction

Ivy gourd (*Coccinia indica* Wight and Arn.) of the family *Cucurbitaceae*. Is one of the important vegetables and medicinal fruits and plant, distributed in Tropical, Asia, Africa Pakistan, India and Sri Lanka (Cooke, 1903; Sastri: 1950). The Ivy gourd fruit is attacked by fungal pathogen. Among them fruit rot disease occurs in our country caused by *Geotrichum candidus.* This disease is managed by applying various systemic and conventional fungicides. But in the present paper on attempt is made to manage the of fruit rot of Ivy gourd by using different agrochemicals.

Materials and Methods

Synergistic effects of various agrochemicals on the development of resistance infected pathogen were studied by mixing different fungicides, insecticides, herbicides, fertilizers, antibiotics and salts. Equal volume of chemicals were mixed in of alone carbendazim were studied (Bhale, 2002).

Effect of various agrochemicals on the development of resistance in *G. candidus* were studied by mixing different Fungicides (mancozeb, copper oxychloride, chlorothalonil), Insecticides (phorate, chlorpyriphos and endosulfan), Herbicides (metribuzin, sodium salt and atrazine), Antibiotics (bacipen, ambistryn-S and penicillin), Salts (mercuric chloride, stannous chloride and barium chloride) and Fertilizers (MOP, urea, 10:26:26). The equal volume of chemicals mixed @4000µg/ml alone carbendazim mixed with equal volume of agro-chemicals (25, 50, 75 and 100µg/ml and 0.1, 0.2, 0.3 and 0.4) were mixed in 4000µg/ml of carbendazim. The resistant and senstive isolates were selected for this study. It was grown on the CzapekDox agar medium (CZA) containing resistant dose of carbendazim4000µg/ml and other chemicals at above said concentrations. Increase in the growth of organisms over control was considered to be increase in the resistance while decrease was considered to be decrease in the resistance growth on carbendazim served as control.

Equal volume of chemicals were mixed in µg/ml of alone of carbendazim resistance isolates of pathogen was incubated on to the plates and its growth was observed at inhibited pathogens and percentage control efficacy (PCE) was calculated (Bhale *et al.*, 2010).

Percentage Control Efficacy (PCE) of Agrochemicals

The percentage control efficacy of each agrochemical was calculated as follows:

$$PCE = 100\left(1 - \frac{X}{Y}\right)$$

where,

X = Is the diameter of the colony on fungicide treated agrochemical.

Y = Is the diameter of the colony on the medium alone.

Results and Discussion

Agrochemicals such as fungicides, insecticides, herbicides, fertilizers, antibiotics and salts were mixed with carbendazim in the Czapek Dox agar medium.

Carbendazimresistant,Gc_6 isolates were inoculated onto the plates and its growth was observed after 8^{th}day's incubation period. Increased growth over carbendazim alone control was considered as increase in the resistance and vice versa.

Fungicides

Results are depicted in Table 37.1. It is observed that Chlorothalonil 100 µg/ml with Carbendazim completely inhibited the growth of the resistant isolate of *G. candidus*.Copper oxychloride and Mancozeb also reduced the growth of *Geotrichum candidus*.

Table 37.1: Synergistic Effects of Fungicides on the Development of Carbendazim Resistance in *G. candidus* with Latent Period of 8 th Days

Sl.No.	*Fungicides with Carbendazim (4000 µg/ml)*	*Linear Growth (mm)*	*PCE*
1.	Copper oxychloride		
	25	40.00	55.55
	50	36.00	60.00
	75	30.00	66.66
	100	25.00	72.22
2.	Chlorothalonil		
	25	17.00	81.11
	50	15.00	83.33
	75	11.00	87.77
	100	9.00*	90.00
3.	Mancozeb		
	25	20.00	77.77
	50	18.00	80.00
	75	17.00	81.11
	100	15.00	83.33
4.	Control (4000 µg/ml) Carbendazim alone	11.00	87.77
	CD (P=0.05)	–	6.53

PCE: Percentage Control Efficacy.

*: Significantly reduced mycelial growth.

Insecticides

Among insecticides,Phorate, chlorpyriphos and endosulphan were mixed with carbendazim. Endosulphan at 75 µg/ml with Carbendazim at 4000 µg/ml was completely inhibited the growth of *G. candidus* (Table 37.2).

Herbicides

Amongherbicides, indicated results in (Table 37.3). Metribuzin at 25 µg/ml with carbendazim at 4000 µg/ml was completely stopped the growth of the *Geotrichum candidus*. Remaining salts also showed inhibition of the pathogens.

Table 37.2: Synergistic Effects of insecticides on the Development of Carbendazim Resistance in *G. candidus* with Latent Period of 8 th Days

Sl.No.	*Fungicides with Carbendazim (4000 µg/ml)*	*Linear Growth (mm)*	*PCE*
1.	Phorate		
	25	0.00	100.00
	50	0.00	100.00
	75	0.00	100.00
	100	0.00	100.00
2.	Chlorpyriphos		
	25	17.00	81.11
	50	15.00	83.33
	75	14.00	84.44
	100	13.00	85.55
3.	Endosulphan		
	25	16.00	82.22
	50	14.00	84.44
	75	12.00*	86.66
	100	0.00	100.00
4.	Control (4000 µg/ml) Carbendazim alone	11.00	87.77
	CD (P=0.05)	–	4.87

PCE: Percentage Control Efficacy.

*: Significantly reduced mycelial growth.

Table 37.3: Synergistic Effects of Herbicides on the Development of Carbendazim Resistance in *G. candidus* with Latent Period of 8 th Days

Sl.No.	*Fungicides with Carbendazim (4000 µg/ml)*	*Linear Growth (mm)*	*PCE*
1.	Metribuzin		
	25	11.00*	87.77
	50	0.00	100.00
	75	0.00	100.00
	100	0.00	100.00
2.	2, 4-D Sodium salt		
	25	17.00	81.11
	50	15.00	83.33
	75	0.00	100.00
	100	0.00	100.00

Contd...

Table 37.3–*Contd...*

Sl.No.	*Fungicides with Carbendazim (4000 μg/ml)*	*Linear Growth (mm)*	*PCE*
3.	Atrazine		
	25	0.00	100.00
	50	0.00	100.00
	75	0.00	100.00
	100	0.00	100.00
4.	Control (4000 μg/ml) Carbendazim alone	11.00	87.77
	CD (P=0.05)	–	4.47

PCE: Percentage Control Efficacy.

*: Significantly reduced mycelial growth.

Antibiotics

Among antibiotics, Bacipen, Ambistryn-S and Penicillin were mixed with carbendazim (Table 37.4). Bacipen @-500 at 0.1 μg/ml with carbendazim at 4000 μg/ml was completely inhibited the growth *G. candidus.*

Table 37.4: Synergistic Effects of Antibiotics on the Development of Carbendazim Resistance in *G. candidus* with Latent Period of 8 th Days

Sl.No.	*Fungicides with Carbendazim (4000 μg/ml)*	*Linear Growth (mm)*	*PCE*
1.	Bacipen @-500		
	0.1	10.00*	88.88
	0.2	0.00	100.00
	0.3	0.00	100.00
	0.4	0.00	100.00
2.	Ambistryn-S		
	0.1	0.00	100.00
	0.2	0.00	100.00
	0.3	0.00	100.00
	0.4	0.00	100.00
3.	Penicillin		
	0.1	0.00	100.00
	0.2	0.00	100.00
	0.3	0.00	100.00
	0.4	0.00	100.00
4.	Control (4000 μg/ml) Carbendazim alone	11.00	87.77
	CD (P=0.05)	–	2.65

PCE: Percentage Control Efficacy.

*: Significantly reduced mycelial growth.

Salts

Among salts, Mercuric chloride, Stannous chloride and Bariu chlorideat (0.1 to 0.4 µg/ml) were mixed with carbendazim. It is observed that Mercuric chloride at 0.4 µg/ml with carbendazim at 4000 µg/ml was completely stopped the growth of *G. candidus* (Table 37.5).

Table 37.5: Synergistic Effects of Salts on the Development of Carbendazim Resistance in *G. candidus* with Latent Period of 8 th Days

Sl.No.	*Fungicides with Carbendazim (4000 µg/ml)*	*Linear Growth (mm)*	*PCE*
1.	Mercuric chloride		
	0.1	23.00	74.44
	0.2	20.00	77.77
	0.3	17.00	81.11
	0.4	15.00*	83.33
2.	Stannous chloride		
	0.1	33.00	63.33
	0.2	27.00	70.00
	0.3	24.00	73.33
	0.4	20.00	77.77
3.	Barium chloride		
	0.1	22.00	75.55
	0.2	20.00	77.77
	0.3	17.00	81.11
	0.4	16.00	82.22
4.	Control (4000 µg/ml) Carbendazim alone	11.00	87.77
	CD (P=0.05)	–	3.80

PCE: Percentage Control Efficacy.

*: Significantly reduced mycelial growth.

Fertilizers

Among fertilizers, MOP, Urea and 10:26:26 were mixed with carbendazim. MOP at 0.3 µg/ml with carbendazim at 4000 µg/ml was completely inhibited the growth of *G. candidus* (Table 37.6).

There are many report's findings to prevent further cases of resistance Ciba Geigy has adopted a basic strategy of the use of prepack mixture of metalaxyl with protective fungicides against foliar oomycetes (Staub and Sozzi, 1981). Benomyl with captan, chlorothalonil and imazilil reduced benomyl resistance in *Venturia inaequalis* (Shabi and Glipatric, 1981), *Sclerotia homococarpic* and *Penicillium digitatium* (Eckert, 1981). Samoucha *et al.* (1987) reported synergistic interaction in the mixture of mancozeb plus symoxanil, mancozeb plus oxadinyl, cymoxanial plus oxydinyl in the control of *P. infestans*on tomato. Dekker (1981) suggested that there was significant delay of resistance build up in the pathogen when mixture of different fungicides was used.

Table 37.6: Synergistic Effects of Fertilizers on the Development of Carbendazim Resistance in *G. candidus* with Latent Period of 8 th Days

Sl.No.	*Fungicides with Carbendazim (4000 µg/ml)*	*Linear Growth (mm)*	*PCE*
1.	MOP		
	0.1	16.00	82.22
	0.2	14.00	84.44
	0.3	13.00*	85.55
	0.4	0.00	100.00
2.	Urea		
	0.1	0.00	100.00
	0.2	0.00	100.00
	0.3	0.00	100.00
	0.4	0.00	100.00
3.	10:26:26		
	0.1	19.00	78.88
	0.2	16.00	82.22
	0.3	0.00	100.00
	0.4	0.00	100.00
4.	Control (4000 µg/ml) Carbendazim alone	11.00	87.77
	CD (P=0.05)	–	5.31

PCE: Percentage Control Efficacy.

*: Significantly reduced mycelial growth.

There are theoretical models suggested by Kable and Jeffiry, 1980. Micronutrients when used in combination with carbendazim reduce the growth of resistant *Aspergillus flavus* causing aflaroot disease of groundnut (Gangawane and Reddy, 1986). Griffin (1981) reported that fungicides and antibiotics having different mode of action are able to control resistant pathogen. Gangawane and Kamble (2001) found that when carbendazim was used in combination with agrochemicals inhibited the growth of resistant isolate of *Macrophomina phaseolina* causing charcoal rot of potato. According to Patil (2009) carbendazim in combination with fungicides (Captan, Zineb, and Mancozeb), insecticides (methomyl, endosulphan and monocrophos) herbicides (2,4-D, Excel mera 71, and zepaclav 500), antibiotics (Griseofulvin and Oflaxacin 400), salt (potassium chloride and magnesium chloride), fertilizers (urea, muriate of potash, 10;26;26, mixture of urea, super phosphate, sufia, nimboli and potash), micronutrients (Mb, Co, Cu, Mn) completely inhibited the growth of *Macrophomina phaseolina* causing charcoal rot of sweet potato.Tilt in combination with fungicides (mancozeb, zineb, captan and rako), insecticides (Thimate and endosulphan), antibiotics (streptomycin and Griseofulvin) completely prevented the infection of *Phakopsora pachyrhizi* to soyabean plants.

The growth of some organisms has been inhibited by herbicides which are meant to destroy weeds while some fungi have been affected by application of insecticides.(Chen *et al.*, 2001; Das *et al.*, 2003). Streptomycin mixed with oxytetracycline is used in control of fire blight of pome fruits and could delay the appearance of resistant strains (Vanneste, 2000) Application of carbendazim in combination with Thiram resulted in the highest seed germination percentage and lowest root rot incidence in chickpea (Prajapati *et al.*, 2003). Synergistic effect of combining biocontrol agents with silicon against postharvest diseases of jujube fruit (Shiping, *et al.*, 2005). Yeasts as biocontrol agents have been widely used to control postharvest diseases in various fruit because they do not produce antibiotics (Droby *et al.*, 1998). The combination of biological control agents with selected chemicals produces a synergistic effect that enhances their efficacy for postharvest disease control (Droby, *et al.*, 2003).

References

Bhale, U. N., Sawant, V. S., Rajkonda, J. N. and Jadhav, D.S. (2010). Influence of different nutrient sources on fungicide resistant and sensitive *Alternaria spinaciae* the incitant of leaf spot of spinach (*Spinacea oleracea* L.). *Ad. Plant Sci.*, **23**(11): 459 - 462.

Bhale, U. N. (2002). Studies on Management of some important diseases of spinach in maharashtra Ph.D. thesis. Dr. Babasaheb Ambedkar Marathwada University, Aurangabad. p. 47.

Chen, S., Edward C. A. and Subler, S. (2001). Effect of the fungicides Benomyl, captan and chlorothalonil on soil microbial activity and N dynamics in laboratory incubation. *Soil Biology and Biochemistry*, **33**(14): 1971-1980.

Cooke, C.*I.E.*T. (1903). "Flora of Presidency of Bombay", Published under the Authority of Secretary of State for Council. **Vol. I:** 571- 572.

Das, A. C., Chakravarty, A., Sukul, P. and Mukherjee, D. (2003). Influence andpersistence of phorate and carbofuran insecticides on microorganism in rice field. *Chemosphere.* (**53**): 1033 – 1037.

Dekker, J.(1981). Counter measures for avoiding fungicides resistance. Fungicides resistance in crop protection (Dekker, J. and S.G. Georopoulos Eds.) *CAPD Wageningen, Netherlands*, pp.128-139.

Droby, S., Cohen, L., Daus, A., Weiss, B., Horev, B., Chalutz, E., Kate, H., KerenTzur, M. and Shachnai, A. (1998). Commercial testing of Aspire: a yeast preparation for the biological control of postharvest decay of citrus. *Biol. Control*, **12:** 97-101.

Droby, S., Wisniewski, M., El Ghaouth, A. and Wilson, C.L. (2003). Influence of food additives on the control of postharvest rots of apple and peach and efficacy of the yeast – based biocontrol product Aspire. *Postharvest Biol. Technol*, **27:** 127- 135.

Eckert, J. W. (1981). Fungicides resistance in *Penicillium*decay of citrus fruits. In fungicides resistance in crop protection (Dekker, J. and S.G. Georopoulos Eds.) *CAPD Wageningen, Netherlands,* pp.231-250.

Gangawane, L.V and S.S.Kamble (2001). Use other agrochemicals in the management of charcoal rot of potato caused by *Macrophomina phaseolina* resistant to carbendazim. *In the proceeding of international conference on frontiers in fungal biotechnology and plant pathogen relation,* pp.58-62.

Gangawane, L.V. and B.B.C. Reddy (1986). Micronutrients reduce the resistance to carbendazim in *Aspergillus flavus. ISPP Chemical control Newsletter,* **7**: 19-21.

Griffin, M.J. (1981). Plant pathology. Notes no.38. *Fungicide resistance. ADAS South Western Region, U.K.*

Kable, R.F. and H. Jaffery (1980). Selection for tolerance in organism exposed to sprays of biocide mixture. A theoretical modal, *Phytopathology,* **70**: 8-12.

Patil, V.B. (2009). Studies on management of charcoal rot of sweet potato.Ph.D. Thesis, Shivaji University, Kolhapur., pp.40-45

Prajapati, R. K. S., Srivastava S. L. and Chaudhary, R.G. (2003). Incidence of chickpea dry root rot in Uttar Pradesh and Uttaranchal and efficacy of seed dressing fungicides on seed germination and seedling infection. *Farm Sci. J.,* **12**(2): 170-171.

Samoucha, Y.U. Hugaishoffer and V. Gisi (1987). Effects of disease intensity and application type on efficacy and synergy of fungicide mixture against *Phytopthora infestans, Phythpathology,* **120**: 44 -52.

Sastri, B. N. (1950). "The Wealth of India, A. Dictionary of Raw Material and Industrial Products", Publication and Information Directorate. CSIR New Delhi, **(2)8**: 257 and 285-293.

Shabi and J.P. Gilpatric(1981). Competition between benomyl resistant and benomyl sensitive strains of *Venturia inaequalis* on apple seedlings treated with benomyl and capton*Neth. J. Plant. Pathol.,* **67**: 250 -251.

Shiping, Tian., Guozheng, Qin. and Yong, Xu. (2005). Synergistic effects of combining biocontrol agents with silicon against postharvest diseases of jujube fruit. *Journal of Food Protection,* **68**(3): 544-550.

Staub, T.and D.Sozzi (1981). First practical experience with metalaxyl resistance *Neth. J. Plant. Pathol.,* **87**: 233-235.

Vanneste, J. L. (2000). Fire blight The Disease and Causative Agent. *Erwinia amylovora* CABI Publications. pp. 1-370.

2015, Major Constrains and Verdict of Crop Productivity *Pages 485–488*
Editor: **Dr. U.N. Bhale**
Published by: **ASTRAL INTERNATIONAL (P) LIMITED, NEW DELHI**

Chapter 38

Use of Algal Biomass in Seed Germination and Growth of Wheat Plant

Milind Jadhav* and Deepika Borkhade

Department of Botany,
Sir Sayyed College, Roshan Gate, Aurangabad – 431 001, M.S., India

ABSTRACT

Algae are known to contain different bioactive compounds. Of these, there are certain growth promoting substances which are found significant in seed germination and plant growth. Present work deals with the study of effect of algal extracts on seed germination and application of algal powder in plant growth. Algal biomass was collected from ponds located near Gogababa hills of Aurangabad in the month of December 2009. Collected algal biomass screened microscopically then washed with sterile distilled water and dried in shade. Fine powder was prepared and stored in acid washed air tight bottles. Algal extracts in cold water, hot water and methanol were prepared separately. Seeds of Wheat (Triticum aestivum) were soaked in algal extracts for four hours. Soaked seeds were placed on moist blotter in sterilized petriplates. Seeds soaked in sterile distilled water served as control. Per cent germination, root and shoot length of seedlings were measured after 7 days of incubation at room temperature. Cold water extract found helpful in seed germination. Algal powder was used in order to understand effects of algal biomass on growth of wheat plant by tray culture method. Algal biomass found useful for the growth of wheat plant and encouraging results were obtained. Present work is one of the important aspects of algal biotechnology which is useful in sustainable agiculture development.

Keywords: *Algal biomass, Seed germination, Plant growth.*

---

* *Corresponding Author.* E-mail: dr.mjjadhav@gmail.com

Introduction

Algae a significant group of plant world, comprises a large heterogenous assemblage of autotrophs. Several of them are useful to human beings in many ways. Algae are known to contain different bioactive compounds. Of these, there are certain growth promoting substances which are found beneficial in seed germination and plant growth. Present research work deals with the study of effect of algal extracts on seed germination and application of algal powder in plant growth. Algal biomass easily available in freshwater habitat has been used for experiments. Gupta (1964) and Gupta and Kusumlata (1964) studied germination of Rice seeds, treated with extracts of different algae. Goyal and Venkataraman (1970) reported an increase in the yield of rice due to blue green algae. Rao and Pattnaik (1974) studied effect of certain blue green algae on flowering and yield of Rice by pot culture method. Bongale (1982) recorded beneficial effect of blue green algae on the growth and yield of rice.Pingle and Abhang (2007) reported promotive effect of hot and cold water extracts of freshwater algae *Nostoc* and *Lyngbya* on seed germination of some vegetable crops. Kamble (2008) studied effect of algal extracts on seed germination of Sorghum, Mothbean and Sesamum.

Materials and Methods

In present study, algal extracts were prepared from freshwater algal biomass. Algal extracts prepared in cold water, hot water and methanol were tested for germination of wheat (*Triticum aestivum*) seeds. Fine algal powder was used in growth of wheat plant.

Algal biomass was collected in plastic jars from ponds located near Gogababa hills of Aurangabad, in the month of December 2009. In laboratory collected algal biomass screened microscopically thoroughly, then washed with sterile distilled water and dried in shade. After drying, fine algal powder was prepared and stored in acid washed air tight bottles. Algal extracts in cold water, hot water and methanol were prepared separately. Surface sterilized seeds of wheat were taken. 10 seeds for each extract were taken and soaked separately for 4 hours. After that, soaked seeds were placed on moist blotter in sterilized petriplates. Seeds soaked in sterile distilled water served as control. Per cent germination, root and shoot length of seedlings were measured after 7 days of incubation at room temperature.

Algal powder was used in order to understand the effect of algal biomass on growth of wheat plant by tray culture method. 50 grams of algal powder was added in 5 kilograms of soil, taken into tray and sufficient water was sprinkled. In another tray only 5 kilograms of soil was taken, sufficient water was sprinkled. This tray is served as control. On second day wheat seeds were sown in both trays. After 30 days, observations on growth and development of wheat plant were recorded.

Results and Discussion

Microscopic observation of algal biomass reveals the dominance of *Spirogyra aequinoctialis, Spirulina gigantea, Oscillatoria subbrevis, Phormidium molle, Microcoleus acutissimus* and *Plectonema gracillimum*. In algal biomass different bioactive compounds are present, of these, some are growth promoting substances. In order to see the effect of these growth promoting substances on seed germination of wheat, algal extracts in cold water, hot water and methanol were prepared and experiments were carried out. It is observed that cold water extract found beneficial in seed germination of wheat. In cold water extract treatment, germination of seeds was 100 per cent, shoot length and root length of wheat seedlings found maximum (Table 38.1). In hot water, the germination percentage was 90 per cent whereas in methanol and in control it was 80 per cent. Gupta (1964) and Gupta and Kusumlata (1964) studied effect of algae on rice seed germination and observed that seed germination is markedly accelerated. Pingle and Abhang (2007) found that cold water and hot water extracts of *Nostoc* and *Lyngbya* were promotive on seed germination on some vegetable crops. Kamble (2008) observed that cold water algal extract was helpful for seed germination.

Table 38.1: Effect of Algal Extracts on Germination of Wheat Seeds

Sl.No.	*Type of Algal Extract*	*Germination (per cent)*	*Shoot Length (cm)*	*Root Length (cm)*
1.	Cold water extract	100	6.04	6.10
2.	Hot water extract	90	4.88	3.10
3.	Methanol extract	80	4.00	5.10
4.	Control	80	4.90	6.00

Algal powder found useful in growth of wheat plant as compared to control (Table 38.2). Wheat plant treated with algal powder showed luxuriant growth, increase in height and number of leaves. Total chlorophyll content and dry matter of treated plant were maximum as compared to control plant. Rao and Pattnaik (1974) reported increase in flowering and yield of rice plant which is treated with blue-green algae. Abhang *et al.* (2010) worked on BGA extract on growth and yield of Spinach. They observed that blue green algal extract is significant in overall growth, development and yield of plant.

Table 38.2: Effect of Algal Fertilizer on Growth and Development of Wheat Plant

Sl.No.	*Parameters*	*Plant Treated with Algal Fertilizer*	*Control Plant*
1.	Growth	Luxuriant	Normal
2.	Colour of Plant	Dark green	Green
3.	Height of Plant (cm)	39	34
4.	Number of leaves	07	05
5.	Total Chlorophyll content (mg/gm)	0.90	0.81
6.	Dry matter (per cent)	19.12	18.00

It is observed that algal extract has definite stimulatory effect on seed germination. Algal biomass can be used as biofertilizer for growth and development of plants. Algal biofertilizers are organic and can be applied in different plants by different ways by farmers as concerned with bioactive supplementary fertilizer and eco-friendly agricultural management.Present work is one of the important aspects of algal biotechnology which is useful in sustainable agiculture development.

References

Abhang, A. R., Pingle, S. D. AND Auti, B. K. (2010), Effect of foliar application of BGA extract on growth and yield of Spinach (*Spinacea oleracea* L.) var. Allgreen. *Flora and Fauna*.Special issue., pp.1-4.

Bongale, U.D.(1982). Effect of *Hapalo siphonwelwitschii* and *Cylindrospermum musicola* on the growth and yield of Jaya paddy.*Phykos.*, **21**: 80-82.

Goyal, S.K. and Venkataraman. (1970).Effect of algalizationon high yields rice varieties, I. Response of rice varieties. *Phykos*, **9**(2): 137-138.

Gupta, A.B. (1964). The algal flora and its importance in the economy of paddy fields. Sci. Tech. Kanpur, (ann. conf.) Labdev, *J.Sci. Tech*, .**2**(1): 84.

Gupta, A.B. and Kusumlata (1964).Effect of algal growth hormones on the germination of paddy seeds.*Hydrobiologia*, **24**: 430-434.

Kamble, S.M. (2008). Studies on effect of bioactive compounds on algae on some fungi. Ph.D. Thesis. Dr. Babasaheb Ambedkar Marathwada University, Aurangabad, Maharashtra.

Pingle, S.D. and Abhang, A.R. (2007). Studies on freshwater algal extract of seed germination some vegetables. Abstract Nat. Conf. on Modern trends in plant science, Dr. Babasaheb Ambedkar Marathwada University, Aurangabad. p. 69.

Rao, S.D. AND Pattnaik (1974).Studies on the effect of certain blue green algae on the flowering and yield of two early varieties of rice.*Phykos*, **13**(1): 84-89.

2015, Major Constrains and Verdict of Crop Productivity *Pages 489–498*
Editor: Dr. U.N. Bhale
Published by: ASTRAL INTERNATIONAL (P) LIMITED, NEW DELHI

Chapter 39

A Conventional Application of Foliar Fertilization for some Selected Crop Plants

C. Maya[1], H.C. Lakshman[2*], Romana M. Mirdhe[2] and Ratna V. Airsang[2]

[1]*Department of Botany, Bangalore University, JB Campus, Bangalore – 560 056, Karnataka, India*
[2]*Microbiology Laboratory, P.G. Department of Studies in Botany, Karnatak University, Pavate Nagar, Dharwad – 580 003, Karnataka, India*

Introduction

To complete life cycle normally, living organism requires a supply of large number of substances from outside. This supply is called nutrition. If the supply needed by the organism is that of inorganic substances only, then the organisms are called autotrophic or self-feeding. All green plants synthesizing their own organic requirements are autotrophs. Under normal conditions of growth, however, all green plants are autotrophic and they require from outside the supply of only inorganic substances. With the exceptions of hydrogen, carbon and oxygen, all other inorganic plant requirements are obtained directly or indirectly from soil. As the sources of these inorganic requirements are mineral elements are known mineral nutrients and the nutrition is called mineral nutrition.

Most of the crop plants require 17 nutrients to complete their life cycle. Essential plant nutrients are divided into macro and micronutrient groups. Macronutrients

* *Corresponding Author.* E-mail: dr.hclakshman@gmail.com

are carbon (C), hydrogen (H), oxygen (O), nitrogen (N), phosphorus (P), potassium (K), calcium (Ca), magnesium (Mg) and sulfur (S). Micronutrients include zinc (An), copper (Cu), iron (Fe), manganese (Mn), boron (B), molybdenum (Mo), chlorine (CI), and nickel (Ni) (Fageria, 1992; Brady and Weil, 2002). Further, the essentiality of silicon (Si), sodium (Na), vanadium (V) and cobalt (Co) has been considered, but is not yet proven (Mengel *et al.*, 2001; Fageria *et al.*, 2002; Epstein and Bloom., 2005). Macronutrients are required in higher amounts compared to micronutrients. However, from the plant essentiality point of view, all the nutrients are equally important for plant growth. First three macronutrients (C, H and O) are supplied to plants by air and water. Hence, their supply to plants is not a problem. Hence, the remaining 14 nutrients should be present in the plant growth medium in adequate amount and proportion for plant growth (Fageria, 2005; 2007; Fageria and Baligar, 2005).

Convenient Classification of Minerals Required by Plants

1. **Structural elements:** Carbon, hydrogen, oxygen, nitrogen, sulphur, phosphorus, which is given below:
2. **Bio-catalytical elements:** The elements forming part of biocatalysts such as iron, copper, manganese and cobalt.
3. **Indispensable elements:** Potassium, calcium, boron and molybdenum.
4. **Radio-active elements:** Uranium, thorium, radium, actinium and rhodium.
5. **Stimulating elements:** Sodium, chlorine, arsenic, titanium, cobalt, nickel, iodine, bromine and rubidium.

Vegetation evolved in the oceans, bathed in a solution containing every imaginable mineral. Seaweed takes food directly from seawater. Like their marine ancestors, land plants, can take in nourishment through the stomatal pores or stomata on their leaf surfaces. They are tiny mouths that breathe in CO_2 and exhale water and oxygen. They also transport nutrients up to ten times more efficiently that of root systems. Foliar fertilization bolsters the nutrients available to each plant, like a regular dose of vitamins and supplements. Foliar fertilization is fast becoming an essential addition to standard cultivation techniques and many growers who have grown up with chemicals; it is a small step to organic fertilization, Practice.

Mechanisms and Uptake of Foliar Applied Nutrients by Plants

Nutrient absorption process by leaves may be different than roots because, leaf cell walls are covered by a cuticle, which are not found in the root structure. Franke (1967), have reported that cuticular membranes are permeable to both organic and inorganic ions and un-dissociated molecules. The penetration of ions is determined by the kind of charge, adsorbability, and ion radius. Under normal conditions uptake of ions constitutes an accumulation against a concentration gradient in leaves as in roots. The energy required for active absorption can be derived from respiratory metabolism, or as in green leaves from photosynthesis proves (Franke, 1967). The light quality and intensity improves the rate of ion absorption by leaves (Franke, 1967). Franke (1967), have also suggested that ion uptake by leaves may be completed in three stages.

If a nutrient is not able to be transported from the sprayed tissues to those developing after the spray treatment, the spray treatment should be repeated every time when a new flush of growth appears (Papadakis *et al.*, 2007). In other words, if a nutrient is immobilized after its foliar application, the positive effects of the spray would be limited only to the sprayed tissues. Hence, deficiency symptoms will appear in the shoots growing after spray. Macronutrients mobility in plant tissue is reasonable, except Ca and S. However, most of the micronutrients mobility in plant tissues is poor. Therefore, foliar fertilization can also be timed to encourage and/or enhance critical points in the plant's growth cycle, such as seedling emergence or rapid growth phases or flowering, fruiting and seed formation. Many crops benefit from 6–8 foliar applications per season. Fish based fertilizers (fish powder) and seaweed soluble seaweed powder or seaweed extract) are among the most common foliar feeding in organic farming, applied either separately or in combination with water. Compost tea has become popular as a foliar spray material because of its nutrient containing disease-suppressive characteristics. Most of the European countries, use soluble organic materials and those from which extracts are easily made include spray- dried blood, bat guano, worm castings, manure teas, humates, molasses, milk, B vitamins, and herbal extracts of plants like stinging nettle and horsetail and other organic foliar fertilizer components.

In recent days some market gardeners spray nutrients on fruit-setting crops such as Tomatoes and cucumbers to increase yields and on leafy greens like lettuce and spinach to speed maturity and increase storage life. Grape growers European use foliar feeds in their vineyards, Chinese farmers use foliar fertilizers on grain crops to increase yields. As, we know that leaves are green factories where the complex chemical processes of photosynthesis produce the compounds plants need for growth. And thus foliar fertilizers are absorbed right at the site where they will be used, and are quite fast acting.

Basic Diagnostic Techniques for Foliar Fertilization

Proper identification of nutritional status of crop species is important to maximize production efficiency. Inadequate knowledge of the crop nutrient status can frequently result in excessive fertilizer applications and nutrient imbalances as well as undetected deficiencies or excess within the crop. Four methods to assess nutrient availability or sufficiency of mineral nutrients for plant growth: *viz.*, 1) visual symptoms; 2) soil testing, 3) plant analysis; and 4) crop growth response. All the four approaches are becoming widely used separately or collectively as nutrient availability or deficiency or sufficiency diagnostic aids. They are extremely useful, without limitations. According to Fageria and Baligar (2005) these techniques as under;

1) Visual Symptoms

Visual symptoms are the cheapest nutritional disorders it needs a lot of experience on the part of the observer, because deficiency symptoms are confused with drought, insects and disease infestation, herbicide damage, soil salinity and inadequate drainage problems. Deficiency symptoms caused by immobile nutrients first appear on the younger or upper leaves. The older leaves do not show any symptoms because immobile nutrients do not move or translocate from older to newer leaves. Immobile nutrients are calcium, zinc, boron, copper, iron, manganese, and molybdenum. In contrast, when there is a deficiency of a mobile nutrient, the symptoms first appear on the older leaves of the plant. This is because the mobile nutrients move out of the older leaves to the younger part of the plant.

2) Soil Test

To develop a sound soil best identifying N deficiency or sufficiency is rather difficult. In the literature some authors have mentioned that N recommendations can be done on the basis of ammonium (NH_4^+) or nitrate (NO_3^-) analysis. However, in the author's opinion it is not a very precise method to identify N nutritional disorder in crop plants. Generally, and samples are taken at 0-20 cm soil depth for food crops. This depth is recommended because about 80 per cent of root systems of food crops remain in this soil volume (Fageria *et al.*, 2006).

3) Plant Analysis

In plant analysis, the different mineral nutrient elements are determined even in micro and semimicro quantities by spectrophotometric, colorimetric, turbidimetric, flame photometric and titrimetric or by other methods. These estimations are done according to needs either in extracts of fresh plant tissue such as organic extracts or in the plant tissue from which organic material has been completely removed either by ashing it at 600° or by digesting it with nitric acid and perchloric acid on a hot plate. Plant tissue tests to be most effective and reliable, it is necessary to have plant analysis results under different agro-ecological regions for each crop species to make interpretation of analytical results. Plant analysis results vary with crop species, plant age and plant part analyzed. Hence, special care should be taken while adopting plant tissue test for identifying nutrient deficiency or sufficiency. Many factors such as soil, climate, plant and their interaction affect absorption of nutrients by growing plants. However, the concentrations of the essential nutrients are maintained within rather narrow limits in plant tissues. Such consistency is thought to arise from the operation.

4) Crop Responses to Foliar Fertilization

Extensive research conducted during the 1970s and 1980 on foliar fertilization of soybean during early vegetative growth stages or during late reproductive growth stages showed inconsistent grain yield increases (Parker and Boswell, 1980; Poole *et al.*, 1983; Haq and Mallarino, 1998; 2000; Mallalrino *et al.*, 2001; Nelson *et al.*, 2005). Garcia and Hanway (1976), have reported yield increases of 27 to 31 per cent when a liquid N-P-K-S fertilizer was sprayed at late reproductive stages (R5 to R6). These authors suggested that root activity decreases during pod fill and that nutrient uptake is not enough to meet the seed demands for nutrients.

Modest soybean yield increase from an addition was obtained by Syverud *et al.* (1980), but little effect of P, K, S, was noted. The authors also reported that seed weight and seed N concentration were also increased by the foliar N treatments. Boote *et al.* (1978) on the other hand failed to increase either soybean yield or seed number and noted slight burn and necrotic spots on the laves. Leaf damage due to foliar applications of N, P, K and S was sufficiently severe in studied by Parker and Boswell (1980), to reduce soybean yield in most cases. Haq and Mallarino (1998), have reported that foliar fertilization with various rates of N, P and K at the R5 growth stage increased soybean grain yield in R7 of R48 field trials and reduced yields slightly in 2 trials, with mean yield increase of 54 kg ha^{-1} across all the trials. Similarly, Poole *et al.* (1983), have reported that foliar applications of fertilizer N, P, K and S to soybean during the pod fill stages (R4 to R7) of growth has been shown to increase yields. An increase in grain protein concentration with foliar fertilization at late growth stage in wheat could prevent price deduction and possibly could result in premium returned to the producer in favorable years (Bly and Woodard, 2003). Studies conducted by Chesnin and Shafer (1953) in Nebraska did not indicate consistent positive response to foliar applied N and P on soybeans. Tennessee Valley Authority (TVA) coordinated 180 foliar fertilizer comparisons on soybeans in 11 states in the USA and in a subsequent report (Peele, 1977) indicated that average yield among states varied from 27 kg ha^{-1} increase to 396 kg ha^{-1} decrease. In Florida, Boote *et al.* (1978) reported that the nutrient concentration of soybean was increased by foliar applications of N, P, K, and S, but yields were not affected significantly and photosynthesis duration was not extended. These authors further reported that control soybeans yielded 3825 kg ha^{-1} as compared to 3617 kg ha^{-1} for foliar treated soybeans.

Iron deficiency chlorosis is a common problem when soybean is grown on calcareous soils. There are two most important management practices to reduce iron chlorosis in soybean grown on calcareous soils. One is use of iron efficient cultivars and second is application of iron as foliar sprays. The application of K through foliar sprays is much lower than N and P. The reasons may be that, K is required in very higher amount by most crop plants and K responses to field crops are limited. However, in Alaska, potato yields were increased from weekly K sprays (Tisdale *et al.* (1985). They reported that, as with P, however, the problem of adding sufficient amounts becomes critical. Further, foliar application of P used less than N largely because most P compounds are damaging to leaves when sprayed on in quantities large enough to make the application beneficial (Tisdale *et al.*, 1985).

Role of AM Fungi on Foliar Fertilized Plants

The current approach of cultivation is following the sustainable agricultural Horticultural plants, using very less of chemical inputs like fertilizers and pesticides having adverse effect on soil health, fertility and environment. Thus, the use of microbial inoculants plays an important role in sustainable plants health. The use of AM fungi in increasing the plant growth is well documented in the last three and half decades. The uses of foliar fertilization on mycorrhizae inoculated plant grow significantly over the control non- mycorrhizal plants (Lakshman, 1996). Arbuscular mycorrhizal (AM) fungi are a ubiquitous group of soil fungi colonizing roots of plant to more than 90 per cent angiosperm families, they enhanced plant growth due to AM association is well documented (Bagyaraj, 2006; Smith and Read, 1997; Lakshman, 2009). This has gained momentum in recent years because of the higher cost and hazardous on natural environment due to the use of chemical fertilizers.

Arbuscular mycorrhizal (AM) fungal association is known to improve plant growth through better uptake of nutrients, tolerance to drought and salinity (Lakshman, 2005). Plants able to withstand transplant shock, resistance to root pathogens and by boosting synergistic interaction with beneficial soil micro-organisms such as N-fixes and P-Solubilisers (Sabannavar and Lakshman, 2008). Further, the mycorrhizal mutualistic symbiosis between plant roots and soil fungi is known to play a significant role in P cycling and uptake of P by plants (Harley and Smith, 1983). Among mycorrhizae, AM fungal inoculants are most promising fertilizer.

Arbuscular mycorrhizal (AM) fungi mutually beneficial symbiotic fungi, is ubiquitous in distribution and physically unspecialized. It is wide spread in plant kingdom. They occurs a non- pathogenic symbiotic association between fungi and roots of the higher plants (Girma *et al.*, 2007). A unique trait of these mycorrhizal fungi is that form direct link between the soil and root, providing for movement of elements in the in the alkaline rhizosphere (Chiramel *et al.*, 2006). These AM fungi are known to improve plant growth through better up-take of nutrients and water resistance to root pathogens. However, no efforts seem to have directed towards the isolation and identification of the AM fungi inhabiting soil of high salinity.

Applications of Foliar Fertilization

Foliar fertilization has played an important role in the yield increases, resistance to diseases and insect pests, improved drought tolerance, and enhanced crop quality.

Foliar spray applications are used to aid plants in recovery from transplant shock, hail damages. Foliar spray help plants compensate for soil deficiencies (low fertility, low soil temperature, etc.) during the growing season.

Foliar spray help in nutrient absorption, foliar fertilization can be from 8 to 20 times as efficient as ground application.

Because of the variability in research results and practical field experience with foliar fertilization, options on its usefulness vary in both conventional and alternative agriculture practice. However, there is general consensus, however, that foliar fertilization should not be considered a substitute for a sound soil- fertility programme.

One of the most benefits from the foliar fertilization is the increased uptake of nutrients from the soil. This is mainly based on the belief that foliar fertilization causes the plant to pump more sugars and other exudates from its roots into the rhizosphere. And there are innumerable beneficial microbial populations in the root zone are stimulated by the increased availability of these exudates. In turn, this enhanced biological activity increase the availability of nutrients, these microbes secret disease- suppressive and other vitamins, and other factors beneficial to the plants.

Its economic value is generally deemed greater for horticultural than for agronomic crops.

Highly concentrated sprays, especially those bearing inorganic salt-based fertilizers, that "burn" plant foliage. This is especially true of chloride salts example potassium chloride therefore; Spray- solution pH should remain in the near-neutral range (pH 5.5-8.5). Sources of water are contamination should not be used at all for foliar feeding. If there is concern specifically some disease organisms, this water can be more effective.Chlorination of water removes harmful bacteria, but it can also kill beneficial organisms, which may be included in some foliar sprays. Therefore, before the use of foliar sprayer, allowing water to stand in an open tank overnight generally renders chlorinated water harmless to beneficial microbial population.Foliar spray should be done, when wind movement is limited and minimum. Now a day's atomized sprays are useful because, they drift readily.Before use of foliar spray, one must undertake preliminary tissue nutrient tests which much more reliable, eye should know the than past history, plant symptoms, or soil tests for establishing whether nutrient deficiencies exist. This could always, work hand in hand to many horticultural plants. One who wants to foliar fertilization, he/she should be accurate foliar analysis depends on good sampling and handling procedures. While doing so instruction should be obtained from each laboratory or consulting service in advance and one should read the instructions entitled leaf/petiole/stem etc.

Precautions before Foliar Spraying

When mixing up your formulation, urea, mineral, organic fertilization or compost tea, use non-chlorinated, well oxygenated water is need of the hour. Bubble air through chlorinated water or leave it to off-gas overnight. One can try using seltzer in your foliar spray to give plants and then added CO_2 boost.One should make sure mineral ingredients must be completely dissolved and the solution is must be very dilute. For example, chemicals in high concentration tend to 'burn' foliage and leave a salt residue. Compost teas need to be diluted more than ten times. While using foliar spray it must be a natural surfactant or wetting agent to help the solution flow over and stick to foliage. For example Yucca, is a natural surfactant and is often a component of compost tea. The use true organic soaps such as Dr. Bronners, Tom's, or Pangea are suitable.Usually transplants plants prefer more alkaline solution (pH 7.0) while, after growth older plants need more acid (pH 6.2) spray. The uses of baking soda to raise pH and apple cider vinegar to lower the pH of spray are more suitable.Enzymes and phytoalexins to lead to diseases, spraying with a fine sprayer for foliar fertilization and low pressure sprayer for compost tea. In fact the microbes in compost tea need large protective water droplets. It is suggested that the early morning or evening when the stomata are open. Should not be during the temperature is over 80° F or in the bright sun. Because harsh ultraviolet rays can kill microbes in compost tea.Foliar spray must cover at least 70 per cent of the foliage, particular attention to the underneath of the leaf surfaces.

Conclusion

Soil application is most common method to supply essential nutrients to plants. In this case applied nutrients are absorbed by plant roots. However, higher plants can also absorb mineral nutrients when applied as foliar sprays in appropriate concentrations. In recent days high yielding cultivars, nutritional requirements macronutrients are met with foliar applications. Furthermore, foliar application of macronutrients requires several sprays, because it will wash off by rain, plant should have sufficient leaf area for absorption and there should not be leaf damage. Despite these drawbacks, under certain circumstances foliar application is most effective methods to correct nutritional disorder. For example, iron deficiency in calcareous soils can be corrected by foliar application of ferrous sulfate or iron chelates solution more efficiently than the soil application of iron sources. Macronutrients, which are required in high amounts by crop plants, are rarely met by foliar application. Hence, so far the most important use of foliar sprays has been in the application of micronutrients. In foliar sprays, macronutrient concentrations of generally less than 2 per cent are used to avoid leaf burning. Macronutrient solution concentrations vary from 0.1 to 1.2 per cent depending on the nutrient. Plant age should also be considered in selecting nutrient concentration. Older plants are more tolerant to higher concentrations of salts compared to younger plants. In foliar fertilization, droplet size and fertilizer solubility should be carefully controlled since it will affect crop response. Foliar fertilization in food crops may not increase yield but may increase protein content of grains, if applied during anthesis or flowering.The yield response of field crops to foliar fertilization of macro and micronutrients is highly variable. Positive results with foliar fertilization of N, P, and K, in soybean have been associated with high yielding environments. Yield responses to foliar fertilization are generally not positive when yield is low or nutrients are at an optimum level in the soil. There may be exceptions to this rule, such as iron. In Brazilian Oxisols, iron content is quite high; however, when soil pH is higher than 6.0, iron deficiency in upland rice is frequently observed. This may happen due to immobilization of iron due to precipitation at higher pH. Leaf damage due to higher concentration of foliar fertilization may be one of the reasons for either yield decreases or the lack of yield increases. Foliar spray of nutrients should be avoided at high temperature during the day to avoid leaf burning. Similarly, windy days may drift the applied nutrient solution and rain immediately after application may washout the sprayed material and reduces its efficiency.

References

Bagyaraj, D. J. (2006). Arbuscular mycorrhizal Fungi in sustainable agriculture. In: *Techniques in mycorrhizae*, Eds., Rodirgues and Bukhari. **1**: 1-8.

Bly, A.G., and Woodard, H. J. (2003). Foliar nitrogen application timing on grain yield and protein concentration of hard red winter and spring wheat. *Agronomy Journal.*, **95**: 335-358.

Boote, K.J., Gallaher, R.N., Robertson, W.K., Hinson, K. and Hammond L.C. (1978). Effect foliar fertilization on photosynthesis, leaf nutrition and yield of soybean. *Agronomy Journal.*, **7**: 787-791.

Brady, N.C. and Weil, R.R. (2002). The nature and properties of soils, 13th ed. Upper Saddle River, N.J. Prentice Hall.

Chesnin, L., and Shafer N. (1953). Foliage applications of urea solutions to grain and forage crops. *Agronomy Journal.*, **45:** 576.

Chiramel, T. Bagyaraj, D. J and Patil, C. S. P. (2006). Response of A*ndrodraphis paniculata* to different Arbuscular mycorrhizal fungi: *J. of Agri. Technology.*, **2**(2): 221-228.

Epstein, E., and Bloom A. J. (2005). Mineral Nutrition of Plants: Principles and Perspectives, 2nd ed. Sunderland, MA: Sinauer Associates.

Fageria, N.K. (2007). Soil fertility and plant nutrition research under field conditions: Basic principles and methodology. *Journal of Plant Nutrition.*, **30**; 2003-223.

Fageria, N.K., and Baligar, V.C. (2005). Nutrient availability. In: Encyclopedia of Soils in the Environment, ed. D. Hillel, . San Diego, CA: Elsevier, pp. 63-71.

Fageria, N.K., Baligar, V.C. and Clark R.B. (2006). Physiology of Crop Production. New York: The Haworth Press.

Fageria, N.K., Baligar, V.C. and Clark R.B. (2002). Micronutrients in crop production. *Advances in Agronomy.*, **77**: 185-268.

Fageria, N.K., Barbosa Filho, M.P., Moreira, A. and Guimaraes, C.M. (2009). Foliar fertilization of crop plants. *Journal of plant nutrition.*, **32**: 1044-1064.

Fageria, N.K. (1992). Maximizing Crop Yields, New York: Marcel Dekker.

Franke, W. (1967). Mechanisms of foliar penetration of solutions. *Annual Review of Plant Physiology.*, **18**: 281-300.

Garcia, R.L. and Hanway, J.J. (1976). Foliar fertilization of soybean during the seed-filling period. *Agronomy Journal.*, **68**: 653-657.

Girma, K., Martin, K.L., Freeman, K. W., Mosali, J., Teal, R.K., Raun, W.R., Moges, S.M. and Arnall, D.B. (2007). Determination of optimum rate and growth for foliar applied phosphorus in corn. *Communications in Soil Science and Plant Analysis.*, **38**: 1137-1154.

Haq, M.U. and Mallarino, A.P. (1998). Foliar fertilization of soybean at early vegetative stages. *Agronomy Journal.*, **90**: 763-769.

Haq, M.U. and Mallarino. A.P. (2000). Soyabean yield and nutrient composition as affected by early season foliar fertilization. *Agronomy Journal.*, **92**: 16-24.

Hardey, J. L and Smith, S. E. (1983). Mycorrhizal Symbiosis, Academic Press New York., p.486.

Lakshman, H. C. (1996). VA Mycorrhizal studies in some important timber tree species. Ph. D thesis Karnatak University Dharwad, India., p.349.

Lakshman, H. C. and Kolkar, K.P. (2009). AM fungal diversity of Forest tree species in dry deciduous forest of Dharwad. In: Forest diversity Eds. K. Muthuchelian, S. Kannaiyan and A. Gopalam. Associated Publishing Company, New Delhi. Vol II., pp. 223-230.

Lakshman, H.C. (2005). Response of Albizzia lebbeck DC., to VAM, Rhizobium and Molybdenum sources. *My forest.*, **41** (2A): 295-300.

Mallarino. A.P., Haq, M.U., Wittry, D., and Bermudez, M. (2001). Variation in soybean esponse to early season foliar fertilization among and within fields. *Agronomy Journal.*, **93**: 1220-1226.

Mengel, K., Kirkby, E.A. Kosegarten, H. and Appel. T. (2001). Principles of Plant Nutrition. 5th ed. Dordreeht: Kluwer Academic Publishers.

Nelson, K.A., Motavalli, P.P. and Nathan, M. (2005). Response of notill soybean (*Glycine max*. L. Merr.) to timing of pre plant and foliar potassium applications in a claypan soil. *Agronomy Journal*., **97:** 832-838.

Papadakis, *I.E.*, Sotiropoulos, T.E. and Therios, I.N. (2007). Mobility of iron and manganese within two citrus genotypes after foliar applications of iron sulfate and manganese. *Journal of Plant Nutrition*., **30**: 1385-1396.

Parker, M.B., and Boswell, F.C. (1980). Foliar injury, nutrient intake, and yield of soybean as influenced by foliar fertilization. *Agronomy Journal*., **72**: 110-113.

Peele, R. (1977). Jury still out in case of soybean foliar. Muscle Shoals, AL: Southeast Farm Press.

Poole, W.D., Randall, G.W. and Ham. G.E. (1983). Foliar fertilization of soybean: I. Effect of fertilizer sources, rates and frequency of application. *Agronomy Journal*., **75**: 195-200.

Sabannavar, S.J. and Lakshman, H.C. (2008). Interaction between Azotobacter, Pseudomonas, and AM fungi on two verities of *Sesamum indicum* L. J. *Agronomy and Crop Sci.*, **194** (6): 472-478.

Smith, S.E. and Read, D.J. (1997). Mycorrhizal symbiosis. Academic press, San Diego., p. 605. Syverud, T.D., Walsh, L.M., Oplinger E.S., and Kelling. K.A. (1980). Foliar fertilization of soybean (*Glycine max* L.). *Communications in Soil Science and Plant Analysis*., **11**: 637-651.

Tisdale, S.L. Nelson, W.L. and Beaton. J.D. (1985). Soil Fertility and Fertilizers, 4th ed. New York: Macmillan.

2015, Major Constrains and Verdict of Crop Productivity *Pages 499–509*
Editor: Dr. U.N. Bhale
Published by: ASTRAL INTERNATIONAL (P) LIMITED, NEW DELHI

Chapter 40

Effect of AM Fungus, Pressmud and Indole Acetic Acid on Two Varieties of *Corchorus capsularis* (L.)

Pushpa K. Kavatagi* and H.C. Lakshman

Microbiology Laboratory, P.G. Department of Studies in Botany, Karnatak University, Pavate Nagar, Dharwad – 580 003, Karnataka, India

Introduction

Arbuscular mycorrhizal fungi (AMF) are an interesting group of soil fungi belong to Glomeromycota and that form a symbiotic association with wide range of plant species. Due to growing concern about environmental protection, research on the possibility of adding or manipulating micro-organisms to enhance plant protection against pathogens have received increasing attention in last three and half decades. The symbiosis has a positive effect on plant water potential, especially for plants under drought stress (Auge, 2001). For sustainable sugar cane production it is neither chemical fertilizers nor organic manures alone, but their integrated use has been observed to be highly beneficial Press mud cake (PMC) is one such sources of organic matter and nutrients of sugarcane based industrial waste which can be profitably utilized for sugarcane production. The microbial activity in the soil improved substantially due to addition of PMC as considered to be as effective sources of carried material for microbial inoculants (Jauhri, 1990). Traditional use of mineral fertilizer in agriculture cannot be over emphasized, but with fertilizer costs going up, these need to be supplemented or substituted with available organic wastes or

* *Corresponding Author.* E-mail: pushpak.k2@gmail.com

manures. They also protect the soils from wind and water erosion thus preventing nutrient losses through runoff and leaching. Arbuscular mycorrhizal fungal inoculation of Tomato plants can lead to increase in fruit yield (Subramanian *et al.*, 2006), and in the content of phosphorus in the fruits (Al-Karaki and Hammad, 2001). The auxin Indole-3-acetic acid (IAA) was the first used as plant hormone to be to stimulate rooting (Cooper, W.C, 1935). The effects of plant growth regulator treatments on root growth may be a function on plant growth regulator and the concentrations applied (Pilet, 1983; Firn, 1986). Further, the beneficial effect on plant growth in presence of free living nitrogen fixing organisms was attributed to phyto-hormone production rather than or in addition to nitrogen fixation (Lakshman and Patil, 2004; Lakshman *et al.*, 2010). AM fungi can protect plants against conditions of suboptimal water potential (drought, flooding, salinity) (Miransari, 2010; Smith *et al.*, 2010; Birhane *et al.*, 2012).

Jute (*Corchorus capsularis* L.) is the least expensive but most important of all bast fibers and occupy second in the production of after cotton, the fifer is obtained from the stem today about 80 per cent of total word's production comes from India and Bangladesh. It thrives on warm humid reach loamy soils where, rain fall raises from 150-250cm.Jute fifers contained 63 per cent of cellulose and they have high proportion of lignin and it finds wide manufacture of drugs and blankets carpets, carpluanies cloth backing, oil cloth, twine, rope, curtains and coarse cloth. The leaves and young shoots are important sources of vegetable food in Egypt, Sudan and Greece. No efforts of inoculating AM Fungi with press mud treatment on this plant. Therefore, to evaluate the growth response by using AM fungus, Press mud and IAA the studies have been undertaken.

Materials and Methods

Soil and Plant Material

The garden soil: sand (3:1 v/v) autoclaved for 12 hr and its (20x20) breadth x length mixture was filled into 17.5 cm diameter pots containing 4 kg of soil in each experimental pot. The seeds of Ca-1 and DH-3 varieties of *Corchorus capsularis* L., were procured from University of Agriculture science Dharwad, India. Seeds were surface sterilized by treating with 1 per cent sodium hypochlorite for 2-3 min before sowing and after germination uniform seedlings were maintained in one per pot.

Inoculum

AM fungus *Rhizophagus fasciculatus* (Thaxt.) Walker and Schußler, were maintained for 12 weeks in green house conditions using Jowar (*Sorghum vulgare* L) as a host for mass multiplication in (30×35 cm) diameter pots containing sterilized sand soil mix (1:1 v/v) and bio-compost were prepared by mixing press mud and water in the ratio of 1:2 were used for composting.

Pot-Experimental Setup

The experimental pots amended 15 grams of the AM fungal inoculum approximately consisted of roots and soil contained 115-172 chlamydospores, as single inoculums placed 4 cm below the surface of experimental pots with 1g pressmud and before sowing the seeds were treated with 2 per cent of IAA hormone, the details as follows.

1. Control
2. *Rhizophagus fasciculatus* (Thaxt.) C.Walker and A.Schußler,
3. Pressmud
4. IAA
5. *Rf* + Pressmud+ IAA

The plants were harvested periodically at every 30, 60 and 90 days. The growth parameters such as, shoot length, fresh weight of root, dry weight of root, fresh weight of shoot, dry weight of shoot, per cent root colonization, spore number and number of leaves, were recorded. The AM fungal spores were counted by wet sieving and decanting technique proposed by (Gerdemann and Nicolson, 1963). The per cent of mycorrhizal infection was evaluated microscopically by clearing of roots in 10 per cent KOH, neutralized in 2 per cent HCL and stained with 0.05 per cent trypan blue in lactophenol according to method described by (Phillips and Hyman, 1970). The following formula was used to calculate the root colonization as per (Giovanneti, Mosse 1980).

$$\text{Per cent of root colonization (\%)} = \frac{\text{No of root bits colonization}}{\text{Total number of root bits observed}} \times 100$$

Statistical Analysis

All the data were subjected to analysis of variance and mean values were compared using the Duncan's multiple range test (at the 0.05 level), as recommended (Sedecor and Cochran, 1980).

Results and Discussion

The inoculation of arbuscular mycorrhizal fungi and press mud with Indole acetic acid augmented plant growth shown in (Figures 40.1 and 40.2), the growth parameters were recorded in all the four Tomato varieties. The results of Ca-1 and DH-3 varieties of *Corchorus capsularis* L., are described as under.

Ca-1 Variety

The (Table 40.1) showed that after 30 days of plant growth in the amendment of pressmud, IAA with inoculation of *Rhizophagus fasciculatus* resulted an increase in shoot length(9.52cm), fresh weight (4.22g) and dry weight of shoot(0.91g), root length(3.09cm), fresh weight (1.52g) and dry weight of root(0.13g), number of leaves(22.00), per cent root colonization(33.00 per cent), spore number(42.00), stem diameter(2.57cm). flowers and fruits were not observed at this stage.

After 60 days with inoculation of *Rhizophagus fasciculatus*, pressmud and IAA showed significant growth in all the treatments like non-inoculated, press mud, *Rhizophagus fasciculatus* and IAA significant increase in shoot length (38.11cm), fresh weight (55.12g) and dry weight of shoot (6.12g), root length (8.52cm), number of

Table 40.1: Showing the Effect of AM Fungi (*Rhizophagus fasciculatus*) on Shoot Length, Fresh Weight of Shoot, Dry Weight of Shoot, Root Length, Fresh Weight of Root, Dry Weight of Root, Root Colonization, Spore Number, Number of Leaves, in Ca-1 Variety. As influenced by different treatment for 30, 60 and 90 days.

Treatments	*Shoot Length (cm)*	*Fresh Weight of Shoot (gm)*	*Dry Weight of Shoot (gm)*	*Root Length (cm)*	*Fresh Weight of Root (gm)*	*Dry Weight of Root (gm)*	*Root Colonization (per cent)*	*Spore Number*	*Number of Leaves*
30 days									
C.N	11.00±0.36e	10.54±0.32d	4.35±0.05e	5.63±0.45c	4.02±0.32e	1.06±0.63e	0.00±0.00e	0.00±0.00e	10.00±0.32c
Rf	29.36±0.31b	31.63±0.15bd	13.25±0.68b	12.54±0.75bc	10.26±0.15b	3.15±0.65b	49.00±0.68b	42.00±0.68b	20.00±0.12b
Pressmud	19.12±0.65c	27.26±0.06cd	11.42±0.62c	11.95±0.48v	7.31±0.42cd	3.09±0.26c	0.00±0.00e	0.00±0.00e	16.00±0.44c
IAA	16.24±0.45d	22.35±0.26d	9.36±0.63e	8.34±0.12de	5.32±0.69d	1.98±0.35d	0.00±0.00e	0.00±0.00e	13.00±0.68d
Rf+Pressmud+IAA	39.45±0.94a	42.84±0.85a	19.52±0.15a	22.48±0.19a	16.52±0.28a	7.45±0.93a	60.00±0.79a	51.00±0.64a	36.00±0.62a
60 days									
C.N	26.36±0.48e	22.46±0.16c	10.36±0.45e	15.63±0.74d	10.06±0.62d	4.05±0.35e	0.00±0.00e	0.00±0.00eb	22.00±0.34e
Rf	35.74±0.67bc	30.22±0.25b	15.42±0.33b	20.64±0.65b	15.85±0.09b	11.32±0.52be	62.00±0.68b	55.00±0.55b	39.00±0.25be
Pressmud	28.69±0.48c	28.42±0.66c	11.62±0.62c	18.46±0.61c	13.26±0.23cd	8.03±0.16c	0.00±0.00e	0.00±0.00e	30.00±0.63ce
IAA	25.52±0.46d	26.54±0.53d	11.02±0.45d	16.75±0.36d	12.54±0.62de	5.42±0.09d	0.00±0.00e	0.00±0.00e	29.00±0.41d
Rf+Pressmud+IAA	48.74±0.36a	55.23±0.48a	25.86±0.62a	38.26±0.63a	24.68±0.36a	9.65±0.65a	78±0.63a	72.00±0.45a	52.00±0.62a
90 days									
C.N	32.65±0.78c	39.61±0.35d	17.36±0.84c	23.69±0.32d	19.45±0.42e	13.62±0.36d	0.00±0.00e	0.00±0.00e	35.00±0.03e
Rf	49.26±0.93b	60.04±0.75b	23.62±0.45be	35.21±0.05b	24.61±0.53bc	15.84±0.64b	84.00±0.45b	52.00±0.48b	59.00±0.99bc
Pressmud	39.52±0.46c	58.52±0.45ce	21.45±0.39c	29.42±0.64cd	21.34±0.62c	9.65±0.26c	0.00±0.00e	0.00±0.00e	43.00±0.74c
IAA	35.53±0.79d	55.24±0.41d	20.46±0.25d	25.36±0.31d	20.03±0.96de	7.52±0.64d	0.00±0.00e	0.00±0.00e	39.00±0.51de
Rf+Pressmud+IAA	61.85±0.62a	79.32±0.33a	31.96±0.63a	50.49±0.16a	35.93±0.36a	22.65±0.51a	91.00±0.68a	80.00±0.62a	75.00±0.95a

CN: Control; *Rf. Rhizophagus fasciculatus*; IAA: Indol acetic acid all the value represents.

Data represents means ± SE of 3 replicates; Each experiment was repeated thrice. Mean separation within column by Duncan's multiple range test at $P < 0$.

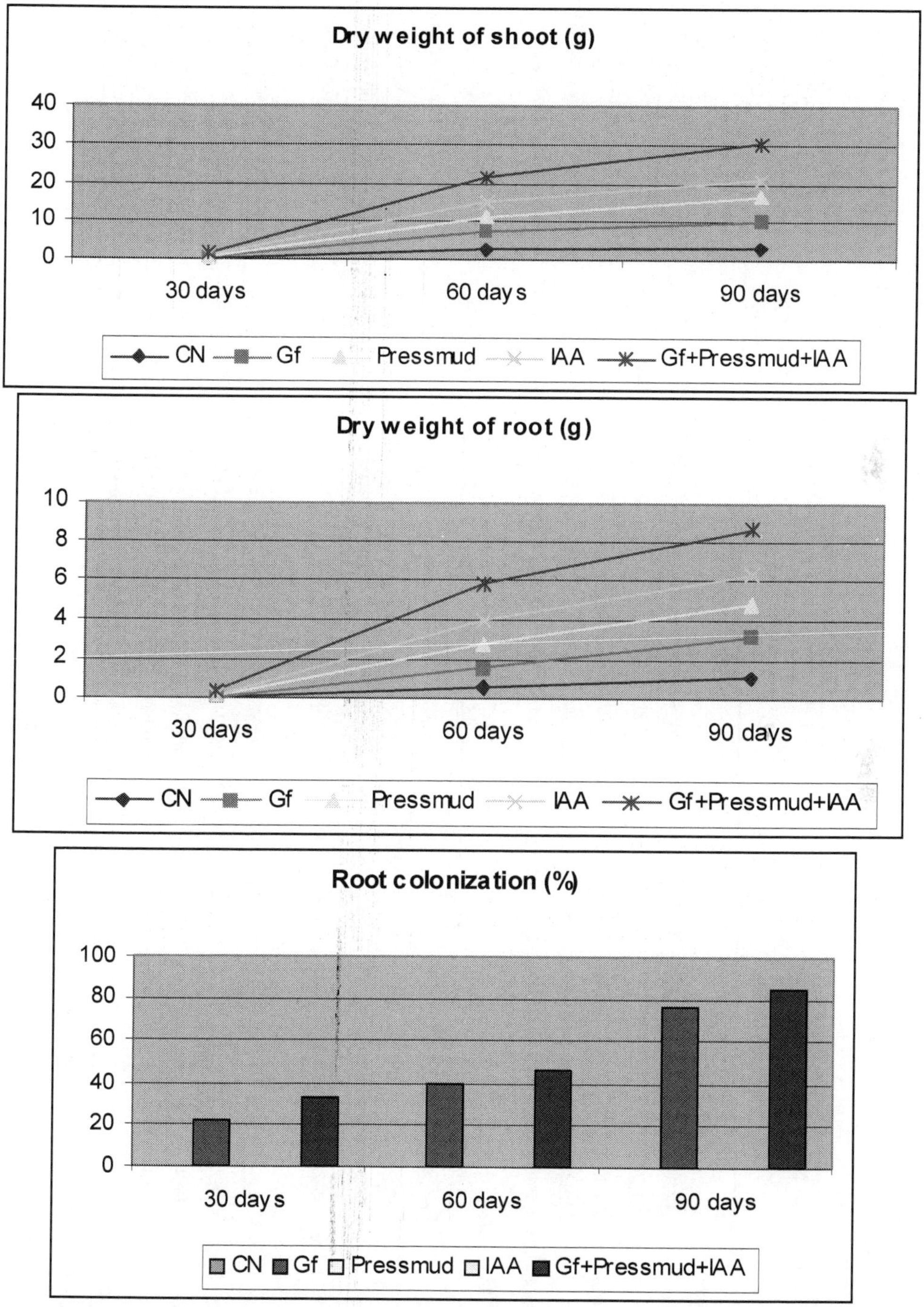

Figure 40.1: Showing Effect of *Rhizophagus fasciculatus*, Pressmud and IAA on Dry Weight of Shoot and Root, and Per cent of Root Colonization in *Corchorus capsularis* L., Variety Ca-1.

Table 40.2: Showing the Effect of AM Fungi (*Rhizophagus fasciculatus*) on Shoot Length, Fresh Weight of Shoot, Dry Weight of Shoot, Root Length, Fresh Weight of Root, Dry Weight of Root, Root Colonization, Spore Number, Number of Leaves, in DH-3 Variety. As influenced by different treatment for 30, 60 and 90 days.

Treatments	*Shoot Length (cm)*	*Fresh Weight of Shoot (gm)*	*Dry Weight of Shoot (gm)*	*Root Length (cm)*	*Fresh Weight of Root (gm)*	*Dry Weight of Root (gm)*	*Root Colonization (per cent)*	*Spore Number*	*Number of Leaves*
30 days									
C.N	13.18±0.57e	12.15±0.24e	7.62±0.55e	6.00±0.45e	5.02±0.45e	2.12±0.42e	0.00±0.00e	0.00±0.00e	8.35±0.54e
Rf	31.26±0.32b	36.25±0.25bc	18.34±0.63b	15.28±0.78bc	13.12±0.15b	6.12±0.61b	45.00±0.36b	35.00±0.26b	16.61±0.29b
Pressmud	25.36±0.84cd	25.85±0.36c	10.36±0.58c	12.32±0.69c	12.16±0.42c	5.15±0.47c	0.00±0.00e	0.00±0.00e	12.36±0.34cd
IAA	22.42±0.24d	20.36±0.75d	8.84±0.59d	10.12±0.54d	8.13±0.87d	3.24±0.26d	0.00±0.00e	0.00±0.00e	11.00±0.69de
Rf+Pressmud+IAA	42.36±0.95a	40.92±0.46a	21.54±0.36a	25.02±0.61a	18.15±0.39a	8.16±0.88a	58.00±0.63a	48.00±0.69a	22.0±0.26a
60 days									
C.N	28.14±0.61e	21.36±0.24e	12.25±0.64b	18.22±0.74e	13.25±0.24e	5.23±0.64e	0.00±0.00e	0.00±0.00e	19.66±0.36e
Rf	41.25±0.36b	42.47±0.64b	25.47±0.74b	22.53±0.62b	19.12±0.31b	12.46±0.84bc	60.00±0.36b	53.00±0.96b	34.36±0.25bc
Pressmud	38.28±0.24c	32.46±0.25c	18.74±0.57c	25.95±0.42cd	17.15±0.74c	9.25±0.26c	0.00±0.00e	0.00±0.00e	26.66±0.65c
IAA	30.64±0.45d	30.95±0.31d	14.85±0.84de	22.34±0.62de	15.25±0.26d	7.15±0.64d	0.00±0.00e	0.00±0.00e	20.00±0.69d
Rf+Pressmud+IAA	55.24±0.75a	58.36±0.41a	30.47±0.62a	40.87±0.43a	45.34±0.34a	20.15±0.34a	75±0.56a	65.00±0.48a	42.61±0.64a
90 days									
C.N	35.29±0.64e	39.25±0.26e	18.34±0.63e	22.65±0.57e	19.24±0.92e	12.25±0.45e	0.00±0.00e	0.00±0.00e	30.69±0.62e
Rf	53.74±0.46b	55.26±0.42b	31.24±0.34bc	38.58±0.24b	23.86±0.12b	17.12±0.96b	78.00±0.69b	69.00±0.62b	41.42±0.43b
Pressmud	46.31±0.31c	48.58±0.34c	22.74±0.34c	31.06±0.24c	21.65±0.34cb	11.12±0.24c	0.00±0.00e	0.00±0.00e	38.63±0.63cd
IAA	41.98±0.34de	39.24±0.63d	16.23±0.62de	29.88±0.56c	35.48±0.34c	12.15±0.36de	0.00±0.00e	0.00±0.00e	32.63±0.56d
Rf+Pressmud+IAA	68.56±0.32a	70.98±0.35a	45.85±0.52a	55.25±0.69a	68.45±0.31a	28.58±0.25a	85.00±0.69a	78.00±0.68a	61.63±0.96a

CN: Control; *Rf*: *Rhizophagus fasciculatus*; IAA: Indol acetic acid all the value represents.

Data represents means ± SE of 3 replicates; Each experiment was repeated thrice. Mean separation within column by Duncan's multiple range test at $P < 0$.

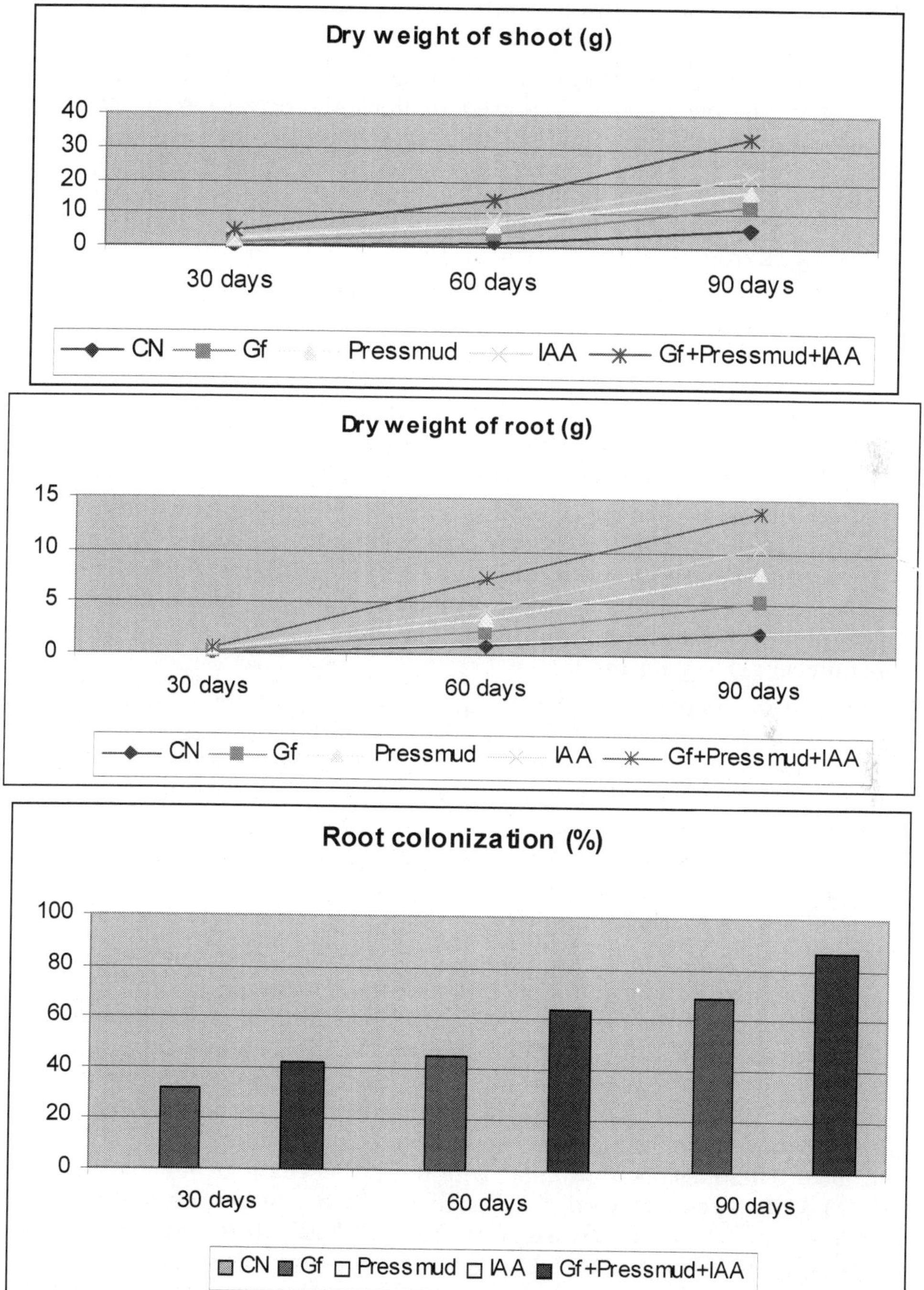

Figure 40.2: Showing Effect of *Rhizophagus fasciculatus*, Pressmud and IAA on Dry Weight of Shoot and Root, and Per cent of Root Colonization in *Corchorus capsularis* L., Variety DH-3.

leaves (14.23), number of flowers (6.00), number of fruits (3.33), per cent root colonization(47.00 per cent) and spores number (65.33) in 50 g of soil.

After 90 days in the treatments inoculation with *Rhizophagus fasciculatus*, pressmud and indole acetic acid showed significantly increase in growth. shoot length (45.08cm), fresh (99.20g) and dry weight of shoot(9.24g), root length(14.53cm), fresh(21.45g) and dry weight of root(2.41g), number of leaves(88.33), number of flowers(2.33), number of fruits(4.33), root colonization(85.00 per cent), spore number (90.00) in 50gm of soil, the fresh and dry weight of root and root colonization (Table 40.1 and Figure 40.1).

DH-3 Variety

The (Table 40.2) have depicts plant growth in the amendment inoculation with *Rhizophagus fasciculatus*, pressmud and IAA resulted in increase in shoot length (12.52 cm), fresh (8.45g) and dry weight of shoot(2.11g), root length(3.50cm), fresh (1.83g) and dry weight of root(0.25g), number of leaves(27.66), root colonization(42.00 per cent), spore number(50.00) in 50 g of soil. There were no records of flowers or fruits.

After 60 days plant growth inoculated with shoot length (40.10cm), fresh (69.53g) and dry weight of shoot *Rhizophagus fasciculatus*, in the pressmud and IAA there was a significant increase in shoot length (5.56g), stem diameter (2.08cm), root length(8.11cm), fresh (13.83g) and dry weight of root(2.89g), number of leaves(71.66), number of flowers(6.66), number of fruits(3.33), root colonization(63.33 per cent), spore number per 50 g soil (67.33).

After 90 days of plant growth to the amendment of press mud, inoculated with *Rhizophagus fasciculatus um* and IAA resulted significant increase in shoot length (45.10cm), fresh (125.91g) and dry weight of shoot (12.14g), root length(15.49cm), stem diameter (3.34cm), fresh weight (24.55g) and dry weight of root(3.23g), number of leaves (102.00), number of flowers (3.66), number of fruits(6.66), root colonization(86.00 per cent), spore number (99.33) in 50 g of soil. The inoculated with *Rhizophagus fasciculatus* alone was significant compare to the treated with alone pressmud and IAA. The dry weight of shoot, root colonization are shown in (Figure 40.2).

An alternative explanation might be that changes in the phytohormone balance of the plant occurred (Hause *et al.*, 2007). AM fungal inoculation of Tomato plants can lead to increase in fruit yield (Subramanian *et al.*, 2006). The auxin indole acetic acid (IAA) was the first plant hormone to the used to stimulate rooting (Cooper, 1935). The effect of plant growth regulator and treatment on root growth may be a function of the type of plant growth regulator and concentration applied (Pilet, 1983; Firn, 1986). These results are similar with the earlier findings of (Vijayalakshmi, 2011). Majority of research work has been directed at expensive pure substrates to a much lesser quantity of solid waste (Hawkes *et al.*, 2002). Pressmud from the sugar mills is a very useful source of fertilizer as well as some chemicals. The major use that has recently been developed in India is in biocomposting (usually trade named as Bioearth) where it is treated with the spent wash from the distillery (Padmanabhan *et al.*, 1993). These results are in line with the work carried out by other researchers

(Nehra and Hooda, 2002; Naik and Rao, 2004), who reported increased plant height in lentil crop due to pressmud application. Mycorrhizae are considered essential for growth and survival of many plant species in disturbed and unproductive soils (Raman and Mahadevan, 1996; Bothe *et al.*, 2001). (Lakshman, 1996) have suggested that AM fungi inoculated plants could be like a stock for transplanting into the field. Mycorrhizal fungi provide direct link between soil and roots and ameliorate heavy metal toxicity (Leyvel *et al.*, 1997). The combined treatments of *G. fasciculatum* along with pressmud were synergistic in reducing the nematode infestation and enhancing plant growth and yield. The increase in plant growth and reduction in nematode infestation that we found may be due to induced systemic resistance or multiple potential defense mechanisms (Wei *et al.*, 1996; Jonathan *et al.*, 2000; Anita and Rajendran, 2002). A similar result was reported by (Rakkiyappan *et al.*, 2000).

The results of the present studies clearly brought out the beneficial effect of inoculation with AM fungi, pressmud and indole acetic acid Ca-1 and DH-3 varieties, suggesting a sustainable way to increasing plant growth.

References

Al-Karaki, G.N., Hammad, R. (2001). Mycorrhizal influence on fruit yield and mineral content of tomato grown under salt stress. *J Plant Nutr.*, **24**: 1311–1323.

Anit, a B. and Rajendran, G. (2002). Nursery application of Pseudomonas fluorescens for the contro1 of Meloidogyne incognita on tomato and brinjal. *Nernatologia Mediterranea*, **30**: 209-210.

Auge, R.M (2001). Water relation, drought and VA mycorrhizal symbiosis. *Mycorrhizal.*, **11**: 3-42.

Birhane, E., F.J. Sterck, M. Fetene, F. Bongers and Kuyper, T.W. (2012). Arbuscular mycorrhizal fungi enhance photosynthesis, water use efficiency, and growth of frankincense seedlings under pulsed water availability conditions. *Oecologia.*, **169**: 895-904.

Bothe, H., Hildebrandt, U., Ouziad, F., Landwehr, M., and Nawrath, K. (2001). The colonization of plants from control European heavy metal soils and salt marshes by arbuscular mycorrhizal fungi. *Minerva Biotech.*, **13**: 65-67.

Cooper, W.C (1935). Hormones in relation to root formation on stem cuttings. *Plant Physiol.*, **10**: 789–794.

Firn, R.D. (1986). Plant growth substance sensitivity: The need for clear ideas, precise terms and purposeful experiments. *Physiol. Plant.*, **67**: 267–272.

Gerdemann, J.W and Nicolson, T.H. (1963). Spores of mycorrhizal endogene species extracted from the soil by wet sieving and decanting. Trans. *Br.Mycol.Soc.*, **46**: 235-241.

Giovannetti, M and Mosse, B. (1980). An evaluation of techniques to measure Vesicular-arbuscular infection in roots. *New Phytol.*, **84**: 489-500.

Hause, B, Mrosk C, Isayenkov S, Strack D. (2007). Jasmonates in arbuscular mycorrhizal interactions. *Phytochem.*, **68**: 101-110.

Hawkes, F.R., Dinsdale, R., Hawkes, D.L. and Hussy, I. (2002) Sustainable fermentative hydrogen production: challenges for process optimization, *Int J Hydrogen Energy,* **27**: 1339-1347.

Jauhri, K.S. (1990). Modified sugarcane pressmud: A potential carrier for commercial production of bacterial inoculants. *Indian J. Agric. Res.,* .**24** (3): 189.

Jonathan, E.I., Barlter K.R., Abdel-Alim F.F., Vrain T.C. and Dickson D.W. (2000). Biological control of Meloidogyne incognita on tomato and banana with rhizobacteria, actinomycetes and *Pasteuria penetrans. Nematropica,* **30**: 23 1-240.

Lakshman, H.C (1996). Vesicular- arbuscular mycorrhizal studies in some important timber tree species. Ph.D, thesis, Karnatak University Dharwad, India., pp.249.

Lakshman, H.C (2009). Arbuscular Mycorrhizal Fungi a promising bioinoculant for sustainable plant growth In: *Proc of ICAR conf., April 14. 16, NAL Bangalore,* pp.117-121.

Lakshman, H.C and Patil, G.B. (2004). Dual inoculation of VAM and *Rhizobium* beneficial to *Acacia nilotica* wild. *Ecology Env. And Conservation.,* **10**(4): 461-464.

Lakshman, H.C. Nitya, K.M. Pushpalata, K.C and Rohini, J. (2010). Biodiversity of endophytic fungi isolted from selected graminae of Mercara region in Karnataka. *Inter. J. of Plant Science.,* **3**(2): 335-341.

Leyvel, C., Turnau, K. and Haselwandter, K. (1997). Effect of heavy metal pollution on mycorrhizal colonization and its function: physiological, ecological and applied aspects. *Mycorrhiza.,* **7**: 139–153.

Miransari, M. (2010). Contribution of arbuscular mycorrhizal symbiosis to plant growth under different types of soil stress. *Plant Biol.,* **12**: 563-569.

Naik, S.K. and Rao, V.S. (2004). Effect of pyrite in combination with organic manures (FYM and Pressmud) on growth and yield of sunflower (*Helianthus annuus* L.) genotypes grown in Alfisols and Vertisols. *J. Interacademicia.,* **8**(3): 383-387.

Nehra, A.S. and I.S. Hooda. (2002). Influence of integrated use of organic manures and inorganic fertilizers on lentil and mung bean yields and soil properties. *Res. Crops.,* **3**(1): 11- 16.

Padmanabhan, P., Rakkiyappan, P., and Alexander, K.C. (1993).Enrichment of Pressmud by Microorganisms, Sugarcane Breeding Institute Extension, Publication No.39.

Phillips, J.M and D.S.Haymann (1970). Improved procedure for clearing roots and vesicular arbuscular mycorrhizal fungi for rapid assessment of infection. *Tran, Br.Mycol.Soc,* . **55**: 158-161.

Pilet, P.E. (1983). Control of root growth by endogenous IAA and ABA. *Monograph British Plant Growth Regulator Group,* **10**: 15–24.

Rakkiyappan, P., Padmanaban, P. and Thangavelu, S. (2000). Effect of organic manure made out of pressmud and distillery effluent on sugarcane. *Proc. International Conference on Managing Natural Resource for Sustainable Agricultural Production in the 21 st Century,* Feb. 14-18, New Delhi pp. 247-248.

Raman, N and Mahadeven, A. (1996). Mycorrhizal research a priority in agriculture. In: concepts of Mycorrhizal Research (Mukerji, K.G., Ed), Kluwer. The Netherlands, pp. 41-75.

Sedecor, G.W and Cochran, W.G. (1980). Statistical methods. Iowa state University press, Ames, Iowa, USA.

Smith, S.E., E. Facelli, S. Pope and F.A. Smith, (2010). Plant performance in stressful environments: Interpreting new and established knowledge of the roles of arbuscular mycorrhizas. *Plant Soil,* **326**: 3-20.

Subramanian, K. S., Santhanakrishan, P., and Balasubramanian, P. (2006). Response of field grown tomato plants to arbuscular mycorrhizal fungal colonization under varying intensities of drought stress. *Sci Hortic,* **107**: 245-253.

Vijayalakshmi, K (2011). Impact of composted pressmud, rhizobium and indole acetic acid on biometric and yield parameters of *Lablab purpureus* (L.) VAR.CO GB-14. *Indian J.Environ and Ecoplan.,* **18**(1): 67-69.

Wei G.J., Kloepper VI and Tuzun S. (1996). Induced systemic resistance to cucumber diseases and increased plant growth by plant growth promoting rhizobacteria under field conditions. *Phytopathology*, **86**: 221-224.

2015, Major Constrains and Verdict of Crop Productivity *Pages 510–523*
Editor: Dr. U.N. Bhale
Published by: ASTRAL INTERNATIONAL (P) LIMITED, NEW DELHI

Chapter 41

Assessment of some Crop Field Weed Residues on Productivity of Black Gram [*Vigna mungo* (L.) Hepper]

Kavitha Duraiswamy* and Prabhakaran Jayaraman

Department of Botany, Botany Wing-DDE,
Annamalai University, Annamalai Nagar – 608 002, Tamil Nadu, India

ABSTRACT

The residues of important crop field weeds viz. Cynodon dactylon (L.) Pers, Echinochloa crus-galli (L.) P.Beauv. and Ammania baccifera L. were evaluated on growth and yield of black gram (Vigna mungo (L.) Hepper) cv.TMV-1. In the field experiments, the weed residues were mixed in the ratio of 1,2,3 and 4 per cent (w/w) of residue/plot. Equal quantities of each weed residue were mixed for combined residues treatment and control plot was maintained without mixing weed residues. The pod yield/plot and ha. was reduced by 85 per cent in the higher concentration (4 per cent) of combined weed residue. This was a cumulative effect of reduction in length, number and weight of pods and seeds. In all cases, the combined weed extracts proved to be more toxic. As an exception, 1 per cent residue of C. dactylon stimulated the growth and yield of black gram. The order of detrimental allelopathic potential of three weed was A. baccifera > E. crus-galli > C. dactylon on the germination, growth and yield responses of black gram cv.TMV-1.

Keywords: *Ammania baccifera, Black gram, Crop field weeds, Cynodon dactylon, Echinochloa crus-galli, Weed residues, Yield.*

* *Corresponding Author:* E-mail: kavipriya2002@gmail.com

Introduction

The concept of weeds is unwanted plant was born when man started to grow plants deliberately for food and other purposes. As cropland is an artificial ecosystem where the chosen plants desired by man are cultivated, the weeds manage to invade the crop fields and compete with the standing crops, ultimately reducing the quantity as well as quality of produce. Allelopathy has received a wide attention in the past three decades in India. Of the total annual loss of agricultural produce due to various pests in India, weeds account for 45 per cent, insects 30 per cent, diseases 20 per cent and other pests, 5 per cent (Rao, 1983). In developed countries, with use of different methods, weed damages has been reduced to 5 per cent and a large number of weeds remain after harvesting and enter the soil and form the most important of soil sources (Jaskuish, 1997). But in developing countries, where weed management is far from adequate, identifying the kind of weed-crop is a prerequisite in fighting the troublesome weeds (Hamid *et al.*, 2012). Weeds have become a part of the agricultural scenario ever since man started cultivating crops. Jethro Tull (1731) was the first scientist to use the word 'weed' for the plants growing in places where they are not desired. A weed is a species which is adapted to conditions of human disturbances (Harlam and de Wet, 1964). Any plant that competes with cultivated plants or which in some way interferes with man's legitimate activities is considered a weed. Weeds compete with crop plants for nutrients, soil moisture and sunlight (Ashton and Crafts, 1981).

The competition of weeds for nutrients may result in such obvious responses as dwarfing in plant size, nutrient starved conditions, wilting and actual dying out of plants (Anderson, *et al.*, 1996). Weed seeds germinate earlier to crops, their seedlings grow faster and aggressive so that they crowd out all other plants which possess more valuable properties and establish a kingdom of their own within a short period of time. Weed species mature ahead of crops so that their seeds are collected with the crop harvest and get distributed to other places (Dangwal *et al.*, 2010).Weeds are the most underestimated crop pests in tropical agriculture although they cause higher reduction/loss in the yields of crops than other pests and disease. Of the total annual loss of agricultural produce from various pests in India, weeds roughly account for 45 per cent, insects for 30 per cent, disease for 20 per cent and other pests for 5 per cent (Das, 2008). Without weed management, the rice yield may be reduced by 16 to 86 per cent or even 100 per cent (Baltazar and De Datta, 1992; Zoschke, 1990).Weeds cause a number of environmental, toxicological, health and agricultural problems and may even clog water ways. Since weeds are so harmful, it is necessary to control them but the cost is heavy (Zimdahl, 1980). Till 1960, weeds were managed with cultural practices such as crop rotation and polyculture which were low input and sustainable strategies. However, during the last decade or so, concern has been raised against synthetic chemicals, in view of their toxicological and environmental impacts. Further, the repeated use of synthetic herbicides has led to the development of resistant ecotypes of weeds. More than 272 herbicide resistant biotypes of weeds belonging to 172 species (98 dicots and 64 monocots) have been identified (Heap, 2003).

The decomposition of plant residues potentially provides the largest quantity of allelochemicals that may be added to the rhizosphere. At plant death, materials compartmentalized in cells are released and recycled. Patrick *et al.* (1964) reported that depending on the decomposing conditions, substances highly toxic, non-toxic or stimulatory to plants may be formed during the decomposition of similar plant residues. In general, more severe and persistent toxicity has been reported from cold and wet soils (Patrick and Koch, 1958; McCalla and Haskins, 1964; Patrick *et al.*, 1964).Since the decomposing plant materials are never equally distributed throughout the soil, the soil adjacent to the decomposing debris may contain more decomposition products than other areas (Patrick *et al.*, 1964; Patrick, 1971; Drost and Doll, 1980). Therefore, as the roots grow through the soil, at some points they may come in contact with decomposing plant residue and suffer. Debris of wheat, oat, barley, rye, corn, sorghum, rice and sugarcane (Patrick and Koch, 1963; Bonner, 1950; Patrick *et al.*, 1963; Toussoun and Patrick, 1963; Guenzi *et al.*, 1967; Toussoun *et al.*, 1968; Kimber 1973b) and weed such as quack grass, canada thistle, johnson grass, yellow nutsedge and giant foxtail (Schreiber and Williams, 1967; Kovacs, 1972; Drost and Doll, 1980; Wilson, 1981) have been implicated in allelopathy during decomposition (Rice, 1974), Yakle and Cruse (1984) reported that the positioning of plant residues in soil might affect allelopathy. They observed that maize residues were allelopathic to maize plants and that the effect was stronger when residues were placed below the seed rather than above the seed. Obviously, the severity was attributed to the roots having direct contact with the decomposing plant material.

A significant portion of the agricultural land in developing countries in the tropics is heavily infested by various native and alien (invasive) weeds and controlling weeds is a big challenge to Asian farmers (Akobundu, 1992). Weeds growing among crop plants adversely affect yield and quality of the harvest and increase production costs, resulting in high economic losses (Alam, 1991). Barnyard grass (*Echinochloa crus-galli*), umbrella sedge (*Cyperus difformis* L.), Monochoria (*Monochoria vaginalis*) and Water plantain (*Alisma plantago-aquatica* L.) are world's worst weeds of the rice (*Oryza sativa* L.) crop, decreasing yield up to 90 per cent through season long competition with rice (Kwesi *et al.*, 1991).

Randhawa *et al.* (2006) reported that the dominant weed species found in upland, paddled direct seeded and transplanted rice, and rain fed lowland rice fields in Tamil Nadu are *Chloris barbata, Croton bonplandianus, Dinebra arabica, Ammania baccifera, Cyperus difformis, Cyperus iria, Digitaria sanguinalis, Eclipta prostrata, Echinochloa crus-galli, Marsilea minuta, Paspalum distichum, Eichornia crassipes, Eclipta alba, Fimbristylis miliacea, Monochoria vaginalis, Scirpus mucronatus* and *Sphaeranthus indicus, Cysnodon dactylon, Cyperus rotundus* and *Salvinia* sp.

It is evident from the foregoing that crop-weed interactions are critical for agricultural productivity, particularly for sustainable agriculture. Therefore, the present work has been designed to evaluate the allelopathic potential of three common paddy field weeds namely, *Cynodon dactylon* (L.) Pers. (Bermuda grass) (Poaceae), *Echinochloa crus-galli* (L.) Beauv. (Barnyard grass) (Poaceae) and *Ammania baccifera* L. (Blistering ammania) (Lythraceae) on growth and productivity of black gram (*Vigna mungo* (L.) Hepper), which is generally cultivated as a follow-up crop after paddy harvest along the coastal districts of Tamil Nadu.

Materials and Methods

Three weed species namely *Cynodon dactylon* (L.) Pers. (Poaceae), *Echinochloa crus-galli* (L.) P.Beauv.(Poaceae) and *Ammania baccifera* L. (Lythraceae) on germination, growth and yield of black gram (*Vigna mungo* (L.) Hepper.), The weed species were selected based on their higher Important Value Index (IVI) out of eighty two weed species, recorded in the pre and post harvest paddy fields of Annamalai Nagar (11°232483N 79°422583E), Cuddalore Dt., Tamil Nadu India. All parts (root, stem, leaves, flowers and seeds) of three weed species were used for experiments.

The seeds of black gram cv.TMV-1 were procured from Tamil Nadu Agricultural University,(TNAU) Coimbatore, Tamil Nadu, India. Seeds were selected uniform in size, colour and weight and stored in metal tins as suggested by Rao (1976).

The study was conducted by pot culture and field experiments. The pot cultures were raised in Botanical garden, Department of Botany and the field experiments were conducted in the premises of Faculty of Agriculture, Annamalai University, Annamalai Nagar during January-2012 to December-2012.

Description of Black Gram cv.TMV-1

Black gram cultivar TMV-1 is a hybrid of Midhi Ulundu and KM-1.The variety was released by Centre for Plant Breeding and Genetics (CPBG), Tamil Nadu Agricultural University, Coimbatore in 1979. It is a rice follow variety sown during the third week of January to second week of February. The duration of the variety was 60-65 days and the grain yield was 950-1200 kg/ha. and it is resistant to yellow mosaic virus (YMV) and tolerant to root rot disease (Agritech,TNAU,2012).

Cynodon dactylon (L.) Pers.

Common Names

Bermuda grass, dog's tooth grass, Bahama grass, devil's grass, couch grass, Indian doab, arugampul, grama, and scutch grass

Systematic Position

Division	:	Phanerogams
Class	:	Monocotyledones
Series	:	Glumaceae
Family	:	Poaceae
Genus	:	*Cynodon*
Species	:	*dactylon*

Echinochloa crus-galli (L.) Beauv.

Common Names

Barnyard grass, barnyard millet, cockspur, common barnyard grass, Japanese millet, water grass.

Systematic Position

Division	:	Phanerogams
Class	:	Monocotyledones
Series	:	Glumaceae
Family	:	Poaceae
Genus	:	*Echinochloa*
Species	:	*crus-galli*

Ammania baccifera Linn.

Common Name

Blistering ammania

Systematic Position

Division	:	Phanerogams
Class	:	Dicotyledones
Sub-Class	:	Polypetalae
Series	:	Calyciflorae
Order	:	Myrtales
Family	:	Lythraceae
Genus	:	*Ammania*
Species	:	*baccifera*

Field Experiments

Field experiments were conducted with weed residues followed by slightly modified methods of Qasem (1995). Residues were made from dried whole parts (root, stem, leaves, flowers and seeds). The experimental field was thoroughly ploughed and leveled. The field was divided into microplots (0.75 x 0.75m and 60cm) as Completely Randomized Block Design (CRBD). Each plot was underlain with 2.50 x 2.50 m polythene sheet (300 guage thickness) in order to prevent the seepage of weed leachates to other plot

The weed residues were mixed in the ratio of 1,2,3 and 4 per cent (w/w) of residue/plot. Equal quantities of each weed residue were mixed for combined residues treatment. Control plot was maintained without mixing weed residues. The plots were left undisturbed for three weeks allowing decomposition of the weed tissues then, seeds of black gram cv.TMV-1/plot were sown @ 20/plot a depth of 2cm at equal distance. Each plot was irrigated with 15 litres of bore-well water (pH.-7.2) at the time of seed sown then the irrigation schedule followed for entire period of the experiments.

Germination percentage, vigour index, seedling length, dry biomass production, number of lateral roots, number of leaves and photosynthetic pigment contents (chlorophyll-a,b and total chlorophyll and carotenoids) were analysed in 7 and 15 day old seedlings in pot cultures. In the field experiments, in addition to the above, number of root nodules, leaf area, biochemical parameters (reducing, non reducing and total sugars, free amino acids, protein, starch, enzymes-catalase and peroxidase, mineral contents- nitrogen, phosphorus and potassium) were analysed in root, stem and leaves of black gram at 15, 30, 45 and 60 DAS. Nitrogenase and leg-haemoglobin contents were estimated in the root nodules. For estimating the yield, the number of pods/plant and seeds/pod, weight of 100 seeds/plot, average yield kg/plot/ha were recorded. The physico-chemical and biological properties of soil were analysed at 15, 30, 45 and 60 DAS.

Seed Treatment

The seed is treated with fungicides Thiram at 2.5gm/kg seed in a seed dressing drum to avoid occurrence of seed borne diseases.

Application of Fertilizers

In field experiments, the recommended doses of organic manures such as farmyard manure (2t/ha) vermicompost (2t/ha) of sugarcane pressmud were applied in single application. 20kg nitrogen, 40kg phosphorus and 30kg potash per hectare were applied. The bio-fertilizers such as N-bio-fertilizer (*Rhizobiun leguminosorum* 2kg/ha) and P-bio-fertilizer (*Bacillus megaterium* 2kg/ha) and k-bio-fertilizer (*Bacillus mucilaginous* 1kg/ha) were applied (5kg/ha) in the experimental soil (ICAR, 2010).

Irrigation Schedule

The plots were irrigated at the interval of three days with due care to avoid excess flooding of water. The plants were allowed to grow up to yield.

Weeding

Two weeds were uprooted from the experimental plots by hand at 20, 35 and 50 days after sowing.

Yield Parameters

The black gram plants were harvested when the pods were dry. The pods were collected during the early hours of the day when the pods are not brittle and are wet with dew, to avoid shattering of grains in the fields.

The yield characteristics such as the number of pods per plant, number of seeds per pod, length weight of pods and seeds per plant, 100 seeds weight and average yield/plot; yield/hectare were calculated.

Number of Pods/Plant

The number of pods harvested from five randomly tagged plants was counted from which total number of pods per plant was calculated.

Pod Length

The length of the pods was measured with thread and scale.

Number of Seeds/Pod

After weighing the pods, were removed from each pod counted manually and the average number of seeds per pod was expressed as number of seeds per pod.

100 Seeds Weight

Weight of 100seeds/plot was recorded and expressed in grams

Seed Yield/Plant

The seeds were extracted from the matured pods and dried under shade for 6-8 days till it reached constant moisture. Seeds obtained from the five tagged plants were weighed and later added to work out the seed yield per plant and expressed in grams.

Seed Yield/Plot

The total seed yield per plot area was recorded from overall pickings and was expressed in grams.

Average Yield per Hectare

Harvested seeds were weighed per plot and converted to hectare following the method of Anene (1987). This was determined by multiplying average yield per plant population per hectare; average yield per hectare (yield/ha) = Average yield/plant X plant population/ha., which was expressed in kilograms.

Statistical Analysis

Five replicates were maintained for all treatments and control. The depth of significance between the treatments could be brought out clearly by multiple range testing programme. Hence, ANOVA followed by Tukey's multiple range test (TMRT) was applied for the experimental data (at 5 per cent level of significance)(Zar, 1984).

Results and Discussion

In agricultural systems allelopathy can be part of the interference between crops and between crops and weeds and thereby affecting the economical outcome of the plant production. The crop growing along with weed species is forced to compete for space, light and nutrition from seed germination to harvest period of the lifecycle. The weeds are causing inhibition on germination and growth of crops as well as reducing the yield of the desirable crops through releasing allelochemicals from the dead or live parts (Narwal, 1994).

The allelopathic nature of weed - crop interaction was tested with *Vigna mungo* (L.) Hepper cv. TMV-1 as a test crop and three most important weeds 1. *C. dactylon* 2. *E. crus-galli* and 3. *A. baccifera* either singly or in combination as the experimental system (Plate 41.1).The per cent germination, the root length and number of lateral roots, were all decreased by the weed residues either singly or in combinations. This indicates an adverse impact on the root system which is the portal for water and nutrients. The shoot development, assessed in terms of shoot length, number of leaves and their area was also suppressed clearly. As the root systems were poorly developed

Plate 41.1: Growth of Black Gram Exposed to Weed Residues.

with increasing concentrations of the residues, their water and nutrient availability should have become a limiting factor for growth. This is further confirmed by the dry biomass data wherein the biomass accumulations of root, stem and leaves were increasingly lowered by increasing doses of weed residues.

This study receives supports on the observations of *Paspalum dilatatum on* wheat varieties (Bansal *et al.*, 1981), *C. dactylon* on corn and soybean (Bhowmik and Doll, 1982), *Chromolaena OdorataL* on tomato (Onwugbuta, 2001),; *Brassica napus on Solanum nigrum* L. and *Amaranthus retroflexus* (Bita and Majd, 2011), *Eupatorium adenophorum, Ageratum conyzoides* and *Lantana camara on* wheat, rice and corn (Katoch *et al.*, 2012).

Allelopathic effects are often due to synergistic activity of several allelochemicals rather than to single compounds (Williamson, 1990). Under eld conditions, additive or synergistic effects become signicant even at low concentrations (Einhellig and Rasmussen,1978).

An indirect relation between lower germination rate and allelopathic inhibition may be the consequential inhibition of water uptake (Tawaha *et al.*, 2003) and alteration in the synthesis or activity of gibberellic acid ($GA_{3)}$ (Olofsdotter, 2001). This may be due to the presence of phenolic compounds inhibit the activity of GA_3 (Einhellig, 1986) or inhibit the synthesis of GA_3 which regulate de novo amylase production during seed germination (Chandler *et al.*, 1984).

Yield is the most important parameter for assessing the performance of any crop species. Though it marks the culmination of growth and reproduction, it is the most economically important aspect of a plant performance. For food crops, it contributes to food security; in addition, any decrease or increase of yield has severe social implications too. Yield was first assessed in plot in terms of length and number of pods, 100 seed weight, number of pods/plant, number of seed/plant. It is evident from Tables 41.1 and 41.2, shown that all the yield components were significantly reduced by weed residues. The reductions range from 67 to 85 per cent. Maximum reductions were 67 per cent by *C. dactylon*, 75 per cent by *E.crus-galli*, 78 per cent by *A.baccifera;* it rose sharply to 85 per cent when the residues were combined. Extrapolated on a ha. basis, the reductions become more alarming and spell danger for the food front.

Before concluding, it is imperative to comment on the deviant behavior of *C. dactylon*. As noted elsewhere, only the higher concentrations were inhibitory; at 1 per cent level it improved the data on all counts. The instances of an allelopathic plant being inhibitory as well as stimulatory are not rare. For example Quayyum, *et al.* (2000) reported that the increasing inhibition of rice growth with increased amount of the soil amending *Cyperus* leaves.

The yield reduction of a crop by the weed residue was reported by number of researchers, *C. dactylon on* barley (Horowitz and Friedman,1971), giant foxtail (*Setaria faberi* Herrm.) on corn (Bell and Koeppe,1972), curly dock (*Rumex crispus* L.) on sorghum and maize (Einhellig and Rasmussen,1973), itch grass (*Rottboellia exaltata* (L.)L.f.) on rice and green gram (Mercado and Manuel, 1978). *Convolvulus arvensis* L. and *Cynodon dactylon* L. on wheat (Yarnia, 2010), *E.crus-galli,; Cyperus difformis, Monochoria vaginalis and Alisma plantago-aquatica* on rice cultivars (Ali Mohadesi *et al.*, 2011) Sorghum bicolor on *Vigna radiata* (L) Wilczek. (Moosavi *et al.*, 2011), *Parthenium* on corn, wheat and soybean, *Trianthema portulacastrum* L., *Dactyloctenium aegyptium* L. and *Eleusine indica* L. on rice (Mubeen *et al.*, 2011), and *Convolvulus arvensis* L. on *Pennisetum* (Esfandiar, 2012). Incidentally, these reports corroborate the present findings too.

Conclusion

The data presented in foregoing and the emerging negative trends are of great concern to sustainable agriculture. Since black gram is a protein supplement important for nutritional security, it portends ill for health security also. The present study may be helpful to the farmers and agriculturists to understand the effects of weeds on the yield of crops and in finding the role of weeds on the yield losses. It may be useful to the scientists involved in the management of weeds. However, detailed analysis of the mechanisms underlying the yield reductions would be more rewarding.

Table 41.1: Yield Responses of (Pods Number/plant, Pod Length (cm/pod) No. Seeds/Pod and No. of Seeds/Plant) Blackgram

Weed Residues (per cent)	C. dactylon				E. crus-galli				A. baccifera				Combined Weeds			
	Pods No./ Plant	Pod L.	No.of Seeds/ Pod	No.of Seeds/ Plant	Pods No./ Plant	Pod L.	No.of Seeds/ Pod	No.of Seeds/ Plant	Pods No./ Plant	Pod L.	No.of Seeds/ Pod	No.of Seeds/ Plant	Pods No./ Plant	Pod L.	No.of Seeds/ Pod	No.of Seeds/ Plant
C	11.2a	5.4a	6.0b	67.2b	11.2a	5.4a	6.0a	67.2a	11.2a	5.4a	6.2a	67.2a	11.2a	5.4a	6.0a	67.2a
1	11.5a	5.6a	7.2a	82.8.a	10.5a	5.2a	6.0a	63.0a	10.3a	5.1b	6.0a	61.8b	10.0a	5.0b	6.0a	60.0b
2	10.0b	5.4a	6.0b	60.0c	10.0ab	5.0a	5.6b	56.0b	9.4c	4.9b	5.2b	48.8c	8.2c	4.7c	5.0b	41.0c
3	9.0c	5.2ab	5.1c	45.9d	8.2c	4.8ab	5.1b	41.82c	8.0d	4.7bc	5.2b	41.6d	7.1d	4.2d	5.0b	35.5d
4	8.2d	5.0c	5.0c	41.0e	7.3d	4.2c	5.1b	37.23d	6.1e	4.0d	5.0b	30.5e	5.2e	3.8e	4.0c	20.8e

Mean with different alphabets in a column differed significantly as per Tukey's Multiple Range Test (TMRT) ($P < 0.05$).

Table 41.2: 100 Seed Weight (g), Average Yield/Plot (g) and Average Yield/ha (kg) of Blackgram

Weed Residues (per cent)	C. dactylon			E. crus-galli			A. baccifera			Combined Weeds		
	100 Seed Weight	Average Yield/ Plot	Average Yield/ Ha.	100 Seed Weight	Average Yield/ Plot	Average Yield/ Ha.	100 Seed Weight	Average Yield/ Plot	Average Yield/ Ha.	100 Seed Weight	Average Yield/ Plot	Average Yield/ Ha.
C	4.64a	61.64b	1105.8a	4.64a	61.64a	1105.8a	4.64a	61.64a	1105.8a	4.64a	61.64a	1105.8a
1	4.71a	64.86a	1163.57a	4.32b	47.98b	860.00b	4.11b	45.02b	807.65b	4.03b	41.76b	749.16b
2	4.22b	45.36c	813.75b	4.04bc	40.32c	722.97c	3.92b	31.12c	558.28c	3.82b	26.21c	470.20c
3	3.63c	31.10d	556.13c	3.22d	22.8d	409.02d	3.23c	19.2d	344.44d	3.2c	15.68d	281.29d
4	3.13d	19.84e	355.00d	3.03d	15.0e	269.09e	3.04c	12.48e	223.88e	3.0c	8.75e	156.97e

Mean with different alphabets in a column differed significantly as per Tukey's Multiple Range Test (TMRT) ($P < 0.05$).

References

Agritech, TNAU, (2012). Crop production Guide. Tamil Nadu Agriculture University, Coimbatore, Tamil Nadu, India.

Akobundu, J.O. (1992). Integrated weed management techniques to reduce soil degradation. In: Combelack, JH.; Parson, J. and Richardson, RG. (eds.), *Proceedings of the First International Weed Control Congress.* Melbourne (Australia): International Weed Science Society, p.341.

Alam, S.M. (1991). Weed Science Problem in Pakistan. *Pak. Gulf. Eco.*, **3**: 25-29.

Ali Mohadesi, Abouzar A., Saeed B., Fatemeh T., Masoud M. S., and Ahad M. (2011). Allelopathy of weed extracts on yield and its components in four cultivars of rice (*Oryza sativa* L.). *J. Central Euro. Agri.*, **12**(1): 70-81.

Anderson, C., Stryhn H., and Streibig J.C. (1996). Decline of the flora in Danish arable fields. *J. Appl. Ecol.*, **33** (3): 619-626.

Anene, C. (1987). The effect of attack by the flea beetle *Podagrica* spp. on seed yield of two okra cultivars at Zaria Nig. *J. Entomol.* (**8**): 95-98.

Ashton, F.M. and Crafts, A.S. (1981). Mode of action of some herbicides, Wiley Interscience Publication, New York.

Baltazar, A.M., and DeDatta S.K. (1992). Weed management in rice. *Weed Abstracts*, **41**: 495–507.

Bansal, G.L., Modgal, S.C. and Singh, K.K. (1981). Allelopathic effect of aqueous extracts of some Weeds on two wheat cultivars. In: *Proceedings, 8th Asian-Pacific Weed Sciences Society Conference*, pp.825-827.

Bell, D.T., and Koeppe, D.E. (1972). Non-competitive effects of giant foxtail on the growth of corn. *Agron. J.*, **64**: 321-325.

Bhowmik, P.C., and Doll, J.D. (1982). Corn and soybean response to allelopathic effects of weed and crop residues. *Agron. J.*, **74**: 601-606.

Bita. Z.i and Majd, A. (2011). Allelopathic Potential of Canola (*Brassica napus* L.) Residues on Weed Suppression and Yield Response of Maize (*Zea mays* L.). *Int. Conf. on Chemical Ecology,* Pattaya, pp.457-460.

Bonner, J. (1950). The role of toxic substances in the interactions of higher plants. *Botanical Review,* **16**: 51-65.

Chandler, P.M., J.A. Zucar, J.V. Jacobson T.J.V. Higgins and A.S. Inglis, (1984). The effect of gibberellic acid and abscisic acid on a-amylase mRNA levels in barley aleurone layers studies using an -amylase cDNA clone. *Plant Mol., Biol.*, **3**: 407-408.

Dangwal, L.R, Singh A, Singh T, Sharma A, and Sharma, C. (2010). Common weeds of rabi (winter) crops of tehsil Nowshera, district Rajouri (Jammu and Kashmir), India. *Pak. J. Weed Sci. Res.*, **16**(1): 39-45.

Das, T.K. (2008). Weed Science, Basics and applications. Jain Brothers, New Delhi.

Drost, S.C. and Doll, J.D. (1980). The allelopathic effect of yellow nut sedge (*Cyperus esculentus*) on corn (*Zea mays*) and soybean (*Glycine max*). *Weed Science,* **28**: 229-233.

Einhellig, F.A, and Rasmussen, J.A. (1979). Effects of three phenolic acids on chlorophyll content and growth of soybean and grain sorghum seedlings. *J. Chem. Ecol.,* **5**: 815-824

Einhellig, F.A. and Rasmussen, J.A. (1973). Allelopathic effects of *Rumex crispus* on *Amaranthus retroflexus*, grain sorghum and corn in field. *American Midland Naturalist,* **90**: 79-86.

Einhellig, F.A. (1986). Mechanisms and modes of action of allelochemicals. In "The Sciences of Allelopathy" (P.R. Putnam and C.S. Tang, eds), pp. 171-188. Wiley Interscience, New York.

Esfandiar F., Sohrabi, S. S. and Gerami, F. (2012). Evaluation the allelopathic effect of bindweed (*Convolvulus arvensis* L.) on germination and seedling growth of millet and basil. *Advances in Environmental Biology,* **6**(3): 940-950.

Guenzi, W.D. and McCalla, T.M. and Norstadt, F. (1967). Presence and persistence of phytotoxic substances in wheat, oat, corn and sorghum residues. *Agronomy Journal,* **59**: 163-165.

Hamid, N., Zabiholah, A.T. and Abolfazl, T. (2012). Effect of weed allelopathic of sorghum (*Sorghum halepense*) on germination and seedling growth of wheat, Alvand cultivar. *Annals of Biol. Res.,* **3** (3): 1283-1293.

Harlam, J.R. and de Met, J.M.J. (1964). Some thoughts about weeds. *Eco.Bot.,* **19**: 16-24.

Heap, I. (2003). International survey of herbicide resistant weeds. Online. Internet www.weedscience.com. Accessed on January **26**, 2003.

Horowitz, M and Friedman, T. (1971). Biological activities of subterranean residues of *Cynodon dactylon* L., *Sorghum halepense* L. and *Cyperus rotundus* L. *Weed Res.,* **11**: 88-93.

ICAR, (2010). Soil biodiversity-biofertilizers. Research Progress 2007-2011. All India network project on soil biodiversity-biofertilizers, Indian Institute of Soil Science

Jaskuish. D. (1997). Zeszyty Naukowe Akademii Rolniczejw Szczecinie Rolnictwo., **65**: 127-133.

Jethro, Tul1. (1731). Horse hoeing husbandry. Berkshire MDCC- 33.

Katoch, R. Anita, S. and Thakur, N. (2012). Effect Of Weed Residues On The Physiology Of Common Cereal Crops. *Int. J. of Engg. Res.* and *Applications,* pp.828-834.

Kimber, R.W.L (1973b). Phytotoxicity from plant residue II. The effects of time of rotting straw from grasses and legumes on the growth of wheat seedling. *Plant and Soil,* **38:** 347-361.

Kovacs Jr., M.F. (1972). Dhurrin: An Allelopathy Identified in Johnson grass Root Exudates Ph. D. Dissertation, University of Maryland, and College Park.

Kwesi, A, Nyarko A.N. and De Datta S.K. (1991). Hand Book of Weed Control in Rice. IRRI. Los Banos, Philippines, p. 100.

McCalla, T.M. and Haskins, F.A. (1964). Phytotoxic substances from soil microorganisms and crop residues. *Bacterial Review.*, **28**: 181 – 207.

Mercado, B.L. and Manuel, J.S. (1978). Biology and control of Paspalum distichum. L.II.Growth behavior and papettern of reproduction Pages 16-23. In *Weed Sci.* Rep. 1976-77. College of Agri., Univ. Philippines, Los Banos, .pp. 64-75.

Moosavi, A., R. Tavakkol Afshari, A. Asadi and Gharineh, M.S. (2011). Allelopathic effects of aqueous extract of leaf stem and root of sorghum bicolor on seed germination and seedling growth of *Vigna radiata* L. *Notulae Scientia Biologicae,* **3**(2): 114-118.

Mubeen, K., Nadeem, M.A., Asif Tanveer and Zahir A.Z. (2011). Allelopathic effect of aqueous extracts of weeds on the germination and seedling growth of rice (*Oryza sativa* L.). *Pak. J. Life Soc. Sci.,* **9**(1): 7-12.

Narwal, S.S. (1994). In: Allelopathy in Crop Production. pp. 2–9. Scientific Publishers.

Olofsdotter, M. (2001). Rice- A step toward use of allelopathy. *Agron. J.*, **93**: 3-8.

Onwugbuta, E.J. (2001). Allelopathic Effects of *Chromolaena odorata* L. (R. M. King and Robinson – (Awolowo Plant') Toxin on Tomatoes (*Lycopersicon esculentum* Mill). *J. Appl. Sci. Environ. Mgt.*, **5**(1): 69-73.

Patrick, Z. A. (1971). Phytotoxic substances associated with the decomposition in soil of plant residues. *Soil Sci.,* **111**: 13-18.

Patrick, Z. A. and Koch. L. W. (1963). The adverse influence of phytotoxic substances from decomposing plant residues on resistance of tobacco to black root rot. *Canad. J. Bot.,* **41**: 747-758.

Patrick, Z.A. and Koch, L. W. (1958). Inhibition of respiration, germination and growth by substances arising during the decomposition of certain plant residues in soil. *Canadian J. Botany,* **36** : 621 – 647.

Patrick, Z.A., Toussoun T.A. and Koch, L.W. (1964.) Effect of crop residue decomposition products on plant roots. *Ann. Rev.Phytopathology,* **2**: 267-93.

Qasem, J.R. and T.R. Hill, (1989). Possible role of allelopathy in competition between tomato, *Senecio vulgaris* L. *Chenopodium album* L. *Weed Res.,* **29**: 349-356.

Quayyam, H.A., Malik, A.V., Leach, D.M. and Gottardo, C. (2000). Growth inhibitory effects of nutgrass (*Cyperus rotundus*) on rice (*Oryza sativa*). *J. Chem. Eco.,* **26**: 2221-2231.

Randhawa, G.J., Verma, D.D., Shashi B., Chalam, V.C. and Tyagil, V. (2006). Document on Biology of rice (*Oryza sativa* L.) in India. National Bureau of Plant Genetic Resources, Alpha Litho graphics Inc., New Delhi-110028.

Rao, R.S.N. (1976). Seed viabilities studies under different storage conditions. Patna Nagar. *J. Res.,* **2**: 253.

Rao, V.S. (1983). *Principles of Weed Science.* Oxford and IBH Publishing Co. New Delhi, India.

Rice, E.L. (1974). Allelopathy. Academic Press, New York, p.353.

Schreiber, M. M. and Williams Jr., J. L. (1967). Toxicity of root residues of weed grass species. *Weeds* **15** : 80 – 81.

Tawaha, A.M, and Turk, M.A. (2003). Allelopathic effects of black mustard (*Brassica nigra*) on germination and growth of wild barley (*Hordeum spontaneum*). *J Agron Crop Sci.,* **189**: 298–303.

Toussoun, T. A., and Patrick, Z. A. (1963). Effect of phytotoxic substances from decomposing plant residues on root rot of bean. *Phytopathology*, **53**: 265-270.

Toussoun, T. A., Weinhold, A. R., Linderman, R.G. and Patrick, Z.A. (1968). Nature of phytotoxic substances during plant residues decomposition in soil *Phytopahtology*, **58** : 41-45.

Williamson, G. B. (1990). Allelopathy, Koch's postulates and the neck riddle, pp. 143–162, in J. B.Grace and D. Tilman (eds.), Perspectives on Plant Competition. Academic Press, New York.

Wilson, Jr. R. G. (1981). Effect of Canada thistle (*Cirsium arvense*) on growth of some crops. *Weed Sci.,* **29**: 159-164.

Yakle, G.A. and Cruse, R.M. (1984). Effects of fresh and decomposing residue extracts on corn seedling development. *Proc. Soil. Soc. Amer.,* **48**: 1143-1146.

Yarnia, M. (2010). Comparison of field bindweed (*Convolvulus arvensis* L.) and bermuda grass (*Cynodon dactylon* L.) organs residues on yield and yield components of bread wheat (*Triticum aestivum* L.). *Advances in Environmental Biology*, **4**(3): 414-421.

Zar, J.H. (1984). Bio-statistical Analysis, Prentice – Hall Inc., Englewood Clliffs, New Jersey. p. 718.

Zimdahl, R. L. (1980). Weed Crop Competition – A Review. *Int. Plant Prot. Centre,* Oregon State Univ. Coravallis.

Zoschke, A. (1990). Yield loss in tropical rice as influenced by the competition of weed flora and the timing of its elimination. In: Grayson B.T., Green M.B., Copping L.G. (eds): Pest Management in Rice. Elsevier Science, London: 301–313.

2015, Major Constrains and Verdict of Crop Productivity Pages 524–535
Editor: Dr. U.N. Bhale
Published by: ASTRAL INTERNATIONAL (P) LIMITED, NEW DELHI

Chapter 42

Degradation of Malathion using *Bacillus polymyxa*

Sarika Kanade[1], Avinash Ade[2]* and Vikram Khilare[1]

[1]*Botany Research Center, Vasantrao Naik Mahavidyalaya, CIDCO, Aurangabad, Maharashtra, India*
[2]*Department of Botany, University of Pune, Pune, Maharashtra, India*

ABSTRACT

Pesticides are the killers of the pest of various kinds. There are several pesticides out of which the category organophosphates are widely used. Malathion is one of the examples of organophosphate which needs to be degraded as it has hazardous effects on the soil microflora. In the present investigation the attempt has been made for the biodegradation of malathion. The Plant Growth Promoting Rhizobacterium (PGPR), Bacillus polymyxa was used for the malathion degradation. The bacterium was isolated from the soil and characterized by following the standard morphological and biochemical criteria. The tolerance of Bacillus polymyxa against the malathion was measured. The subtolerant (1000 μg/ml) concentration was taken in the LB broth medium and inoculated with Bacillus polymyxa. There was 86.4 per cent degradation of malathion was reported. The degradation products were identified through GCMS technique. There were two products named as succinic acid mercapto diethyl ester, s-ester with o,s-dimethy phosphorodithioate and butanedioic acid [(dimethoxy phosphinothiol)thio]-, diethyl ester.

Keywords: *Bacillus polymyxa, Malathion, PGPR, GCMS technique.*

Introduction

Pesticides are those substances which are used to control, destroy, repel or attract pests in order to minimize their detrimental effects. Pests are the organisms like

* *Corresponding Author.* E-mail: avinashade@unipune.ac.in

weeds, insects, bacteria, fungi, viruses and animals which adversely affect our way of life. Pesticides are therefore used in many situations such as livestock farming, cropping, horticulture, forestry, home gardening, homes, hospitals, kitchens, roadsides including recreational and industrial areas. Pesticides are derived from inorganic sources (copper, sulphate), natural organic sources (paints) or organic compounds synthesized in the laboratory. Most of these are synthesized products which mimic the activity of natural organic compounds (Kent, 1991), these includes chlorinated hydrocarbons, organophosphates, carbamates, cyclodines, phenylurease, thiocarbamates, phenylcarbamates, triazines, etc.

At present these are about 35,000 pesticidal compounds in the world market involving 1,500 basic ingredients. The number of pesticide registered in India is 221. Synthetic organophosphorus compounds are used especially as insecticides accounting for an estimated 34 per cent of worldwide pesticide sales. These compounds possess high mammalian toxicity about 2,00,000 of nerve agents have to be destroyed worldwide under chemical weapons convention, 1993.

Malathion is one of the largest organophosphorus insecticides in the world and is a product that has a wide variety of application in the agriculture industry around the world. Its chemical name is 'diethyl mercaptosuccinate' or 'o, o-dimethyl phosphorodithioate'. Malathion is also known as carbophos, maldison and mercaptothion.

The toxic effects of Malathion are well known on the wide range of animals. The effects on mammals especially on human being were noted as leukemia, kidney damage, brain damage, lung damage, etc. Its carcinogenic effects include the chromosome defects which were reported in human blood cells and gene loss from the human DNA. The birth defects in birds, turtles and frogs are well known. Like fishes the toxicity was also scored in terms of annelids, crustaceans, echinoderms, insects, molluscs, nematodes, flatworms and many zooplanktons (PAN pesticide-chemicals, Sinclair).

There are enough evidences where the application of chemicals to soil has resulted change in soil microflora (Gangawane and Saler, 1988). Most of the pesticides are administered as spray or dust on plants which ultimately reaches the soil as run off or drift where they influence the microbial balance of the soil. Moreover the residual effect on the plant and life is known (Alexander, 1969). Therefore degradation of pesticide in the environment and soil is desirable as soon as the pests or diseases are controlled.

Pesticide degradation is the breaking down of the toxic pesticide into the non-toxic compounds, in some cases, down to the original elements from where they were derived (Vargas, 1975). There are three types of pesticide degradation, photodegradation, chemical degradation and microbial degradation. Photodegradation is carried out by light, particularly sunlight and can destroy the pesticides on foliage of the soil surface and even in the air (Kiss and Virag, 2009) Chemical degradation is the process which does not involve microorganisms but due to temperature, moisture, pH and adsorption degradation is done. Microbial degradation is the breakdown of pesticide by fungi, bacteria and other microorganisms that use pesticides as food source. Soil conditions such as moisture, temperature, aeration, pH and amount of organic matter affects the rate of microbial degradation because of their direct influence on the microbial growth and activity (Foster *et al.*, 2006).

Microbial process plays an important role in the biological transformation or degradation of pesticides (Matsumura, 1974). The major aspect of pesticide degradation is being done with the various types of saprophytic microbes (Mostafa *et al.*, 1972a).

Bacillus polymyxa is the nitrogen fixing bacterium. It also synthesizes antibiotics, polymyxin which inhibits the growth of *Bacillus subtilis, Micrococcus luteus, Pseudomonas aeruginosa, Staphylococcus aureus* and *Streptococcus bovis* (Quinn *et al.*, 2012). It is also considered as Plant Growth Promoting Rhizobacterium (PGPR). In the present investigation attempts have been made to find out the ability of *Bacillus polymyxa* to degrade the organophosphorus insecticides, malathion.

Materials and Methods

Isolation and Characterization of *Bacillus polymyxa*

One gram of soil was serially diluted and spreaded on nutrient agar plates. The well grown colonies were subjected to grow and spreaded on nutrient agar plates (Peptone 10 g, Beef extract 2 g, Sodium chloride 2 g, Distilled Water 1000 ml, Agar 15 g) and incubated for 48 hours. The well grown colonies were subjected to Gram's iodine test and biochemical tests for identification.

Cultural characteristics were seen on the basis of morphological features of the bacterial colonies such as colour, texture, outline, margin, elevation of the colonies. Gram staining was performed according to the method of Gram (1984). The identification of the strain was done through biochemical tests described in Bergey's manual (Bergey and Holt, 1994). These were reaction on litmus milk, 3-ketolactose production, indole production from tryptophan, Voges Proskauer test, production of hydrogen sulphide, hydrolysis of gelatine, hydrolysis of starch, reduction of nitrate and the test for catalase.

Testing Malathion Tolerance Limit of *Bacillus polymyxa*

To check the malathion tolerance limit of *Bacillus polymyxa* it was inoculated in the nutrient broth medium with various concentrations of malathion. The growth was measured in terms of turbidity (O.D. at 600 nm) using spectrophotometer. The concentrations used for testing tolerance limit were 25, 50, 75, 100, 200, 1000 and 2000 µg/ml.

Degradation of Malathion

Sub-lethal concentration of Malathion (1000 µg/ml) was used for the degradation. The sterilized nutrient broth medium was inoculated with the *Bacillus polymyxa* aseptically. The incubation was done at room temperature with regular shaking condition at the time interval of one day each for 10 days. The residual Malathion was measured according to the method of Getz and Watts (1964).

Two ml of broth from each tube was centrifuged and extracted five times with equal volume of chloroform. The chloroform extracts were allowed to evaporate to dryness. To the dry tubes 0.2 ml of 4-(p) nitrobenzyl pyridine solution and 0.2 ml of cyclohexylamine solution were added. The tubes were attached to the column and immersed in pre-heated oil bath (175-180°C) for 3 minutes. The tubes were cooled with cold water. These were diluted to 3 ml with ethyl acetate and absorbance was recorded at 520 nm against blank reagent. The above procedure was repeated for the chloroform extract from the inoculated and un-inoculated flasks. A calibration curve for standard Malathion was prepared by dissolving various concentrations of Malathion in acetone and by following the same methodology.

Analysis of Pesticide Degradation Products

Analysis of Malathion degradation products was done by gas chromatography and mass spectrometry (GCMS) from the same assay of degradation. The mixture was centrifuged to separate the bacterial cells. The bacterial cells were lysed through sonication. The metabolites were detected and their possible functional groups were determined by GCMS analysis.

Results and Discussion

The *Bacillus polymyxa* colonies were whitish in colour, rough in texture, flat with undulate margin. In liquid medium the yellow colour was reported. For the identification out of 8 biochemical tests (Gram's iodine test, ketolactose production, indole production from tryptophan, production of hydrogen sulphide, hydrolysis of gelatine, hydrolysis of starch, reduction of nitrate and the test for catalase) *Bacillus polymyxa* has shown positive results for 4 tests (Gram's iodine test, starch hydrolysis, catalase production and nitrate reduction).

The tolerance limit of *Bacillus polymyxa* against malathion was found at 1000 µg/ml (Table 42.1). The estimated residual malathion after 10 days was found 136 µg/ml which was taken initially as 1000 µg/ml. Therefore 86.40 per cent malathion was found degraded (Table 42.2).

Table 42.1: Tolerance Limit of Malathion by *Bacillus polymyxa*

Growth (mm) at following Concentrations (µg/ml)							
0	25	50	75	100	200	1000	2000
35	30	21	16	18	18	05	00

Table 42.2: Residual Malathion after 10 Days Degradation by *Bacillus polymyxa* (Spectrophotometric analysis)

Sl.No.	*Treatment*	*Concentration of Malathion Before Treatment (µg/ml)*	*Concentration of Malathion After Treatment (µg/ml)*	*Per cent Degradation of Malathion*
1.	*Bacillus polymyxa*	1000	136	86.40
2.	Control	1000	1000	–

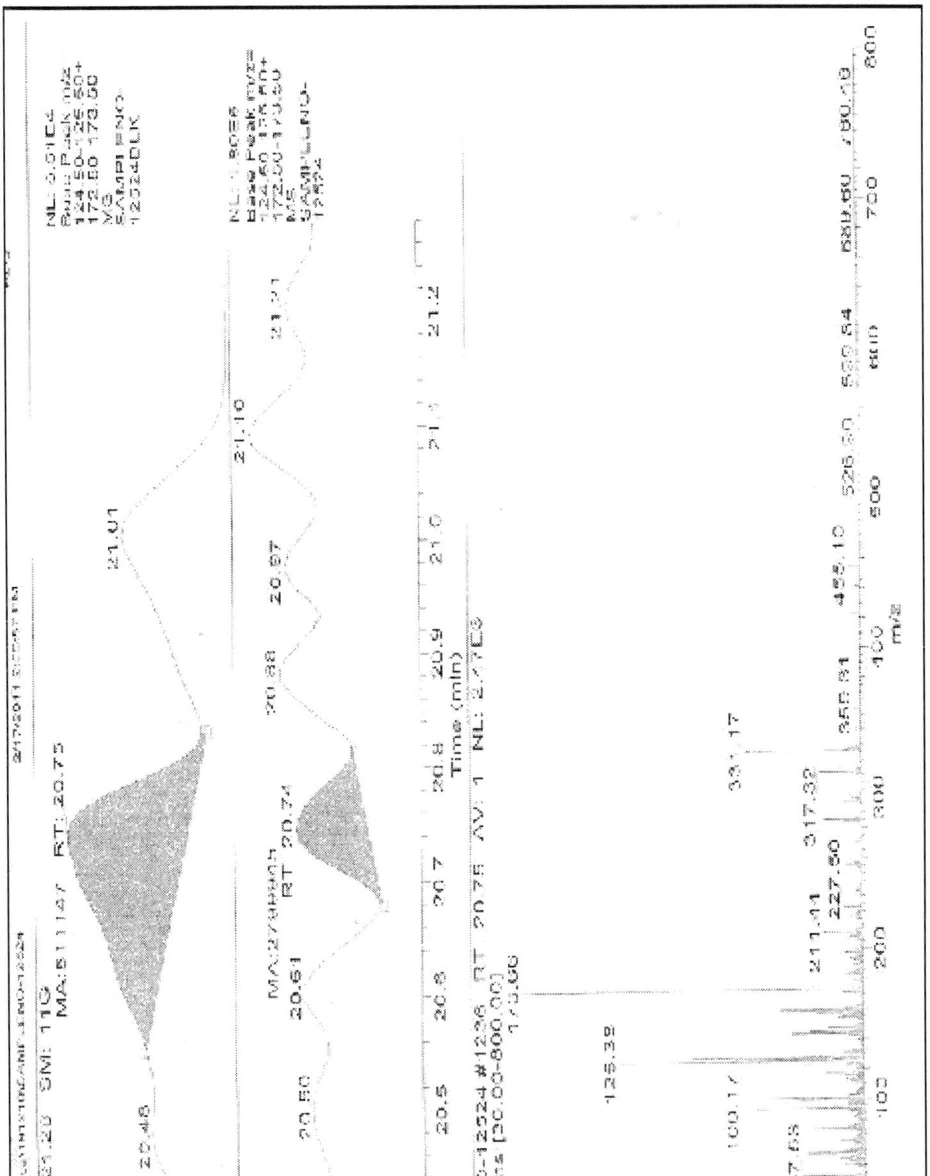

Figure 42.1: Chromatogram Area Covered by Residuial Malathion After Degradation by *Bacillus polymyxa.*

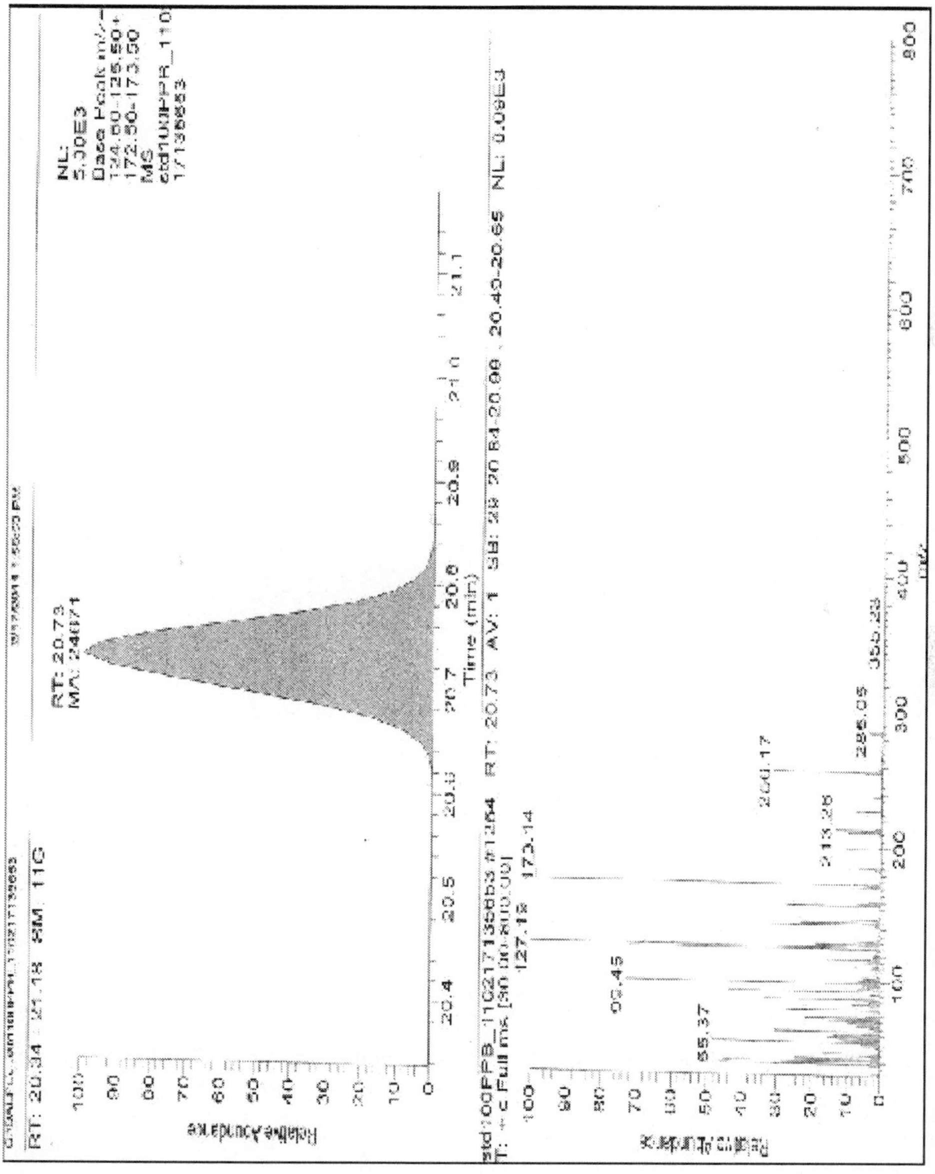

Figure 42.2: Chromatogram Area Covered by Standard Malathion.

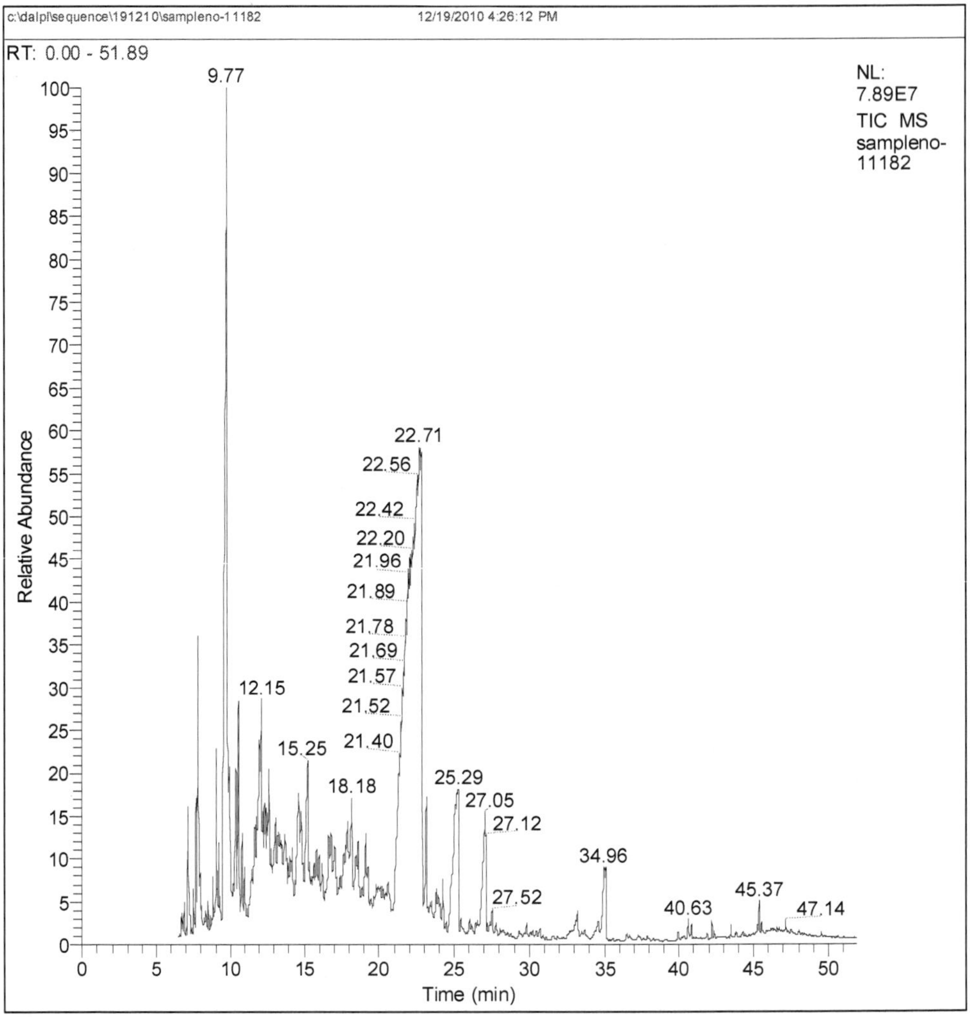

Figure 42.3: Gas Chromatogram of Malathion Degradation Products by *Bacillus polymyxa.*

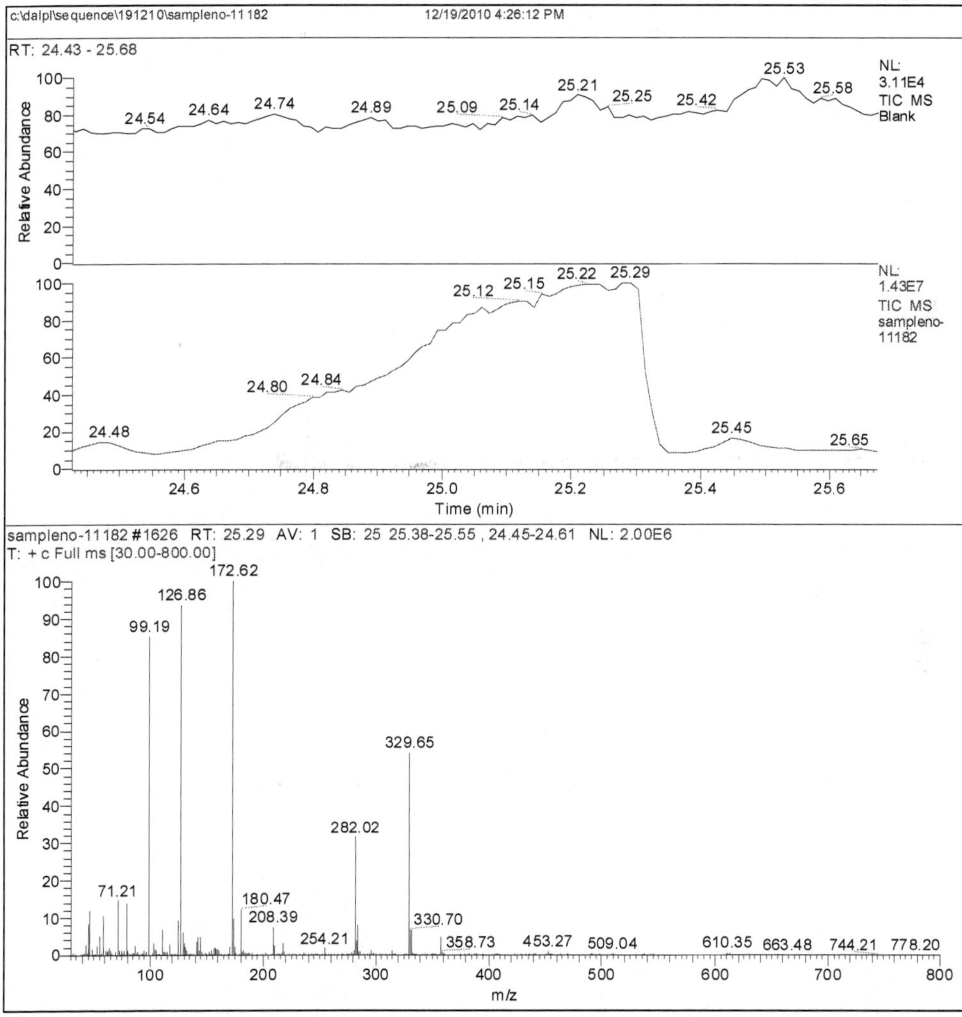

Figure 42.4: GCMS Analysis of Malathion Degradation Product, Succinic Acid Mercapto Diethyl Ester, s-ester with o,s-dimethyl Phosphorodithioate by *Bacillus polymyxa.*

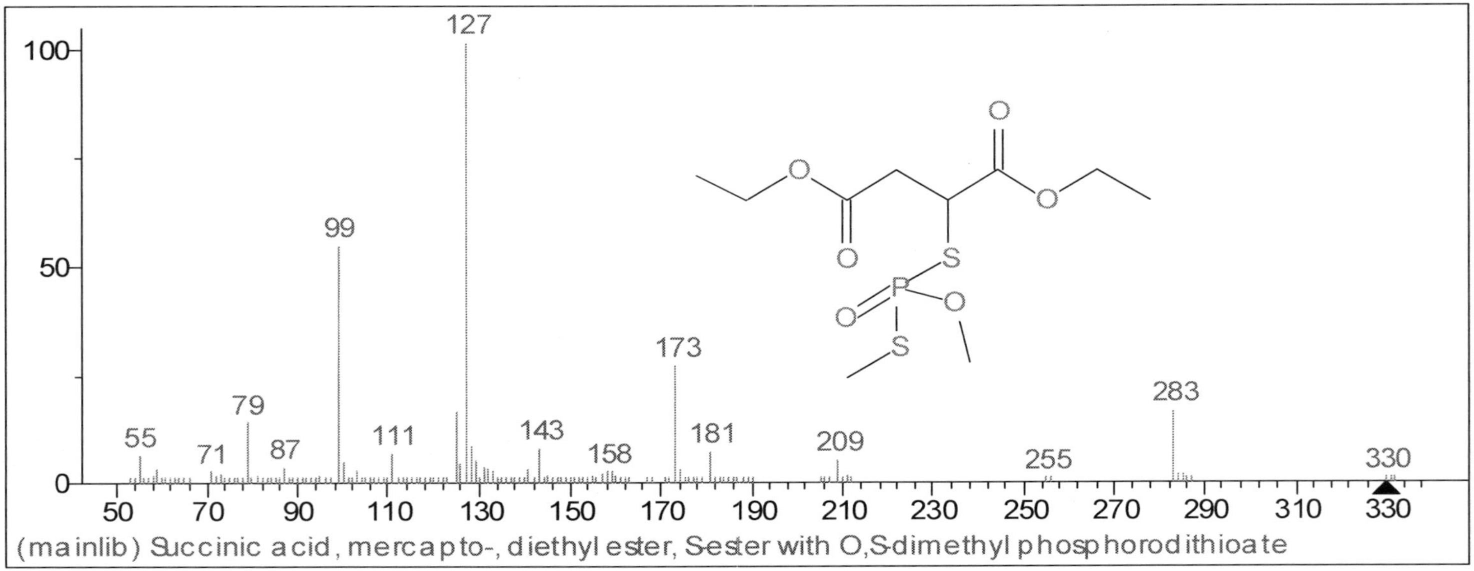

Figure 42.5: GCMS Analysis of Malathion Degradation Product, Succinic Acid, Mercapto-, Diethyl ester, S-ester with O,S-dimethy Phosphorodithioate by *Bacillus polymyxa.*

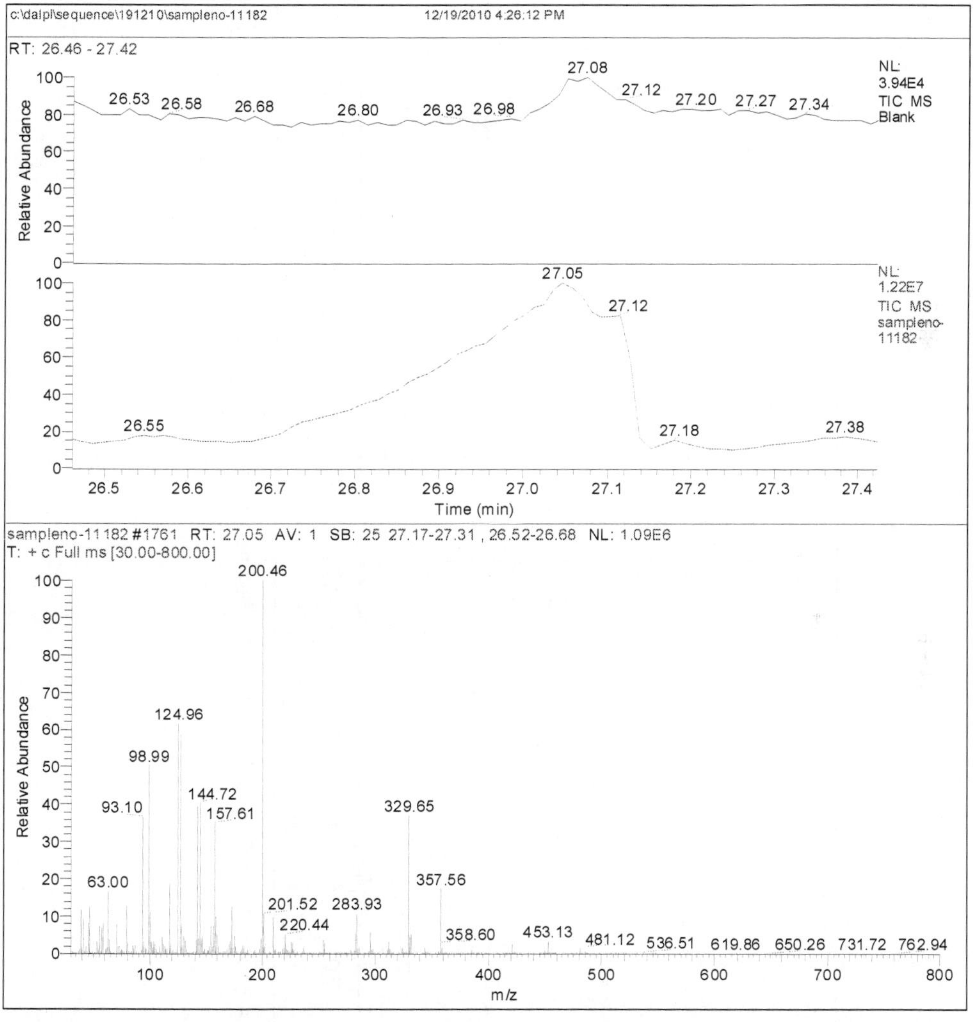

Figure 42.6: GCMS Analysis of Malathion Degradation Product, Butanedioic Acid [(dimethoxy phosphinothiol)thio]-, Diethyl Ester by *Bacillus polymyxa.*

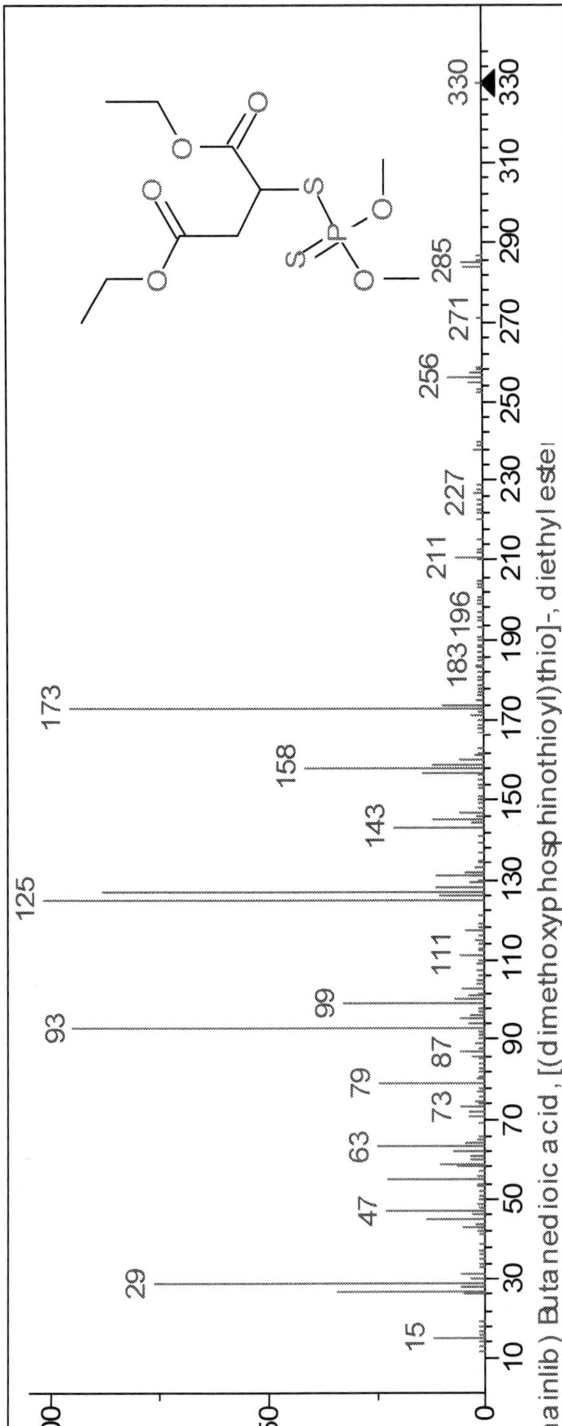

Figure 42.7: GCMS Analysis of Malathion Degradation Product, Butanedioic Acid, [(dimethoxy phosphinothioyl)thio]-, Diethyl Ester by *Bacillus polymyxa*.

The products formed after degradation of malathion was analyzed by GCMS by taking control as standard malathion (Figures 42.1 and 42.2). The gas chromatogram was obtained (Figure 42.3). The products were analyzed through mass spectra. There were two compounds detected. These were succinic acid mercapto diethyl ester, s-ester with o,s-dimethy phosphorodithioate and butanedioic acid [(dimethoxy phosphinothiol)thio]-, diethyl ester (Figures 42.4–42.7).

References

Alexander, M. (1969). Microbial degradation and biological effects of pesticides in soil. *In: Soil Biol*. Reviews of research, UNESCO, Paris, p.209-240.

Bergey, D.H. and Holt, J.H. (1994). Bergey's Manual of Determinative bacteriology Williams and Wilkins, Baltimore, Maryland, USA.

Foster, L., John, R., Kwan, B.H. and Vancov, T. (2006). Microbial degradation of the organophosphate pesticide ethion. *FEMS Microbiology Letters*. **240**: 49-53.

Gangawane, L.V. and Saler, R.S. (1988). A comparative study of tolerance of fungicides by *Rhizoctonia batalicola* and other microfungi in the rhizosphere of groundnut. *Pesticides*. **22**: 27-37.

Getz, M.E. and Watts, R.R. (1964). Application of 4-(p-nitrobenzyl) pyridine as a rapid quantitative reagent for organophosphate pesticides. *J.A.O.A.C.* **47**: 1094-1096.

Gram, C. (1984). The differential staining of Schizomycetes in tissue sections and in dried preparations. *Firtschitte der Medicin*., **2**: 185-189.

Kent, J. (1991). Education and training in farm chemical management. *Proc.Conf. Agriculture, Education and Information Transfer*, Murumbridgee College of Agriculture.

Kiss, A. and Virag, D. (2009). Photostability and photodegradation pathways of distinctive pesticides. *J. Environ. Qua*., **38**: 157-163.

Matsumara, F. (1974). Microbial degradation of pesticides. In: Survival in ToxicEnvironments. (M.A.O. Khan, J.P. Bederka Eds.) Academic press, NewYork. pp. 129-154.

Mostafa, N.I.Y., Bahig, M.R.E., Fakhr, I.M.I. and Adam, Y. (1972a). Malathion breakdown by soil fungi. *Z. Naturforsch*., **27 b**: 1115-1116.

Quinn, GA; Maloy, AP; McClean, S; Carney, B; Slater, JW (2012). "Lipopeptide biosurfactants from *Paenibacillus polymyxa* inhibit single and mixed species biofilms.". *Biofouling*., **28** (10): 1151–66.

Vargas, J.M. (1975). Pesticide degradation. *J. Arboriculture*., pp.232-233.

2015, Major Constrains and Verdict of Crop Productivity Pages 536–549
Editor: Dr. U.N. Bhale
Published by: ASTRAL INTERNATIONAL (P) LIMITED, NEW DELHI

Chapter 43

Assessment of Floral Quality Index in Aquatic Ecosystem of Selected Freshwater Ponds in Ariyalur District, Tamil Nadu, India

Saravanakumar Kaliyaperumal*
and Prabhakaran Jayaraman

Botany Wing-DDE, Annamalai University,
Annamalai Nagar – 608 002, Tamil Nadu, India

ABSTRACT

The Floristic Quality Index was used as a tool for evaluating the floristic and conservation value based on the abundance of some freshwater ponds in Ariyalur District of Tamil Nadu, India. The aquatic plant species were identified with available literature source and species list (70 species) were prepared. All plant species were clearly exhibited varying degrees of tolerance to disturbances of both natural and anthropogenic in addition to varying degrees of fidelity to specific habitats. This can express quantitatively by Coefficient of Conservatism (C of C). Using C of C on floristic quality assessment of various sites was calculated. The following four parameters discussed with the findings of the present study, (1) the identification of remnant native habitats of floristic significance, (2) the comparison of Floristic Quality Index (FQI) among different sites, (3) long-term monitoring of floristic quality in natural areas, and (4) monitoring of habitat restoration using a standard ecological sampling design for comparison between sites.

Keywords: *Floristic quality index, Freshwater ponds, Ariyalur district.*

---

* *Corresponding Author.* E-mail: drsaravanak2007@gmail.com

Introduction

Plant communities are often qualitatively described with unquantiable attributes, which vary depending on the surveyor, type of habitat, etc. The Floristic Quality Assessment Index (FQI), rst designed by Swink and Wilhelm (1994) offers a quantitative characterization of plant communities. This index is based upon the degree of conservatism of species, which depends on the specicity a certain plant has toward its habitat. The FQI of plant communities in selected ecosystems reported by various workers (Ladd, 1993; Oldham *et al.*, 1995; Herman *et al.*, 2001; Nichols, 1998; Mathews, 2003; Andreas *et al.*, 2004 and Cohen *et al.*, 2004).

FQI has been used in a variety of habitats: marshes (Cohen *et al.*, 2004), prairies (Klips, 2003), restored wetlands (Mushet *et al.*, 2002), wetlands (Lopez and Fennessy, 2002, and Matthews 2003), and woodlands (Francis *et al.*, 2000). We demonstrate the use of FQI and modied FQI (Rooney and Rogers, 2002) to compare habitat quality in freshwater ponds and its neighbourhood. We also document the oristic diversity and assess the oristic quality of aquatic angiosperm diversity in Ariyalur district.

Floristic Quality Assessment can also be used to assist in making permitting decisions and developing performance standards and mitigation criteria (Andreas and Lichvar, 1995, and Wilhelm and Masters, 1995). Good botanical skills are particularly needed when identifying species that are in vegetative state only *i.e.*, not in fruit or flower, (Rooney and Rogers, 2002). In the present study has been planned to assess the FQI of some selected water pond ecosystems in Ariyalur district of Tamil Nadu, India.

Materials and Methods

Study Area

Ariyalur (11°.8′14′N,79°.4′40′ E) is a capital of Ariyalur District which is situated about 292 km south of Chennai city, is a capital of Tamil Nadu state, India. There are four freshwater ponds selected for the present study *viz.* 1. Soorakuzhi, 2.Agaram, 3.Silambur and 4.Idyakurichi which are from nearly 36 to 41 km away from Ariyalur. A total of 104 sites were selected in the four ponds for the present study to assess the FQI.

Floristic Quality Assessment Method

The method of Coefcients of conservatism were adopted from Freeman and Morse (2002). Floristic Quality Assessment was applied by calculating a mean coefficient of conservatism (C of C) and a Floristic Quality Index (FQI) from a comprehensive list of plant species obtained from a particular site. Using the C of C, a floristic quality assessment of a site can be calculated using two separate measures: 1) the average or Mean C of C, and 2) the FQI.

The Mean C is calculated:

$$\text{Mean C} = \Sigma (C_1 + C_2 + C_3 + \ldots C_n)/N$$

FQI was calculated according to the following formula from Swink and Wilhelm (1994).

Map Depicting Study Area

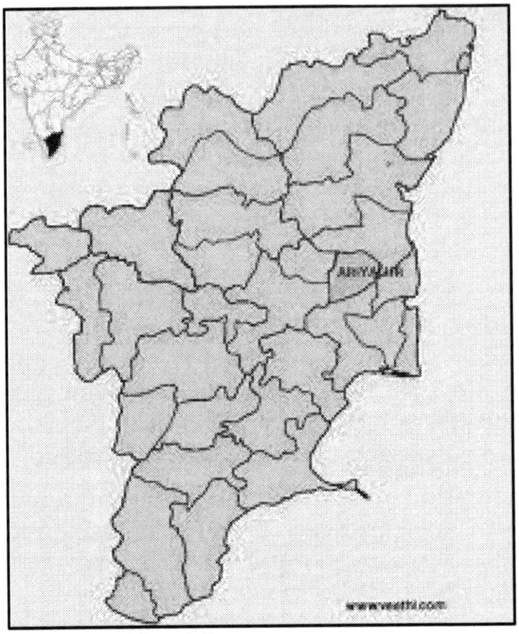

Tamil Nadu State

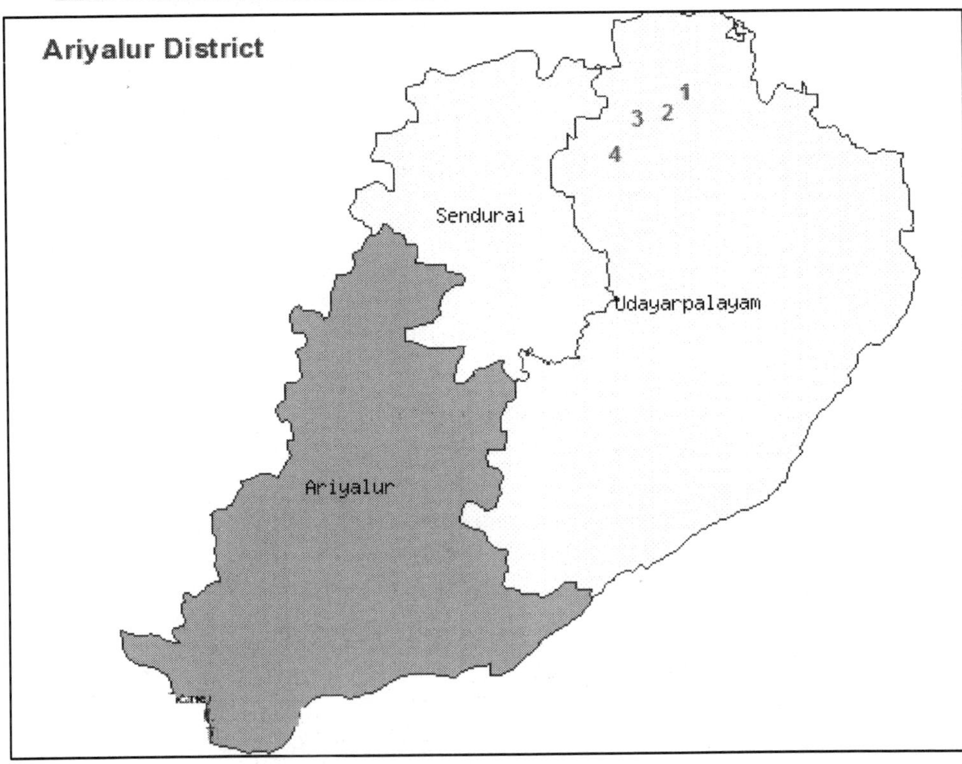

Ariyalur District

1. Soorakuzhi Pond

2. Agaram Pond

3. Silambur Pond

4. Idyakurichi Pond

I=R/√N,

where,

I= Floristic quality assessment index,

R= Sum of coefcients of conservatism for all plants recorded in the area, and

N = Number of different native species recorded.

Species that did not have coefcients assigned to them, such as all non-native taxa, were not considered in the equation. The modied oristic quality assessment index for each community type was also calculated as modiûed FQI=C, where C is the mean coefcient of conservatism (Rooney and Rogers, 2002).

Results and Discussion

A complete species list were (Table 43.1) prepared based on plot sampling and compiled for each site within each management type. C of C were assigned to each species and it ranges from 0-10, with higher values, such as 10, which indicating plants with higher delity to specic habitats and occupy a small ecological niche, cannot tolerate disturbance within surroundings, are habitat specic, and tend to perish easily with changes in habitat. Low C of C plants are tolerant of many different conditions, and are typically weedy generalists. Exotic taxa/invasive natives are all assigned a 0. Plants with a C of C of 1-3 occur in a variety of habitats. Those with a C of C of 4-6 are associated with a specic plant community but tolerate moderate disturbance, while those with scores of 7-8 are associated with a community of advanced successional stage. High delity plants have scores of 9-10.

The FQI of individual quadrats established along a transect may be used to establish baseline data serving as a benchmark for future monitoring of changes in floristic quality during site restoration or site rehabilitation efforts. If one is comparing floristic quality data among similar communities, *i.e.* fen to fen, or bog to bog, data must be obtained using a standard ecological sampling design for comparison between sites. Details of survey methods and effort should accompany any reporting of inventory or sampling results derived from applying the FQI because indiscriminate comparisons of floristic quality with dissimilar methods used in evaluation can be misleading (Taft *et al.*, 1997). Keddy *et al.* (1993) noticed that the species diversity as an essential indicator of integrity, but also recommends assessing guild diversity.

The coefficient of conservatism is applied to a species based upon its fidelity to a pre-settlement landscape, not its rarity or legal status. Although many plant species listed as endangered, threatened, or special concern by the Ariyalur district, Tamil Nadu are highly conservative, C = 8, 9, or 10, many are not, C = 4 - 7 (Figure 43.1). Many conservative, listed species are faithful to rare, high quality natural communities in ponds. Such communities include prairies, where species rarity is largely attributable to habitat loss, and the ponds in study area that have always been relatively rare. Losses of rare, high quality communities can have a serious impact on native biodiversity. Conversely, a number of rare species do not have a high fidelity to specific communities, or the communities to which they are faithful are relatively frequent.

Table 43.1: Floral Quality Index of Freshwater Ponds in Ariyalur District

Sl.No.	*Botanical Name/Family*	*Habit*	*Group*	*Nature*	*C Value*
1.	*Alisma plantago-aquatica* L. (Alismataceae)	MONO	AN	Rooted emergent	3
2.	*Alternanthera philoxeroides*(Mart)Griseb (Amaranthaceae)	DI	AN	Amphibious	1
3.	*Alternanthera sessilis* (L.) R.Br. ex DC. (Amaranthaceae)	DI	AN	Floating	1
4.	*Ammannia baccifera* L.(Lythraceae)	DI	AN	Rooted emergent	2
5.	*Aponogeton natans* (L.) Engler(Aponogetonaceae)	MONO	AN	Anchored floating	1
6.	*Brachiaria mutica* (Forssk.) Stapf.(Poaceae)	MONO	AN	Rooted emergent	7
7.	*Ceratophyllum demersum* L.(Ceratophyllaceae)	DI	AN	Floating	1
8.	*Chloris barbata* Sw.(Poaceae)	MONO	AN	Rooted emergent	1
9.	*Commelina diffusa* Burm.f.(Commelinaceae)	MONO	AN	Rooted emergent	1
10.	*Cynodon dactylon* (L.) Pers.(Poaceae)	MONO	AN	Rooted emergent	1
11.	*Cyperus bulbosus* Vahl(Cyperaceae)	MONO	PE	Rooted emergent	2
12.	*Cyperus difformis* L.(Cyperaceae)	MONO	PE	Rooted emergent	2
13.	*Cyperus rotundus* L.(Cyperaceae)	MONO	PE	Rooted emergent	2
14.	*Dactyloctenium aegyptium* (L.) Willd.(Poaceae)	MONO	AN	Rooted emergent	3
15.	*Echinochloa colona* (L.) Link(Poaceae)	MONO	AN	Rooted emergent	3
16.	*Eclipta prostrata* (L.)L.(Asteraceae)	DI	PE	Rooted emergent	4
17.	*Eichhornia crassipes* (Mart) Solms(Pontederiaceae)	MONO	PE	Floating	2
18.	*Elodea canadensis* Michx.(Hydrocharitaceae)	MONO	PE	Rooted submersed	7
19.	*Enicostemma axillare* (Lam.) Raynal (Gentianaceae)	DI	PE	Rooted emergent	9
20.	*Heteranthera limosa* (Sw.) Wild.(Pontederiaceae)	MONO	AN	Amphibious	9
21.	*Hydrilla verticillata* (L.f.) Royle (Hydrocharitaceae)	MONO	AN	Rooted submersed	2
22.	*Hydrocotyle umbellata* L.(Apiaceae)	DI	PE	Amphibious	8

Contd...

Table 43.1–*Contd...*

Sl.No.	Botanical Name/Family	Habit	Group	Nature	C Value
23.	*Hygrophila auriculata* (Schum.) Heine (Acanthaceae)	DI	PE	Rooted emergent	7
24.	*Ipomoea aquatica* L.(Convolvulaceae)	DI	PE	Anchored floatingAmphibious	2
25.	*Ipomoea carnea* Jacq.(Convolvulaceae)	DI	PE	Rooted emergent	2
26.	*Kyllinga bulbosa* P. Beavu.(Cyperaceae)	MONO	AN	Rooted emergent	2
27.	*Lemna gibba* L.(Lemnaceae)	MONO	AN	Floating	6
28.	*Lemna minor* L.(Lemnaceae)	MONO	AN	Floating	5
29.	*Ludwigia adscendens* (L.) H. Hara(Onagraceae)	DI	PE	Rooted emergent	7
30.	*Ludwigia perennis* L.(Onagraceae)	DI	PE	Rooted emergent	6
31.	*Limnophila aromatica* (Lam.) Merr.(Scrophulariaceae)	DI	PE	Rooted emergent	10
32.	*Limnophila heterophylla* (Roxb.) Benth.(Scrophulariaceae)	DI	PE	Rooted emergent	9
33.	*Limnophyton obtusifolium* (L.) Miq.(Alismataceae)	MONO	AN	Rooted emergent	8
34.	*Monochoria vaginalis* (Burm. F.) Presl(Pontederiaceae)	MONO	PE	Floating,Amphibious	6
35.	*Myriophyllum spicatum* L.(Haloragaceae)	DI	PE	Rooted submersed	3
36.	*Najas graminea* Del.(Naiadaceae)	MONO	AN	Rooted submersed	7
37.	*Najas indica* (Willd.) Cham.(Naiadaceae)	MONO	AN	Rooted submersed	6
38.	*Najas minor* All.(Naiadaceae)	MONO	AN	Rooted submersed	7
39.	*Nasturtium indicum* DC.(Brassicaceae)	DI	AN	Rooted emergent	6
40.	*Nelumbo nucifera* Gaertn.(Nymphaeaceae)	DI	PE	Anchored floating	5
41.	*Neptunia oleracea* Lour.(Mimosacceae)	DI	PE	Floating	5
42.	*Nymphaea alba* L.(Nymphaeaceae)	DI	PE	Anchored floating	6
43.	*Nymphaea nouchali* Burm. f.(Nymphaeaceae)	DI	PE	Anchored floating	5
44.	*Nymphaea rubra* Roxb. ex Andrews(Nymphaeaceae)	DI	PE	Anchored floating	6

Contd...

Table 43.1–*Contd...*

Sl.No.	Botanical Name/Family	Habit	Group	Nature	C Value
45.	*Nymphoides cristata* (Roxb.) Kuntze(Gentianaceae)	DI	AN	Floating anchored	8
46.	*Nymphoides hydrophylla* (Lour.) Kuntze(Gentianaceae)	DI	PE	Floating anchored	9
47.	*Ottelia alismoides* (L.) Pers.(Hydrocharitaceae)	MONO	PE	Rooted submersed	8
48.	*Paspalum repens* P.J.Bergius.(Poaceae)	MONO	AN	Amphibious	7
49.	*Persicaria amphibia* (L.) Delarbre.(Polygonaceae)	DI	PE	Floating	6
50.	*Persicaria glabra* (Willd.) M.Gómez.(Polygonaceae)	DI	PE	Rooted emergent	5
51.	*Phragmites communis* Trin.(Poaceae)	MONO	AN	Rooted emergent	4
52.	*Pistia stratiotes* L.(Araceae)	MONO	AN	Floating	3
53.	*Polygala arvensis* Willd.(Polygalaceae)	DI	AN	Rooted emergent	7
54.	*Pontederia cordata* L.(Pontederiaceae)	MONO	PE	Rooted emergent	8
55.	*Sagittaria guayanensis* Kunth(Alismataceae)	MONO	AN	Anchored floating	8
56.	*Schoenoplectiella articulata* (L.) Lye.(Cyperaceae)	MONO	PE	Rooted emergent	7
57.	*Schoenoplectus lacustris* (L.) Palla(Cyperaceae)	MONO	PE	Rooted emergent	6
58.	*Schoenoplectus subterminalis* (Torr.) Soják(Cyperaceae)	MONO	PE	Rooted emergent	5
59.	*Schoenoplectus tabernaemontani* (C.C.Gmel.) Palla(Cyperaceae)	MONO	PE	Rooted emergent	5
60.	*Sphenoclea zeylanica* Gaertn.(Campanulaceae)	DI	PE	Rooted emergent	7
61.	*Spirodela polyrrhiza* (L.) Schleid.(Lemnaceae)	MONO	PE	Floating	6
62.	*Sporobolus coromandelianus* (Retz.) Kunth.(Poaceae)	MONO	PE	Rooted emergent	6
63.	*Striga angustifolia* (D.Don) Saldanha(Scrophulariaceae)	DI	AN	Rooted emergent	4
64.	*Trapa natans* var. *bispinosa* (Roxb.) Makino.(Trapaceae)	DI	AN	Rooted emergent	7
65.	*Typha domingensis* Pers.(Typhaceae)	MONO	AN	Amphibious	6
66.	*Utricularia caerulea* L.(Lentibulariaceae)	DI	AN	Rooted emergent	4

Contd...

Table 43.1–*Contd...*

Sl.No.	*Botanical Name/Family*	*Habit*	*Group*	*Nature*	*C Value*
67.	*Utricularia aurea* Lour.(Lentibulariaceae)	DI	AN	Submersed	5
68.	*Utricularia polygaloides* Edgew.(Lentibulariaceae)	DI	AN	Submersed	4
69.	*Vallisneria natans* (Lour.) H. Hara(Hydrocharitaceae)	MONO	AN	Rooted submersed	3
70.	*Wolffia globosa* (Roxb.) Hartog and Plas (Lemnaceae)	MONO	AN	Floating	2

MONO: Monocotyledons; DI: Dicotyledons; AN: Annual; PE: Perennial.

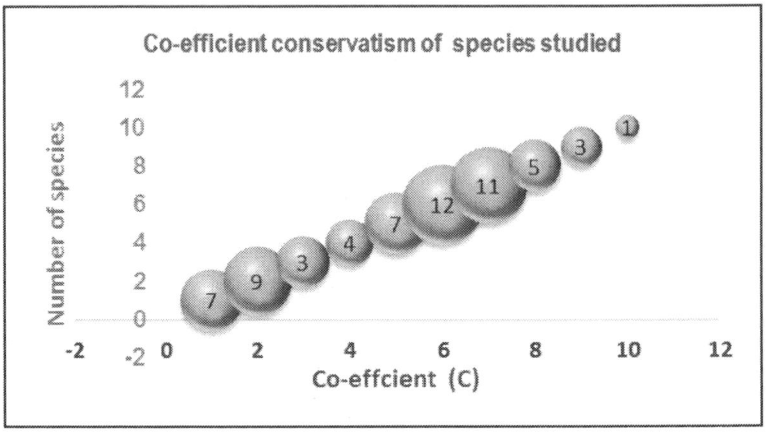

Figure 43.1: Co-efficient of Conservatism of Species.

FQA measures the extent and proportion to which constellations of conservative plants are present on a site (Swink and Wilhelm 1994, Wilhelm and Masters 1995). Areas with a greater proportion of conservative species will have a greater C and higher FQI, alerting us to its or any community's floristic quality, potentially restorable floristic quality, and implicitly natural quality.

The present data correlates with the study of Francis *et al.* (2000), and Matthews (2003) those examined data from over 600 wetland samples in Illinois province and found that mean FQI and N (but not Mean C) were influence by seasonal fluctuations. According to Lopez and Fennessy (2002) found that combined summer and autumn samples in the study areas increased FQI by an average of three points over a summer sample.

In this study a total 70 species were recorded which were represented by 53 genera under 30 families (Table 43.1). Of these, 66 (82.5 per cent) were native while 4 (17.5 per cent) taxa were non-native. The invasive alien species were *Eicchornia crassipes, Hydrilla verticillata, Ceratophyllum demersum* and *Phragmites communis.* Families represented by the most number of species were: Poaceae (8), Cyperaceae (8), Hydrocharitaceae (6), Pontederiaceae, Lemnaceae and Nymphaeaceae (4), and Alismataceae, Naiadaceae and Lentibulariaceae (3). The genera, *Cyperus, Lemna, Nymphaea, Shoenoplectus* and *Utricularia* with 3 species followed by *Limnophila, Nymphoides* with 2 species.

Plant species were found with high C of C, *Limnophila aromatica* and *Nymphoides hydrophylla* species was 10 and 9 respectively.

Plants with a wide range of ecological tolerances, and non-native taxa, were assigned a rank 0-10. These species are often opportunistic invaders of natural areas, due to natural or human disturbances and frequently create extensive monocultures.

1. Rankings of 1 and 2 were applied to taxa that are widespread and not typical of (or only marginally typical of) a particular community, *e.g.* Rigid hornwort (*Ceratophyllum demersum*) and Umbrella plant (*Cyperus difformis*).

2. Plants with an intermediate range of ecological tolerances were assigned average coefficients (3, 4, or 5). These taxa may be found in a stable phase of some native community, but persist under some disturbance. Examples include Asiatic witchweed (*Striga angustifolia*) C of C = 4, Spiked water milfoil (*Myriophyllum spicatum*) C of C = 3, and Water bulrush (*Shoenoplectus subterminalis*) C of C = 5.
3. Rankings of 6 to 8 are applied to those plants that typify stable or near "climax" conditions. Examples include Fat duckweed (*Lemna gibba*) C of C = 6, Ricefield waternymph (*Najas graminea*) C of C =7, and Duck lettuce (*Ottelia alismoides*) C of C = 8.
4. Plants with a narrow range of ecological tolerances were assigned high coefficients. Rankings of 9 and 10 were reserved for those plants that exhibit relatively high degrees of fidelity to a narrow range of habitat requirements. Examples include (*Limnophila aromatica*) C of C = 10 and white Duck salad (*Heteranthera limosa*) C of C = 9.

The coefficients of conservatism provide a useful surrogate for experimentally derived species tolerances. To the extent a species has a narrow or broad range of natural ecological tolerances, this tolerance (or sensitivity) as ranked by the C of C's is a useful tool for evaluating its ability to tolerate human disturbance. For example, the relative area cover of plants with low C of C's can be summed to measure the dominance of "tolerant" species.

Floristic Quality Assessment has been suggested as being a problematic indicator of the quality of a habitat, and modied FQI provides advantages over FQI (Rooney and Rogers, 2002). However, the present study depict an identical signicant differences in FQI values for elds across different management types and the corresponding modied FQI values for these ponds. FQI values for these ponds indicate the true nature of the poor quality of these freshwater ecosystem. These four ponds, had one or two species each with C of C greater than 4, but the predominant vegetation within each individual eld was that of non-native grasses and ruderal species. However, interpreting values of modied FQI alone in this case would lead to the idea that these ponds are of higher superior oristic quality than they actually are.

Conclusion

The present observations of four pond ecosystems will provide most appropriate picture of the condition of a particular pond and it will be helpful to calculate both FQI and modied FQI values. These data shows the most oristically diverse and are very high quality habitats. However, the detailed study needed for managing and conserving these freshwater ecosystems.

References

Andreas, B. K. and Lichvar, R. W. (1995). Floristic index for establishing assessment standards: A case study for northern Ohio. Technical Report WRP-DE-8, U. S. Army Waterways Experiment Station, Vicksburg, MS. 16 pp.

Andreas, B. K., Mack, J. J. and McCormac, J.S.(2004). Floristic Quality Assessment Index (FQAI) for Vascular Plants and Mosses for the State of Ohio. Ohio Environmental Protection Agency, Division of Surface Water, Wetland Ecology Group, Columbus, Ohio, pp. 219.

Cohen, M.J., Carstenn, S. and Lane, C.R. (2004). Floristic quality indices for biotic assessment of depressional marsh condition in Florida. *Ecological Applications,* **14**: 784-794.

Francis, C. M., Austen, M. J. W., Bowles, J. M. and Draper, W. B. (2000). Assessing floristic quality in southern Ontario woodlands. *Natural Areas Journal*, **20**: 66-77.

Freeman, C.C., and C.A. Morse. (2002). Kansas oristic quality assessment: coefcients of conservatism for plants in southeast Kansas. Kansas Department of Health and Environment, Lawrence.

Herman, K.D., Masters, L.A. Penskar, M. R. Reznicek, A. A. Wilhelm, G. S. Brodowicz, W. M. and Gardiner, K.P. (2001). Floristic Assessment with Wetland Categories and Examples of Computer Applications for the State of Michigan (FQA), Revised 2nd Edition. Natural Heritage Program, Michigan Department of Natural Resources, Lansing, Michigan. p. 19.

Keddy, P.A., H. T. Lee, and I. C. Wisheu. 1993. Choosing indicators of ecosystem integrity: wetlands as a model system. in S. Woodley *et al.* (eds.) Ecological Integrity and the Management of Ecosystems, St. Lucie Press, Delray Beach, FL.

Klips, R.A. (2003). Vegetation of Claridon railroad prairie, a remnant of the Sanducsky Plains of Central Ohio. *Castanea*, **68**: 135-142.

Ladd, D.M. (1993). Coefcients of conservatism for Missouri vascular ora. The Nature Conservancy, St. Louis, Mo.

Lopez, R. D. and Fennessy, M. S. (2002).Testing the floristic quality assessment index as an indicator of wetland condition. *Ecological Applications*, **12**: 487-497.

Matthews, J.W. (2003). Assessment of the floristic quality index for use in Illinois, USA, wetlands. *Natural Areas Journal*, **23**: 53-60.

Mushet, D.M., Euliss, N.H. Jr., Shaffer, T L. (2002). Floristic quality assessment of one natural and three restored wetland complexes in North Dakota, USA. *Wetlands,* **22**: 126-138.

Nichols, S.A. 1998. Floristic quality assessment of Wisconsin lake plant communities with example applications. *J. Lake and Reservoir Management,* **15**: 133-141.

Oldham, M., Bakowsky, W. and Sutherland, D. (1995). Floristic quality assessment for southern Ontario. Natural Heritage Information Centre, Ontario Ministry of Natural Resources, Peterborough, Ontario.

Rooney, T.P. and D.A. Rogers. (2002). The modified floristic quality index. *Natural Areas J.*, **22**: 340-344.

Swink, F. and Wilhelm, G. (1994). Plants of the Chicago Region. Fourth Edition. Indiana Academy of Science, Indianapolis, IN., p. 921.

Taft, J. B., G. S. Wilhelm, D. M. Ladd, and L. A. Masters. 1997. Floristic quality assessment for vegetation in Illinois, a method for assessing vegetation integrity. *Erigenia.***15**: 3-95.

Wilhelm, G. and Masters. L. A. (1995). Floristic quality assessment in the Chicago Region and application computer programs. Morton Arboretum, Lisle, IL., p. 17.

2015, Major Constrains and Verdict of Crop Productivity *Pages 550–557*
Editor: **Dr. U.N. Bhale**
Published by: **ASTRAL INTERNATIONAL (P) LIMITED, NEW DELHI**

Chapter 44

Strategies of Deficient Soil for Improvement of AMF Inoculum and Biomass Production of Pigeonpea

S.A. Bansode and U.N. Bhale*

Research Laboratory, Deparment of Botany, Arts, Science and Commerce College, Naldurg, Tq. Tuljapur., Dist. Osmanabad – 413 602, M.S.

ABSTRACT

Pigeonpea is the second most important pulse crop after chickpea. It is belongs to family Fabaceae. As a result, most of the deficient soils are poor in the essential plant nutrients required for optimum crop growth leading, to low crop yield. The mycorrhizas are beneficial fungi growing in association with plant roots and exist by taking sugars from plants in exchange for moisture and nutrients gathered from the soil by the fungal strands. The mycorrhizas are belongs to phylum Glomeromycota. The present study revealed that eight parameters for biomass production and assessment of root colonization and spore density of pigeonpea (Cajanus cajan). The percentage of germination of seeds is more than that of control pigeonpea plants. Biomass was observed after 12 weeks viz. height of stem (18.4 cm), width of stem (1.0 cm), number of leaves (9), fresh shoot weight (1.33 gm), fresh root weight (0.19 gm), total leaf area (282.110 $cm^{2)}$, dry shoot weight (0.56 gm) and dry root weight (0.11 gm). In AMF studies, percentage of root colonization (90.62 per cent) and AMF spore density (257/100 gm soil) was also observed. Among AMF spore viz. Glomus, Gigaspora, Acaulospora and Sclerocystis genus was found but Acaulospora and Glomus was found dominant and Glomus aggrigatum and Acaulospora scrobiculata species were recorded.

Keywords: *Deficient soil, Pigeonpea, Biomass production, Root colonization, AMF Spore density.*

* *Corresponding Author.* E-mail: unbhale2007@rediffmail.com, bansodeshree@gmail.com

Introduction

Soil fertility decline is a major problem confronting crop production in India. This is caused by crop nutrient removal and losses through soil erosion. Soil is natural resources with crucial ecological, economic and social functions. Soil is the essential component of the terrestrial environment and forms the interface between geosphere, atomosphere, hydrosphere and biosphere (Doran and Parkin, 1994). As soils become more acid therefore some plant nutrients become deficient while others increase in availability to toxic levels. These changes in availability of nutrients in acidic soils and cause most of the effects on plant growth. In addition, soil acidity may decrease microbial activity affecting nitrogen fixation by some legumes and mineralisation of organic matter (Helyar and Porter, 1989). In an ecosystem that remained closed there would be no change to the acid/alkali balance of the system, only redistribution over time through the system. However loss of organic anions with removal of plant material and through leaching of nitrate nitrogen before the plant can utilise it leaves the ecosystem a little more acid. If these losses are small then the small amount of acid added is sufficient to maintain a fertile soil. Grant *et al.* (2002) stated that intensification and diversification of cropping systems influence soil physical, chemical and microbiological characteristics.

In natural ecosystem, plant roots are colonized by numerous species of fungi, with extensive symptomless, nonpathogenic, endophytic or biotrophic phases in their life cycle, these includes, associations formed by mycorrhizal and non-mycorrhizal fungi. The AM Fungi is ubiquitous group of fungi (Smith and Read, 1997). AM Fungi belongs to phylum Glomeromycota (Schenck and Perez, 1988). AMF help plants to capture nutrients such as phosphorous, sulfur, nitrogen and micronutrients from the soil. Enhanced plant nutrition in AMF plant is achieved by increasing the percentage of germination of seeds and biomass of the plant.

Therefore the present investigation was made to use deficient soil along with rhizospheric soil for improvement of biomass and AMF status.

Materials and Methods

Deficient Soil Sample Collected Site

Deficient soil samples were collected from the hill region at Naldurg (17.82°N 76.30°E) Osmanabad districts of Marathwada region of Maharashtra for pot culture. Naldurg place is located at an altitude of 566m and receives an average annual rainfall of 760 mm.

The deficient soil samples were collected in separate pot for culture of pigeonpea plants and these soil samples were used for Isolation, Identification and diversity of AM fungal spores.

Pot Experiment

Deficient soil was collected from study site. In earthen pot (1.5kg) capacity was used. The pots were 0.22 cm in height and 0.21 cm in diameter with a small hole at bottom to drain excess water. In pot 1 kg of deficient soil was incorporated and 100 gm of rhizospheric soil as an AMF inoculum was used at upper layer. Sterilized (1 per cent $HgCl_2$) seeds of pigeonpea were used. In pot, 8 seeds were sown during the crop season of 2013. Without rhizospheric soil is considered as control. After one week to check the per cent seed germination. Thoroughly measuring biomass *i.e.* height of stem, width of stem and number of leaves. After 12 week, it was checked Fresh weight and Dry weight of shoot and root, total leaf area and also calculated AMF spore density and root colonization. In these present investigation three replications was used.

AMF Study Inoculum

AMF Spore Isolation, Identification and Density

The treated pot culture soil one hundred grams of soil was dissolved in 1000 ml of water and decanted through a series of 355 to 35 μm sieves (Gerdemann and Nicolson, 1963). Residues were filtered through Whatman filter paper No 1 and all spores were counted under the stereo-zoom dissection microscope. Intact and healthy AM fungal spores were mounted in PVLG (Polyvinyl alcohol-lacto glycerol) with and without Melzer's reagent for identification using keys11 and INVAM (http://www.invam.caf.wvu.edu). Some spores were identified only up to genus level.

Assessment of AM Fungal Root Colonization

The treated pot culture plant roots were fixed in to FAA. Fixed roots were washed to free FAA cleared in water. Roots are placed in glass vial with 10 per cent KOH and stained in 0.05 per cent trypan blue-lactophenol to determine mycorrhizal colonization, these VAM fungal colonized segment studied9. Then transferred the sieving on to a gridded petriplate and observed it under the binocular microscope 400X (Lawrence and Mayo LM-52-3521). The per cent root colonization was measured by using the formula (Giovannetti and Mosse, 1980).

Biomass Production

Pigeonpea plants were observed until 12 weeks Fresh weight of shoot and root samples were recorded. Shoots and roots were separated and oven dried at 60°C for 48 h for the determination of dry mass after recording their lengths (Muthukumar and Udaiyan 2000). Leaf area was measured after 12 week post sowing harvest by disc method. Fifty leaf disc of known size was taken from randomly selected leaves of the plant, discs and remaining leaf blades were oven dried and leaf area was calculated by using formula (Vivekanandan *et al.*, 1972).

Results and Discussion

Biomass Production

The present study the result revealed that, after 12 week of pigeonpea, studies percentage of germination, eight parameters for biomass production and assessment of root colonization and spore density described in Table 44.1. In this study the percentage of germination is more and earlier than that of control pigeonpea plants. In biomass measour the eight parameters up to the 12 Weeks *viz.* Height of stem (18.4 cm), Width of stem (1.0 cm), Number of leaves (9), Fresh shoot weight (1.33 gm), fresh root weight (0.19 gm), total leaf area (282.110 cm ^{2}), dry shoot weight (0.56 gm) and dry root weight (0.11 gm) (Figure 44.1).

Table 44.1: Percentage of Germination, Biomass and Status of AMF in *Cajanus cajan*

Sl.No.	*Parameters/Crops*	*Treated Plant*	*Untreated Plant*	*Week*
1.	Number of Seeds	8	8	One
2.	Germination of Seeds	8	3	Two
3.	Percentage(per cent) of germination	100	37.5	Two
4.	Height of Stem(cm)	5.4	5.8	Two
5.	Width of Stem (mm)	2	2	Two
6.	Number of Leaves	2	2	Two
7.	Height of Stem(cm)	9.2	11.2	Four
8.	Width of Stem (mm)	5	5	Four
9.	Number of Leaves	3	3	Four
10.	Height of Stem(cm)	13.7	17.5	Six
11.	Width of Stem (mm)	6	7	Six
12.	Number of Leaves	6	7	Six
13.	Height of Stem(cm)	14.8	22.7	Eight
14.	Width of Stem (mm)	7	7	Eight
15.	Number of Leaves	8	8	Eight
16.	Height of Stem(cm)	18.4	26.5	Ten
17.	Width of Stem (cm)	1.0	1.4	Ten
18.	Number of Leaves	10	14	Ten
19.	Fresh weight of shoot (gm)	1.33	4.00	Ten
20.	Fresh weight of root (gm)	0.19	0.45	Ten
21.	Dry weight of shoot (gm)	0.56	1.63	Twelve
22.	Dry weight of root (gm)	0.11	0.22	Twelve
23.	Total Leaf area (cm ^{2})	282.110	375.329	Ten
24.	Percentage(per cent) of root colonization	90.62	69.31	Ten
25.	AMF Spore density (100 gm soil)	257	158	Ten

AMF Study

In AMF studies, 32 segments of roots were assayed and found percentage of root colonization (90.62 per cent) and AMF spore density (257/100 gm soil) was recorded. Among AMF spore *viz. Glomus, Gigaspora, Acaulospora* and *Sclerocystis* genus was found but *Acaulospora* and *Glomus* genus was found dominant. *Glomus aggrigatum* and *Acaulospora scrobiculata* species were recorded from tested pigeonpea containing pot soil (Figures 44.2 and 44.3).

Treated Pot **Untreated Pot**

Figure 44.1: Treated (Rhizospheric Soil) and Untreated (Control) of Pigeonpea.

Howarth (2005) stated that management of nutrients to maintain productivity and quality of cropping systems is a challenge that must be met through a combination of organic amendments and management of soil organic matter (SOM). The SOM is the source of nutrients for crops that maintain soil fertility and productivity in farming systems (Powell *et al.,* 1999). Mycorrhizal plants have also been shown to better tolerate to the environmental stresses such as nutrient deficient soils, drought conditions, salinity and pathogens than non-mycorrhizal plant (McArthur and Knowles, 1993; Sylvia *et al.,* 1993; West, 1995). Positive effects of AMF on host-plant growth and development were already observed in low soil fertility conditions and also in drought environments (Sylvia *et al.,* 1992, Picone, 2003 and Liu *et al.,* 2007). AMF is of interest for the reclamation and revegetation of degraded lands (Miller *et al.,* 1992) AMF has been taken into account and has had an impact in landscape regeneration, horticulture, alleviation of desertification andin the bioremediation of contaminated soils (Peter *et al.,* 2003).Pond soil when amended with renovated land soil increased the root length of *Glycine max* (26.66 cm), increased Arbuscular type of root colonization and spore density (Bansode *et al.,* 2014).

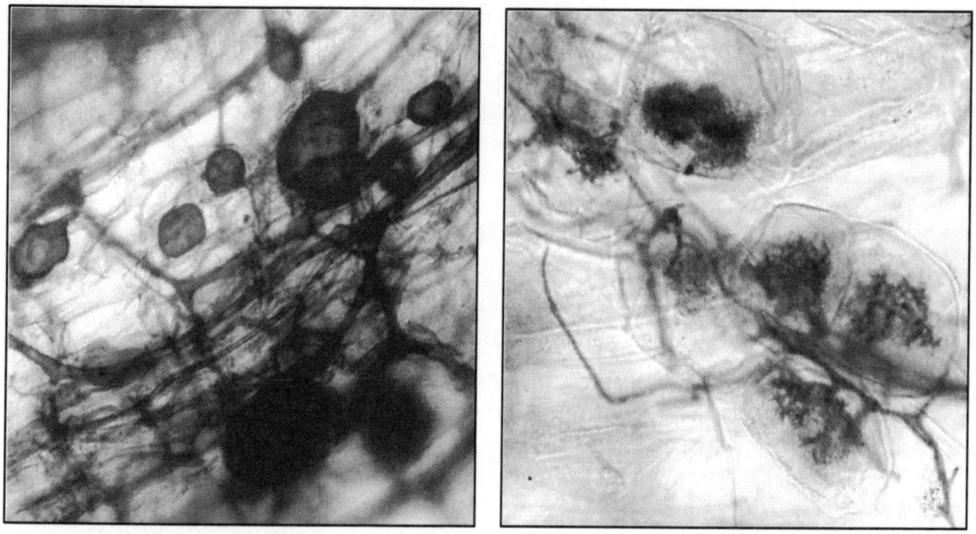

Intraradical-Vesicles **Hyphal and Arbuscular**

Figure 44.2: AMF Root Colonization in Pigeonpea Plant.

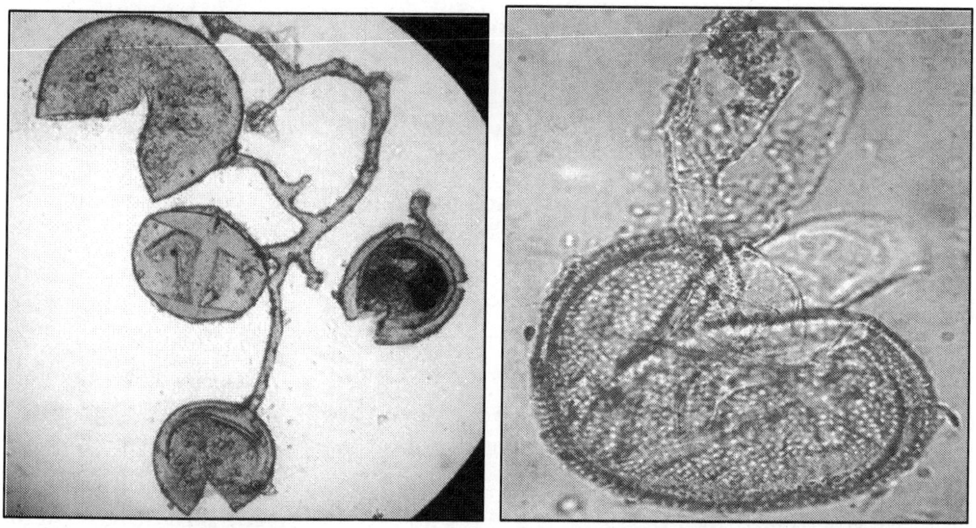

Glomus aggrigatum ***Acaulospora scrobiculata***

Figure 44.3: Studied Arbuscular Mycorrhizal Fungal (AMF) Genera.

Conclusion

Agricultural practice is more acidifying than the natural eco-systems. To establish an agricultural eco-system that will continue indefinitely will mean planning. This plan will be based on an understanding the causes of the acidity. It will mean changing some practices and most likely applying finely ground limestone as the source of alkalinity to maintain this balance.This investigation was concluded that in the deficient soil and rhizospheric soil used in mix then the percentage of germination, biomass parameters and AMF spore density. Deficient soil is also converted into fertile soil when come in a contact with fertilizers and microbes (AMF).

References

Bansode, SA., Sawant, V.S. and Bhale, U.N. (2014). Reclamation of degraded land through pond sedimentary soil And impact of biomass production and arbuscular mycorrhizal Fungal (amf) status of soybean field. *International Journal of Biotechnology and Allied fields.*, **2** (1): 33-41.

Doran, J. W. and Parkin, T. B. (1994). Defining and assessing soil quality. *Soil Science Society of America*, **35**: 3-21.

Gerdemann, J.W. and Nicolson T.H.(1963). Spores of mycorrhizal Endogone species extracted from soil bywet-sieving and decanting. *Trans. Br. Mycol. Soc.*, **46** : 235-244.

Giovannetti, M. and. Mosse, B. (1980). An evaluation of techniques for measuring vesicular-arbuscular infection in roots. *New Phytol.*, **84:** 489-500.

Grant, C.A., A. B. Gary and Campbell, C.A. (2002). Nutrient considerations for diversified cropping systems in the Northern Great Plains. *Am. Soc. of Agronomy. Agronomy Journal.*, **94**: 186- 198.

Helyar, K.R., Porter, W.M. (1989). Soil acidification, its measurement and the processes involved. In: Robson, A.D. (ed.). *Soil Acidity and Plant Growth*. Australia, Academic Press., p.61-101.

Howarth, R.B. (2005). "Against High Discount Rates." In Perspectives on Climate Change: Science, Economics, Politics, Ethics (W. Sinnott-Armstrong and R.B. Howarth, eds.). Amsterdam: Elsevier.

Liu, A, C. Plenchette and Hamel C. Soil nutrient and water providers: how arbuscular mycorrhizal myceliasupport plant performance in are source limited world, 2007.

McArthur, D. A. J and Knowles, N. R. (1993). Influence of species of vesicular-arbuscular mycorrhizal fungi and phosphorous nutrition on growth, development and mineral nutrition of potato (*Solanum tuberosum* L.). *Plant Physiology*, **102**: 771-782.

Miller, R.M. and Jastrow, J.D. (1992). The application of VA mycorrhizae toecosystem restoration andreclamation, 438-467. *In* In M. F. Allen (ed.), Mycorrhizal functioning, Chapman and Hall, Ltd., London, England.

Muthukumar, T. and Udaiyan, K. (2000). The role of seed reserves in arbuscular mycorrhizal formation and growth of *Leucaena leucocephala* (Lam.) de Wit. and *Zea mays* L., *Mycorrhiza*, **9** : 323 -330.

Peter, J, S. Gianinazzi, S, Perotto, K. Turnau and Barea JM, (2003) Thecontribution of arbuscularmycorrhizal fungi in sustainable maintenance of plant health and soil fertility, *Biol. Fertili. Soils*, **37**: 1-16.

Picone, C, (2003). Managing mycorrhizae forsustainable agriculture in the tropics. *In*: Vandermeer, J.H. (ed.), *Tropical Agro ecosystems*, CRC Press, Boca, p.95-132.

Powell, J.M., F.N. Ikpe and Somda, Z.C. (1999). Crop yield and the fate of nitrogen and phosphorus following application of plant materials and feces to soil. *Nutrient Cycling in Agrosystems*, **54**: 215 – 226.

Schenck, N.C. and Perez, Y.(1988). Manual for the Identification of VA Mycorrhizal Fungi. 2nd Edn., University of Florida, Gainesville, Florida, pp.167-223.

Smith, S. E. and Read, S.E. (1997). Mycorrhizal symbiosis. Academic San Diego, CA.

Sylvia, D. M. and Williams S. E. (1992) Vesicular arbuscular mycorrhizal and Environmental stress. *In*: Bethlenfalvay, G.J. and R.G.Linderman (eds.), *Mycorrhizae in Sustainable Agriculture*, Am. Society of Agronomy, pp.101-124.

Sylvia, D. M., L. C. Hammond, J. M Bennett., J. H. Haas and Linda, S. B. (1993). Field response of maize to a VAM fungus and water management. *Agronomy Journal*, **85**: 193-198.

Vivekanandan, A.S., Ganasena, H.P.M.and Shivanayagan, T.(1972). Statistical evaluation of the accuracy of three techniques used in the estimation of leaf area of crop plant. *Indian J. Agril. Sci.*, **42**: 457 - 860.

West, H. M. (1995). Soil phosphate status modifies response of mycorrhizal and non-mycorrhizal *Senecio vulgaris* L. to infection by the rust, *Puccinia lagenophorae* Cooke. *New phytologist*, **129**: 107-116.

2015, Major Constrains and Verdict of Crop Productivity *Pages 558–563*
Editor: **Dr. U.N. Bhale**
Published by: **ASTRAL INTERNATIONAL (P) LIMITED, NEW DELHI**

Chapter 45

Micropropagation Studies Inplant *Uraria picta* (Jacq.)Dc

D.S. Jadhav[1] and N.B. Pandhure[2]*

[1]Department of Botany,
DMSM College, Kalam, Dist. Osmanabad, M.S., India
[2]Department of Botany,
Dr. Babasaheb Ambedkar Marathwada University,
Aurangabad – 431 001, M.S., India

ABSTRACT

Micropropagation of valuable medicinal plant Uraria picta has been carried out during present piece of work. During micropropagation studies various plant parts were used as an explant. Explants showed regeneration of callus and shoots with restricted growth within 25 days on modified Murashige and Skoog's (MS) medium supplemented with various concentration of growth hormones like 1.0 mg/L, 1.2 mg/L, 1.4mg/L, 1.6mg/L, 1.8mg/L, 2.0mg/Lof 6-benzylaminopurine. These hormones were added incombination with 0.1mg/L, 0.2mg/L, 0.3mg/L, and indole-3-acetic acid. Good proliferation rate was achieved by reducing the concentrations of 6-benzylaminopurine and indole-3-acetic acid to 0.1 mg/l(-1) each and incorporating 0.5 mg/l(-1) gibberellic acid in the medium. Maximum number of shoots per explant wasobtained. Isolated shoots were rooted in 0.5 mg/lindole-3-butyric acid. In vitro-raised plants after hardening in inorganic salt solution grew normally in soil.

Keywords: *Uraria picta, Nodal explant, Multiplication.*

* *Corresponding Author.* E-mail: drnarayan14872@gmail.com

Introduction

Uraria picta is valuable medicinal plant belongs to family Leguminosae. It is a perennial herb reaching about 1 to 2.5 m height widely distributed throughout India, Australia, Africa and all most all parts of Asia. In India it is found in dry grasslands and open forests in the sub-Himalayan tract from Kashmir to west Bengal, Assam and wide distribution.In Maharashtra it ranges from Western Ghats, Nanded and Gondia districts of Vidarbha. It is one of the important constituents in ten herb formulation called 'Dashmula, a well-established Ayurvedic drug of the Indian system of medicine. The whole plant is use in medicine. Roots are aphrodisiac in nature. They are mainly used for healing of fracture.

Decoction is useful in cough, chills and fevers. Leaves are antiseptic in nature these are mainly useful in gonorrhea and also for ascaricidal properties in human and animal. *Uraria picta* is also used for its analgesic, anti-inflammatory, antioxidant, different parts of the plants are reported for different activities such as, decoction of the entire plants are useful for cough, chill and fever. *Uraria picta* contains isoflavanones it is isolated from root bark of this plant have anti-microbial activity. Flavonoids are known to exhibit a range of biological activities including Anti-inflammatory, Anti-thrombotic, anti- viral, hepatoprotective and vasoactive properties. Thus the major bioactive flavonoid rhoifolin has been considered as marker compound for qualitative and quantitative standardization of *Uraria picta* based on liquid chromatographic analysis(Akhilesh *et al.*, 2009). Therefore it is need to conservation of this valuable medicinal plant and evaluation standard in vitro protocol of *Uraria picta* for human welfare. In present work it is proposed to carry out in vitro study and multiplication of *Uraria picta*.

Materials and Methods

Initially *Uraria Picta* grown in the greenhouse (Botanical Garden, Dr. Babasaheb Ambedkar Marathwada University, Aurangabad), these plant was used as experimental material. Plants were grown in the beds. The temperature in the greenhouse varied from 28 to 32°C. No artificial light was provided. Various explants were used for establishing callus, including axillary leaves, axillary shoots, hypocotyls etc. Axillary leaves and nodal segments were taken from six month-old plants while hypocotyl and root segment explantswere removed from 10–15 day-old seedlings.

Axillary Leaves and Axillary Shoots

Axillary leaves (0.5–0.8 cm) were taken from 1st to 3rd node from the apical region of lateral branches and surface-sterilized with 0.01 per cent (w/v) mercuric chloride for 2–3 min and washed 3–4 times with sterile double distilled water The explants were inoculated on agar-solidified MS (Murashige and Skoog, 1962) medium supplemented with different concentrations of 2, 4-D and KN, BAP either alone or in combination. The pH of the medium was adjusted to 5.8 before sterilization. Cultures were maintained at 25±1°C with a 16-h photoperiod with 40 mol m^2/s provided by cool white fluorescent tubes light. Callus was sub-cultured after 25 days on the original callus-inducing medium.

For the growth of Axillary shoots arise from using the nodes of nodal segments as an explant. Nodal segments (1–2 cm) were taken from lateral branches and surface-sterilized as for leaves with 0.2 per cent mercuric chloride and inoculated on agar-solidified MS medium supplemented with different concentrations of BA and KN (alone or in combination).

Seeds and Root Segments

Seeds were collected and stored at 4°C for about 2 months were used. These were soaked in water for 24 h, washed with 5 per cent (v/v) teepol solution (5 min), surface-sterilized with 0.3 per cent mercuric chloride (12–13 min), washed 3–4 times with sterile double-distilled water, and inoculated on MS medium with 3 per cent (w/v) sucrose without any growth regulators. Seeds were incubated at 25°C (dark) for 1 week and then transferred to white fluorescent lights with a 16-h day length. After the 3 weeks these seeds were germinated. Hypocotyls and root segments of 10–15 day old seedlings were cultured on the MS-based agar solidified medium supplemented with 2, 4-D (2 mg/l) and KN (0.2mg/l) for callus formation. Tubes inoculated with root segments (0.8–1.0 cm) were incubated both in light and dark at 25°C.

Results and Discussion

Callus obtained from various explants were used for regeneration assessments on MS medium supplemented with 2, 4-D, KN or BA, either alone or in combination. For each treatment, 25 vessels were inoculated with desired explants, *i.e.* axillary leaves, axillary shoots, hypocotyls and root segments for callus induction and incubated under optimal conditions as defined above. The experiment was terminated after an interval of 30 days.

In another set of experiments where the shoot regeneration capacity was scored, the shoot obtained from each nodal explant was subculture on MS medium supplemented with various concentration of, KIN or BA, IAA, IBA either alone or in combination. For each treatment, at least 25 vessels were inoculated and incubated under optimal conditions of temperature, light and photoperiod as defined above. After an interval of 30 days, the experiment was terminated, and the data with respect to cultures producing shoots, number of shoots in each such culture, height of shoots, and number of shoots were recorded.

Formation of Callus

Various explant were inoculated on MS medium contains 3 per cent sucrose, 5 gm/L agar solidified agent with supplemented different concentration of growth regulators as shown Table 45.1. High frequencies of callus induction were found in 1.0 mg/L, 1.2 mg/L2, 4-D and 2.0 mg/L BAP either alone or in combination after the 30 days, this callus was creamish in colour and friable were observed.

Effect of Growth Regulator on Shoot Multiplication

Shoot regeneration for using axillary shoots, hypocotyls and nodal segment. These explant were inoculated on MS medium containing 3 per cent sucrose,

Table 45.1: Effect of different Growth Hormone on Callus Initiation Using Axillary Leaves and Apical Shoot Explant

Source of Explant	*Concentration of Growth Regulator (mg/l)*				*Frequency of Callus Formation*	*Number of Shoot/ Callus*
	BAP	*2,4-D*	*IAA*	*IBA*		
Axillary leaves	1.0	0.2	–	–	++	–
	1.5	0.4	–	–	+++	–
	2.0	0.6	–	–	++++	–
	1.0	–	0.1	–	+	–
	1.5	–	0.2	–	+++	–
	2.0	–	0.3	–	++	–
Apical shoot	1.0	–	–	0.1	+	3
	1.5	–	–	0.2	++	2
	2.0	–	–	0.3	+++	2
	–	0.2	–	–	++	–
	–	0.4	–	–	++	–
	–	0.6	–	–	++++	–

5 gm/L solidified agent agar with supplemented various concentration of growth regulator as shown Table 45.2 with 1.0 mg/L, 1.2 mg/L, 1.4 mg/L, 1.6 mg/L, 1.8 mg/L, 2.0 mg/L BAP and KIN alone with 0.1 mg/L, 0.2 mg/L IAA, IBA or either. The maximum number of multiple shoots were found in 1.2 mg/L BAP with combination 0.1 mg/L IBA, whereas maximum average shoot length were recorded in 1.0 mg/L BAP without combination of any auxins. Similar kind of results were recorded in *Withania somnifera* (Sen and Sharma, 1991).

Table 45.2: Effect of different Growth Hormone on Shoot Multiplication using Nodal and Apical Shoot Explant

Source of Explant	*Concentration of Growth Hormones (mg/l)*		*Per cent of Shoot Formation*	*Shoot Length (Mean± SE) (cm)*
	BAP	*IAA*		
Nodal	1.0	0.2	60	4.12±0.115
	1.2	0.2	85	3.64±0.256
	1.4	0.2	80	2.8±0.300
Apical shoot	1.0	0.2	65	6.64±0.129
	1.2	0.2	85	4.1±0.113
	1.4	0.2	75	4.52±0.278

* After 30 days mean ± SE of 5 replicate.

Shoots and Root Formation

Shoots were initiated, from nodal explant, only on medium supplemented with BAP (1.0 mg/l) and KN (0.5 mg/l). Callus obtained from the axillary shoot base showed multiple shoot formation with BA (.2 mg/l) alone. Shoot buds were initiated from hypocotyl-derived callus on medium supplemented with BAP (1.2 mg/l) and IBA (0.2 mg/l). Shoots were excised and transferred to medium supplemented with BAP (1.0 mg/l) alone to foster multiplication growth of culture.

To increase rooting, shoots derived from culture or axillary shoot base callus were excised and cultured on MS medium with different concentrations of auxins. After 12–15 days post-transfer to rooting medium, roots appeared and, by 30days, many were found to be 4.5–5.5 cm long. Regenerated shoots obtained from the axillary shoot base callus were rooted best on MS medium containing IBA (0.5 mg/L and IAA(0.5 mg/l). These are the embryonic shoots derived from callus. Such kind of results was recorded in Phalaenopsis (Tokuhare and Mii, 2001). As well as sucrose, light regime, and temperature, can play critical roles in favoring bud development and early embryogenesis in vitro plant Propagation of Higher Plants (Zimmerman, 1993), These Plantlets were transferred to pots containing a sand/soil mixture (1:1)

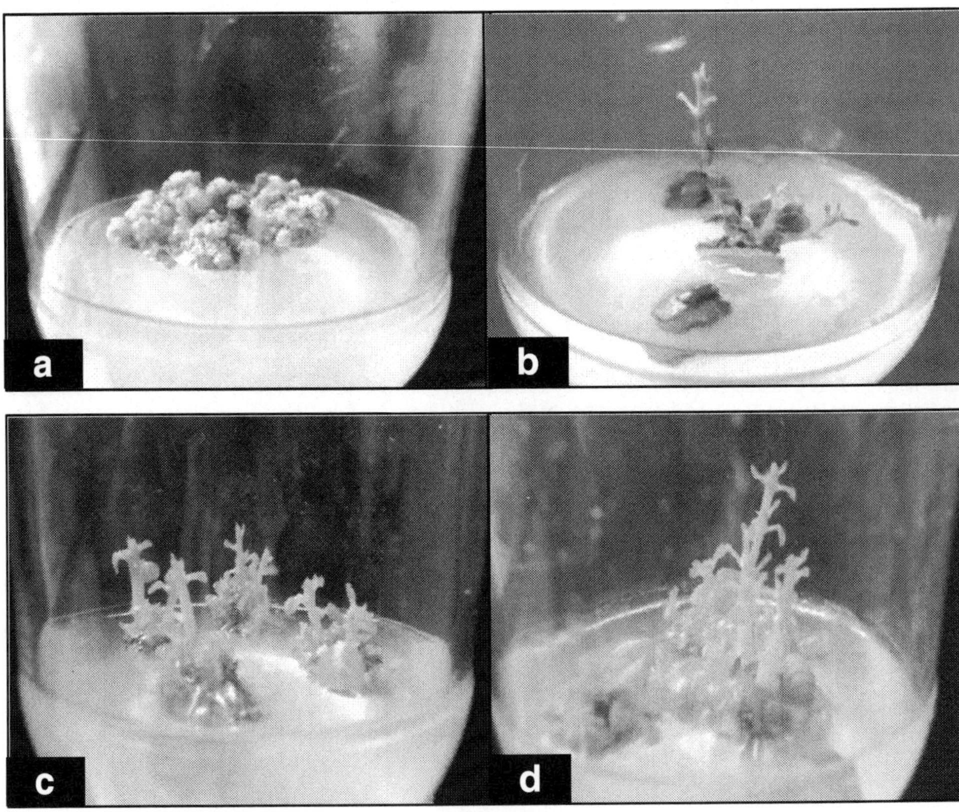

Plate 45.1: (a) Showing crimish colour calus; (b) Shoot regenaration from nodal explant; (c) Multiple; d) Shoot formation using apical shoot tip explant.

initially covered with beakers, and kept at 25C (Plate 45.1) and transferred (25– 30 days) to the glasshouse. The survival rates were as high as 65 per cent.

The comparison of callusing and shoot regeneration potential of different explants showed that nodal explant is the best one, as its callusing capacity was 83 per cent and the number of shoot average was 8 per callus. Moreover, the *Hemidesmus indicus* height of shoots within 30 days was approximately twice that of hypocotyls and 6-fold that of axillary leaves (Patnaik and Debata, 1996).Though root segment was best in callusing potential, no shoots could be regenerated from this source of callus even by using different combinations of growth regulators. Similarly, callus initiation occurred most readily in root segments of *Capsicum frutescens*, but shoot regeneration capacity from root segments was found to be low (Subhash and Christopher, 1988).

As we know that *Uraria picta* contains is oflavanones it is isolated from root bark of this plant has anti-microbial activity. Flavonoids are known to exhibit a range of biological activities including Anti-inflammatory, Anti-thrombotic, anti- viral, hepatoprotective and vasoactive properties. Thus the major bioactive flavonoid rhoifolin has been considered as marker compound for qualitative and quantitative standardization of *Uraria picta*. Therefore such studies are important for conservation of this valuable medicinal plant and evaluation standard in vitro protocol of *Uraria picta* for human welfare.

Acknowledgement

Authors are grateful to Professor and Head, Department of Botany, Dr. Babasaheb Ambedkar Marathwada University, Aurangabad (M.S.), India for providing all necessary facilities and encouragement for the present research work.

References

Akhilesh, KY, Deepti Y, Karuna S, Ram KV, Ajit KS, Madan MG, Flavone glycoside Based Validated RP-LC Method for Quality Evaluation of Prishniparni (*Uraria picta*), *Chromatographia*, **69**: 653-658.

Murashige, T and Skoog F (1962). A revised medium for rapid growth and bioassays with tobacco tissue cultures. *Physiol. Plant.*, **15**: 473–497.

Sen, J and Sharma AK (1991). Micropropagation of *Withania somnifera* from germinating seeds and shoot tips. *Plant Cell Tiss. Org. Cult.* **26**: 71–73.

Subhash, K and Christopher T (1988). Direct plantlet formation in cotyledon cultures of *Capsicum frutescens*. *Curr. Sci.* **57**: 99–100.

Tokuhara, K.; Mii, M. (2001). Induction of embryogenic callus and cell suspension culture from shoot tips excised from flower stalk buds of Phalaenopsis (Orchidaceae). In Vitro Cell. *Dev. Biol. Plant*, **37**: 457-461.

Patnaik, J. and Debata, B.K. (1996). Micropropagation of *Hemidesmus indicus* (L) R. Br. through axillary bud culture. *Plant Cell Rep.*, **15**: 427-430.

Zimmerman, J. L. (1993). Somatic embryogenesis: a model for early development in higher plants. *Plant Cell*, **5**: 1411-1423.

2015, Major Constrains and Verdict of Crop Productivity *Pages 564–595*
Editor: Dr. U.N. Bhale
Published by: ASTRAL INTERNATIONAL (P) LIMITED, NEW DELHI

Chapter 46

Application of Biotechnological Tools in Production of Plant Secondary Metabolites

Srikanth Kagithoju[1], Vikram Godishala[1], Madusudhan Kairamkonda[1], Ugandhar Thirunahari*[2], Anitha Uppu[3] and Rama Swamy Nanna[1]

[1]Department of Biotechnology,
Kakatiya University, Warangal – 506 009, A.P., India
[2]Department of Botany,
S.R.R. Govt. Arts and Science College, Karimnagar, A.P., India
[3]Department of Botany,
Govt. Degree College for Women, Karimnagar, A.P., India

ABSTRACT

Plants represent an immense biodiversity and serves as the chief source of secondary metabolites (SMs) or allelochemicals which are important pharmaceutically and industrially. Due to over and unscientific exploitation, natural disasters and lack of suitable cultivation strategies, such valuable medicinal plant species have become vulnerable and endangered and hence there is a great need to conservation and multiplication. Biotechnological tools like plant tissue culture techniques, genetic and metabolic engineering, biotransformation, combinatorial biosynthesis, use of elicitors, immobilization, cryopreservation etc are opened new avenues and useful in this regard. Efforts are ongoing on production of important SMs through these techniques. In vitro production of secondary

* *Corresponding Author.* E-mail: t.ugandharbiotech@gmail.com

metabolites in plant cell suspension cultures has been reported from various medicinal plants. Bioreactors are the key step towards commercial production of secondary metabolites by plant biotechnology. Biotransformation, combinatorial biosynthesis, immobilized plant cell cultures and genetic transformation may be a powerful tool for enhancing the productivity of novel secondary metabolites, especially by Agrobacterium rhizogenes induced hairy roots. The present review highlights the nature, applications and perspective for the production of valuable secondary metabolites in vitro.

Keywords: *Secondary metabolites, In vitro culture, Genetic engineering, Metabolic engineering, Biotransformation, Combinatorial biosynthesis, Elicitors, Immobilization.*

Introduction

Plants represent an immense biodiversity with more than 2, 50,000 known species existing within the plant kingdom. These plants interact with their environment by producing huge diverse array of organic compounds called secondary metabolites (SMs) (Harborne, 1996). These are the most functionally and structurally diverse molecules in nature, of which only a small fraction has been isolated and characterized. These are also known as allelochemicals. It was estimated that only 1, 00, 000 plant secondary metabolites or natural products (20-30 per cent) have been identified (Table 46.1). These compounds include alkaloids, flavonoids, glycopeptides, volatile oils, phenolics, peptides, terpinoids, steroids etc (Wink, 1999; Verpoorte, 1999). For the vast majority of plant secondary metabolites or natural products, their function for the plants as well as their role in human health care or human welfare is

Table 46.1: Number of known Plant Secondary Metabolites from different Species (Wink, 1999 and Verpoorte, 1999)

Type of Secondary Metabolite	*Number*
Nitrogen-containing	
Alkaloids	15765
Nonprotein amino acids (NPAAs)	600
Amines	100
Cyanogenic glycosides	100
Glucosinolates	100
Without nitrogen	
Sesquiterpenes	8650
Monoterpenes	1946
Diterpenes	7834
Triterpenes, steroids, saponins	4952
Tetraterpenes	352
Flavonoids	8128
Polyacetylenes	1000
Polyketides	2442
Phenylpropanoids	500

still to be elucidated. These compounds are concerned with some important traits of plants like color or fragrance of flowers (helps in attracting pollinators), taste and color of stored food and plants, resistance to disease and pests, detoxification of substances accumulated in primary metabolism, acts as physiological elicitors, gives provision for chemical signals to coordinate the metabolism of organism, coordination activities of different individuals of the same species and development of ecological relationships (Luckner, 1986) which plays important and major role in the survival of the plant in its own environment. These molecules play critical role in a wide variety of processes and fields like pharmaceuticals (as drugs or their precursors), agrochemicals (as fungicides, bactericides, insecticides), nutrition (flavoring agents, coloring foods), cosmetics (fragrances, oils) and industrial products (alkaloids, phenols, paper, resins, gums, terpenoids, latex, oils, dyes, pigments, plastics) etc and for this reason plants are considered as living factories for biosynthesis of a wide array of important molecules (Figure 46.1) which help in producing life saving drugs. Plant secondary metabolites (PSMs) can boost the immune system, protect the body from free radicals, kills pathogenic organisms and much more.

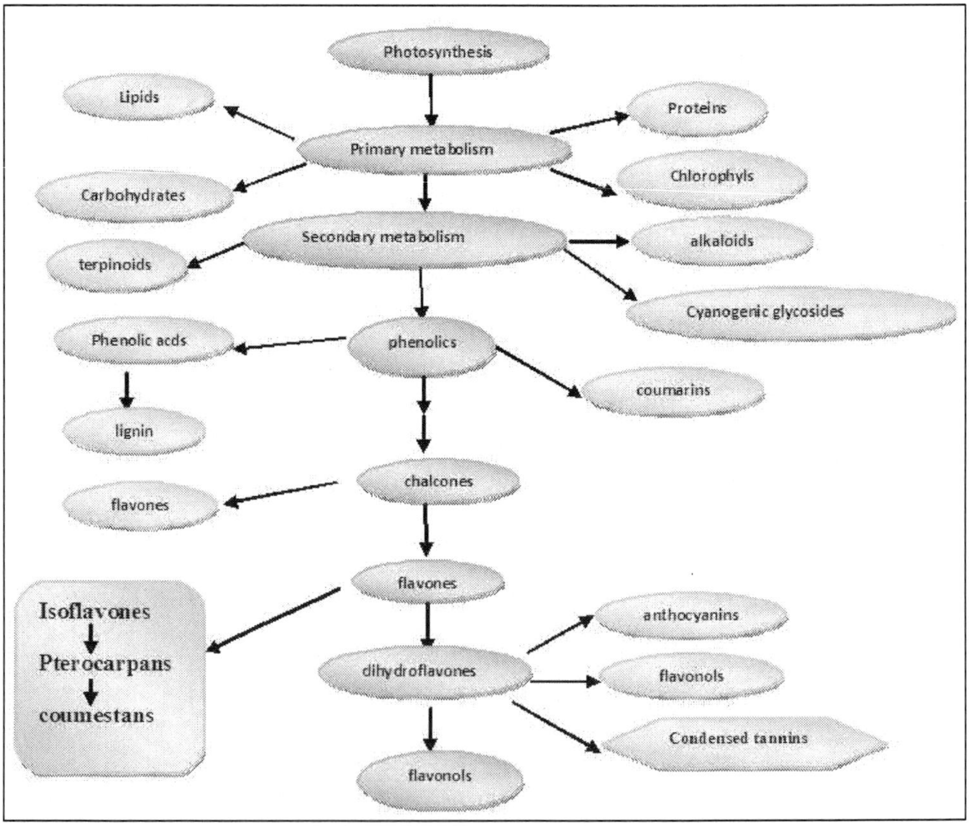

Figure 46.1: Synthesis of Various Secondary Metabolites in Plant Systems.

Materials and Methods

The production of PSMs is due to the expression of cell specialization which either triggered by the process of cell differentiation (Luckner and Nover, 1977, Yeoman *et al.*, 1982) or represents an aspect of the process of plant development. PSMs are such products that occur only in special, differentiated cells and are not necessary for the cells themselves but may be useful for the plant as a whole. Cell differentiation itself is a basic component of metabolic regulation of an organism. which is under the control of basic genetic composition, the expression of which is in turn under the control of environmental factors like biotic and abiotic stimuli, carbon - nutrition balance (Kliebenstein, 2004; Laitinen *et al.*, 2005; Lerdau, 2002; Mary Ann Lila, 2006).

Thus, the biosynthesis of secondary metabolites is usually restricted to a particular developmental stages and/or specialized cells and their production can be controlled by providing specialized or controlled conditions or techniques.

Occurrence of valuable metabolites in plants has stimulated immense interest on the part of industry in the field of pharmaceuticals, agrochemicals, cosmetics, nutrition and industrial products. For this, secondary metabolites are collected from plants growing in wild or from field cultivated sources. These secondary metabolites (SMs), with some exception, occur in amounts less than 0.01 per cent of the dry weight of whole plant material. These compounds may be unstable and present as part of a complex mixture. The identification, separation, isolation and purification of these PSMs require considerable skill. But, the availability of PSMs depends on number of factors which may be either biotic or abiotic like seasonal period of growth, age or maturity, climatic conditions, disease, insect predation and high costs for labor and machinery during harvesting and even political situation.

On the other hand, total chemical synthesis of several compounds is also possible, but economically not feasible and is not eco-friendly. Moreover, in most cases, for SMs isolation whole plants are destroyed, as different SMs are distributed throughout the plant or in parts of plants (bark, leaves, flowers, roots, fruits, seeds etc.) which leads to the loss of plant and loss of availability in near future (Gordon and David, 2001).

Therefore, when plant material is not available throughout the year and chemical synthesis is not possible, particularly for large complex molecules, an alternative, economically feasible and ecofriendly production source for desired SMs is of great interest. Hence, for conservation, multiplication and genetic enhancement of the SMs yielding plants, Biotechnological tools are important which include cell, tissue (*i.e.* dedifferentiated cultures), organ cultures(*i.e.* differentiated cultures); genetic transformation methods, biotransformation, use of elicitors, metabolic engineering, adopting cocultivation methods, immobilization, combinatorial biosynthesis, cryopreservation etc (Figure 46.2). Principle involved in all these methods more or less is:

1. By increasing enzymatic activities which are limiting in normal conditions.
2. Inducing expression of regulatory genes.
3. Increasing total carbon flux through metabolic pathways.

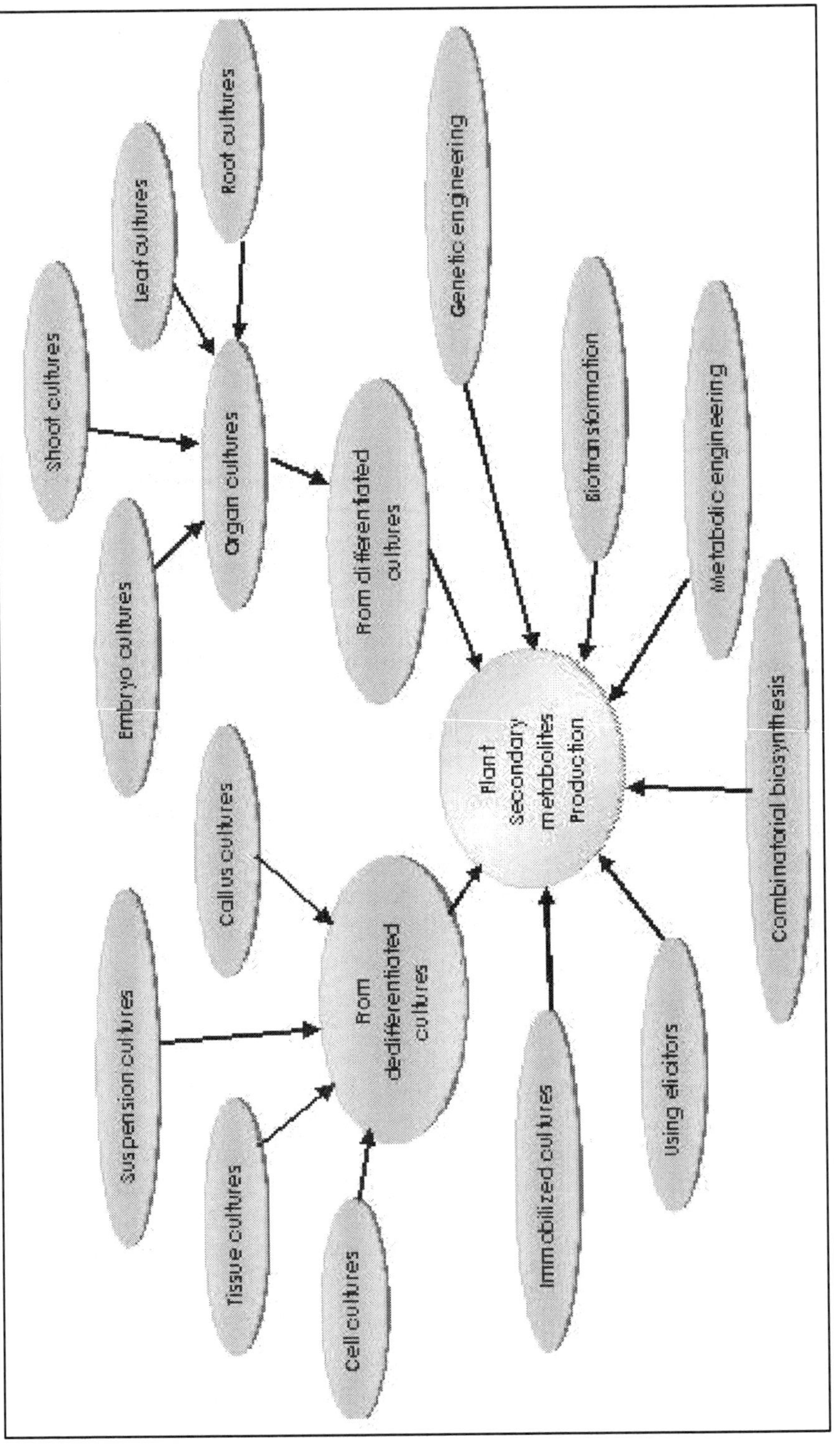

Figure 46.2: Possible Ways for Production of PSMs through Biotechnology.

4. Blocking the competitive pathways.
5. Blocking the catabolism.

The first three cases require the expression of genes yielding active enzymes, the latter two cases require adopting of new biotechnological methods like antisense technology. In all cases the respective biosynthetic pathway has to be known on the level of substrate, product, enzymes and genes, as well as the regulation on all these levels, including aspects as compartmentalization and transport.

Advance in plant biotechnology, particularly in methods for plant cell, tissue and organ cultures; genetic engineering, metabolic engineering, biotransformation, germplasm conservation etc provides new means for the commercial processing of even rare and endangered plants and chemicals given by them. These new technologies will extend and enhance the usefulness of plants as renewable resources of valuable compounds.

SMs Production through Plant Tissue Culture Techniques

Medicinal plants are the chief source of most of the life saving drugs. It is estimated that nearly 70-80 per cent of the world population is depending on the traditional medicines which are derived from plant based compounds. But the availability of these compounds depends on number of biotic and abiotic factors; more over it is difficult to increase the area under plantation for a particular veriety to meet the ever increasing demand from the natural resources. To overcome all these hurdles, an alternative method of assured supply of uniform material throughout the year is required and is fulfilled by plant tissue culture (Ramachandra Rao and Ravishanker, 2002).

Plant tissue culture refers to *in vitro* culture of isolated cells, tissues, or organs under aseptic and appropriate environmental conditions on defined semi solid or in liquid media. The ability of plant cell, tissue and organ cultures to produce and accumulate many of the same valuable chemical compounds as the parent in nature has been recognized almost since the inception of *in vitro* technology. The strong and growing demand for natural, renewable products has refocused attention on *in vitro* plant material as potential factories for SMs, and has paved the way for new research exploring SMs expression *in vitro*. Discoveries of plant tissue cultures capable of producing specific SMs compounds at a rate similar or superior to that of intact plans have accelerated in the last few years (Table 46.2).

Table 46.2: Production of PSMs through Callus Cultures

Plant Name	*Active Ingredient*	*Reference*
Agave amaniensis	Saponins	Andrijany *et al.*, 1999
Allium sativum L.	Alliin	Malpathak and David, 1986
Ambrosia tenuifolia	Altamisine	Goleniowski and Trippi, 1999
Bupleurum falcatum	Saikosaponins	Wang and Huang, 1982
Canavalia ensiformis	L-Canavanine	Ramirez *et al.*, 1992

Contd...

Table 46.2–*Contd...*

Plant Name	*Active Ingredient*	*Reference*
Chrysanthemum cinerariaefolium	Pyrethrins	Rajasekaran *et al.*, 1991
Chrysanthemum cinerariaefolium Citrus sp.	Naringin, Limonin	Barthe *et al.*, 1987
Coffea arabica L.	Caffeine	Waller *et al.*, 1983
Corydalis ophiocarpa	Isoquinoline alkaloids	Iwasa and Takao, 1982
Croton sublyratus Kurz	Plaunotol	Morimoto and Murai., 1989
Cryptolepis buchanani Roem and Schult	Cryptosin	Venkateswara *et al.*, 1987
Duboisia leichhardtii	Tropane alkaloids	Yamada and Endo, 1984
Eriobotrya japonica	Triterpenes	Taniguchi *et al.*, 2002
Eucalyptus tereticornis SM.	Sterols and Phenolic compounds	Venkateswara *et al.*, 1986
Gentiana sp.	Secoiridoid glucosides	Skrzypczak *et al.*, 1993
Glycyrrhiza echinata	Flavanoids	Ayabe *et al.*, 1986
Glycyrrhiza glabra var. *Glandulifera*	Triterpenes	Ayabe *et al.*, 1990
Hyoscyamus niger	Tropane alkaloids	Yamada and Hashimoto, 1982
Mucuna pruriens	L-DOPA	Brain, 1976
Nandina domestica	Alkaloids	Ikuta and Itokawa., 1988
Nicotiana rustica	Alkaloids	Tabata and Hiraoka, 9-1976
Nothapodytes foetida	Camptothecin	Thengane *et al.*, 2003
Ophiorrhiza pumila	Camptothecin related alkaloids	Kitajima *et al.*, 1998
Panax ginseng	Saponins & Sapogenins	Furuya *et al.*, 1973
Papaver bracteatum	Thebaine	Day *et al.*, 1986
Papaver somniferum L.	Alkaloids	Furuya *et al.*, 1972
Polygala amarella	Saponins	Desbene *et al.*, 1999
Portulaca grandiflora	Betacyanin	Schroder and Bohm, 1984
Ptelea trifoliata L.	Dihydrofuro [2,3-b] quinolinium alkaloids	Petit-Paly *et al.*, 1987
Rauvolfia serpentina x *Rhazya stricta* Hybrid plant	3-Oxo-rhazinilam	Gerasimenko *et al.*, 2001
Ruta sp.	Acridone and Furoquinoline alkaloids and coumarins	Baumert *et al.*, 1992
Salvia miltiorrhiza	Lithospermic acid B and Rosmarinic acid	Morimoto *et al.*, 1994
Scopolia parviflora	Alkaloids	Tabata *et al.*, 1972
Scutellaria columnae	Phenolics	Stojakowska and Kisiel., 1999
Tecoma sambucifolium	Phenylpropanoid glycosides	Pletsch *et al.*, 1993
Withaina somnifera	Withaferin A	Ray and Jha, 2001*

* Callus and shoots.

In plant tissue culture, more often a portion of plant called as an explant is used to initiate their growth in culture. The non-dividing, differentiated, quiescent cells of the explant when grown *in vitro* first undergo changes to achieve the meristematic state to form undifferentiated mass of tissue called callus and the phenomenon is called *dedifferentiation*. If cells or tissues formed from such phenomenon are cultured *in vitro* then it is called *dedifferentiated* cultures (cell cultures, suspension cultures, callus cultures). On other hand, if completely organized structures (like roots, leafs, shoots, embryos etc) or cells from them were *in vitro* cultured then it is called *differentiated* cultures. Both, these culture techniques can be used to produce SMs effectively but the SMs production is oftenly restricted to specialized or differentiated cells and hence, it is better to adopt differentiated cell culture techniques. For this, one should have knowledge about optimization of growth and media composition, proper cell line selection, metabolic pathway and its affecters, and stage or age of cultures. Once all these have been came to be known, various techniques of Biotechnology like biotransformation, genetic engineering, metabolic engineering, combinatorial biosynthesis, use of elicitors, immobilization etc can be adopted to enhance the SMs production.

There is a series of distinct advantages to produce valuable SMs by using *in vitro* culture techniques rather than *in vivo* in the whole plant which includes:

1. Production is more reliable, simpler and more predictable.
2. Independence from various environmental factors like climate, pests, disease, geographical and seasonal constraints etc.
3. Useful compounds could be produced under controlled conditions and geared more accurately to market demands.
4. A consistent product quality could be assured with the use of specific cell lines.
5. Isolation of the compound can be rapid and efficient, as compared to extraction from complex whole plants.
6. Production of SMs in chemically controlled environment facilitates later processing and product recovery steps.
7. Compounds produced *in vitro* can directly parallel compounds in the whole plant.
8. Interfering compounds that occur in the *in vivo* can be avoided.
9. This method can yield a source of defined standard SMs in large volumes.
10. Tissue and cell cultures are a potential model to test elicitation.
11. Cell cultures can be labeled, such that the accumulated SMs, when provided as feed to laboratory animals, can be traced metabolically.

SMs from Cell, Callus and Suspension Cultures or *dedifferentiated* Cultures

Plants are the traditional source of many pharmaceutical chemicals which are produced by secondary metabolism and hence the production of SMs *in vitro* can be possible through plant cell cultures (Barz and Ellis, 1981; Deus and Zenk, 1982). Successful establishment of plant cell lines capable of producing high yields of SMs in cell suspension cultures has been reported (Zenk, 1978). The accumulation of SMs in plant cell cultures depends on number of factors like medium composition, phytohormones, pH, temperature, aeration, agitation, light, stage or age of culture, genotype and environmental conditions (Fett-Neto *et al.*, 1995; Wang *et al.*, 1999; Goleniowski and Trippi, 1999; Lee and Shuler, 2000). For plant cell culture techniques to become economically viable, it is important to develop methods that would allow for consistent generation of high yields of products from cultured cells (Berlin and Sasse, 1985). Careful selection of productive cells and cultural conditions resulted in accumulation of several products in cultured cells. In order to obtain yields in higher concentrations for commercial exploitation, different strategies for improving SMs production in suspension cultures using different media for different species have been reported (Robins, 1994) and efforts have been focused on the stimulation of biosynthetic activities of cultured cells using various methods (Buitelaar and Tramper, 1992; Ravishanker and Venkataraman 1993; Dixon, 1999; Ramachandra Rao, 2000). Manipulation of physical aspects and nutritional elements in a culture is perhaps the most fundamental approach for optimization of culture productivity. The production of different SMs in plant cell suspension cultures has been reported from various medicinal plants (Table 46.3).

Table 46.3: Production of PSMs through Cell Suspension Cultures

Plant Name	*Active Ingredient*	*Reference*
Ailanthus altissima	Alkaloids	Anderson *et al.*, 1987
Ailanthus altissima	Canthinone alkaloids	Anderson *et al.*, 1986
Aloe saponaria	Tetrahydroanthracene glucosides	Yagi *et al.*, 1983
Anchusa officinalis	Rosmarinic acid	De-Eknamkul and Ellis, 1985
Brucea javanica (L.) Merr	Canthinone alkaloids	Liu *et al.*, 1990
Camellia Sinensis	Theamine, γ-glutamyl derivatives	Orihara and Furuya., 1990
Capsicum annuum L.	Capsaicin	Johnson *et al.*, 1990
Cassia acutifolia	Anthraquinones	Nazif *et al.*, 2000
Catharanthus roseus	Indole alkaloids	Moreno *et al.*, 1993
Catharanthus roseus	Catharanthine	Zhao *et al.*, 2001
Choisya ternata	Furoquinoline alkaloids	Sejourne *et al.*, 1981
Chrysanthimum cinerariaefolium	Chrysanthemic acid and pyrethrins	Kueh *et al.*, 1985
Cinchona L.	Alkaloids	Koblitz *et al.*, 1983
Cinchona robusta	Robustaquinones	Schripsema *et al.*, 1999
Cinchona sp.	Anthraquinones	Wijnsma *et al.*, 1985
Cinchona succirubra	Anthraquinones	Khouri *et al.*, 1986

Contd...

Table 46.3–*Contd...*

Plant Name	*Active Ingredient*	*Reference*
Cornus kousa	Polyphenols	Ishimaru *et al.*, 1993
Cruciata glabra	Anthraquinones	Dornenburg and Knorr., 1996
Cryptolepis buchanani Roem. and Schult *Digitalis purpurea* L.	Cardenolides	Hagimori *et al.*, 1982
Dioscorea deltoidea	Diosgenin	Heble and Staba., 1980
Dioscorea doryophora Hance	Diosgenin	Huang *et al.*, 1993
Ephedra spp.	L-Ephedrine-D-pseudoephedrine	O'Dowd *et al.*, 1993
Eucommia ulmoides	Chlorogenic acid	Wang *et al.*, 2003
Fumaria capreolata	Isoquinoline alkaloids	Tanahashi and Zenk., 1985
Ginkgo biloba	Ginkgolide A	Carrier *et al.*, 1991
Glehnia littoralis	Furanocoumarin	Kitamura *et al.*, 1998
Isoplexis isabellina	Anthraquinones	Arrebola *et al.*, 1999
Linum flavum L.	5-Methoxypodophyllotoxin	Uden *et al.*, 1990
Lithospermum erythrorhizon	Shikonin derivatives	Fujita *et al.*, 1981
Lithospermum erythrorhizon	Shikonin derivatives	Fukui *et al.*, 1990
Lycium chinense	Cerebroside	Jang *et al.*, 1998
Morinda citrifolia	Anthraquinones	Zenk *et al.*, 1975
Morinda citrifolia	Anthraquinones	Bassetti *et al.*, 1995
Mucuna pruriens	L-DOPA	Wichers *et al.*, 1993
Nicotiana tabacum L.	Nicotine	Mantell *et al.*, 1983
Panax notoginseng	Ginsenosides	Zhong and Zhu, 1995
Papaver somniferum	Morphine, Codeine	Siah and Doran, 1991
Peganum harmala L.	β-Carboline alkaloids	Sasse *et al.*, 1982
Phytolacca americana	Betacyanin	Sakuta *et al.*, 1987
Picrasma quassioides	Bennett Quassin	Scragg and Allan, 1986
Podophyllum hexandrum royle	Podophyllotoxin	Uden *et al.*, 1989; Chattopadhyay *et al.*, 2002
Polygonum hydropiper	Flavanoids	Nakao *et al.*, 1999
Rauwolfia sellowii	Alkaloids	Rech *et al.*, 1998
Rauwolfia serpentina Benth.	Reserpine	Yamamoto and Yamada, 1986
Salvia fruticosa	Rosmarinic acid	Karam *et al.*, 2003*
Salvia miltiorrhiza	Cryptotanshinone	Miyasaka *et al.*, 1989
Sapium sebiferum	Tannin	Neera and Ishimaru., 1992*
Solanum chrysotrichum (Schldl.)	Spirostanol saponin	Villarreal *et al.*, 1997
Solanum laciniatum Ait	Solasodine	Chandler and Dodds., 1983
Solanum paludosum	Solamargine	Badaoui *et al.*, 1996

Contd...

Table 46.3–*Contd...*

Plant Name	*Active Ingredient*	*Reference*
Stizolobium hassjoo	L-DOPA	Huang *et al.*, 2002
Tabernaemontana divaricata	Alkaloids	Sierra *et al.*, 1992
Taxus spp.	Taxol	Wu *et al.*, 2001
Taxus baccata	Taxol, baccatin III	Cusido *et al.*, 1999
Taxus cuspidata	Taxoids	Ketchum *et al.*, 2003
Thalictrum minus	Berberin	Kobayashi *et al.*, 1987
Thalictrum minus	Berberin	Nakagawa *et al.*, 1986
Torreya nucifera var. *radicans*	Diterpenoids	Orihara *et al.*, 2002
Trigonella foenumgraecum	Saponins	Brain and Williams, 1983

* Callus and suspension

SMs from Organ Cultures or differentiated Cultures

Secondary metabolites are products of differentiation and are produced in special or differentiated cells only. Formation of such type of cells or compounds is not possible in the case of cell suspension cultures. For that reason quite some research has been done on the *in vitro* culture of differentiated cells or organs. Roots, shoots and embryos have been thus cultured *in vitro* for the production and isolation of PSMs (Table 46.4).

Table 46.4: Production of PSMs through Organ Culture

Plant Name	*Active Ingredient*	*Culture Type*	*Reference*
Aconitum heterophyllum	Aconites	Hairy root	Giri *et al.*, 1997
Adhatoda vasica	Vasine	Shoot culture	Shalaka and Sandhya., 2009
Agastache rugosa	Rosmarinic acid	Hairy root	Lee *et al.*, 2007
Ajuga reptans	Phytoecdysteroids	Hairy root	Matsumoto and Tanaka, 1991
Ammi majus	Umbelliferone	Shootlet	Krolicka *et al.*, 2006
Amsonia elliptica	Indole alkaloids	Hairy root	Sauerwein *et al.*, 1991
Angelica gigas	Deoursin	Hairy root	Xu *et al.*, 2008
Anisodus luridus	Tropane alkaloids	Hairy root	Jobanovic *et al.*, 1991
Arachys hypogaea	Resveratol	Hairy root	Kim *et al.*, 2008
Armoracia laphthifolia	Fisicoccin	Hairy root	Babakov *et al.*, 1995
Artemisia absinthum	Essential oil	Hairy root	Nin *et al.*, 1997
Artemisia annua	Artemisinin	Kinetin Hairy root	Rao *et al.*, 1998
Astragalus mongholicus	Cycloartane saponin	Hairy root	Ionkova *et al.*, 1997
Astragalus mongholicus	Cycloartane	Hairy root	Ionkova *et al.*, 1997
Beeta vulgaris	Betalain pigments	Hairy root	Taya *et al.*, 1992

Contd...

Table 46.4–*Contd...*

Plant Name	*Active Ingredient*	*Culture Type*	*Reference*
Brugmansia candida	Tropane	Hairy root	Marconi *et al.*, 2008
Brugmansia candida	Tropane alkaloid	Hairy root	Giulietti *et al.*, 1993
Bupleurum falcatum	Saikosaponins	Root	Kusakari *et al.*, 2000
Calystegia sepium	Cuscohygrine	Hairy root	Jung and Tepfer. 1987
Campanula medium	Polyacetylenes	Hairy root	Tada *et al.*, 1996
Cassia obtusifolia	Anthraquinone	Hairy root	Ko *et al.*, 1995
Centella asiatica	Asiaticoside	Hairy root	Kim *et al.*, 2007
Centella asitica	Asiaticoside	Hairy root	Paek *et al.*, 1996
Centranthes ruber	Valepotriates	Hairy root	Granicher *et al.*, 1995
Cephaelis ipecacuanha	Alkaloids	Root	Teshima *et al.*, 1988
Chaenatis douglasei	Thiarbrins	Hairy root	Constabel and Towers, 1988
Cinchona ledgeriana	Quinine	Hairy root	Hamill *et al.*, 1989
Coleus forskohlii	Forskolin	Hairy root	Sasaki *et al.*, 1998
Corydalis ambigua	Corydaline	Embryo	Hiraoka *et al.*, 2004
Cymbopogon citratus	Essential oil	Shoot	Quiala *et al.*, 2006
Datura stramonium	Hyocyamine	Hairy root	Hilton and Rhodes, 1993
Digitalis purpurea	Cardioactive glycosides	Hairy root	Saito *et al.*, 1990
Drosera rotundifolia	7-Methyljuglone	Shoot	Hohtola *et al.*, 2005
Duboisia leichhardtti	Scopalamine	Hairy root	Muranaka *et al.*, 1992
Fagopyrum esculentum	Flavono	Hairy root	Trotin *et al.*, 1993
Fagopyrum esculentum	Rutin	Hairy root	Lee *et al.*, 2007
Fritillaria unibracteata	Alkaloids	Multiple shoot	Gao *et al.*, 2004
Gentiana macrophylla	Glucoside	Hairy root	Tiwari *et al.*, 2007
Gentianella austriaca	Xanthone	Multiple shoot	Vinterhalter *et al.*, 2008
Geranium thunbergii	Tannin	Hairy root	Ishimaru and Shimomura., 1991
Glycyrrhiza glabra	Glycyrrhizin	Hairy root	Mehrotra *et al.*, 2008
Glycyrrhiza glabra	Flavonoid	Hairy root	Asada *et al.*, 1998
Gynostemma pentaphyllum	Saponin	Hairy root	Fei *et al.*, 1993
Gypsophila paniculata	Saponin	Root suspension	Fulcheri *et al.*, 1998
Hemidesmus indicus	Lupeol, Rutin	Shoot culture	Misra *et al.*, 2005
Hyocyamus niger	Tropane alkaloid	Hairy root	Jaziri *et al.*, 1988
Hyoscyamus albus	Phytolexins	Hairy root	Kuroyanagi *et al.*, 1998
Hyoscyamus muticus	Hyoscyamine	Hairy root	Halperin and Flores, 1997
Hypericum perforatum	Hypericins	Multiple shoot	Kornfeld *et al.*, 2007
Hypericum perforatum	Hypericin	Multiple shoot	Santarem and Astarita., 2003

Contd...

Table 46.4–*Contd...*

Plant Name	*Active Ingredient*	*Culture Type*	*Reference*
Hypericum perforatum	Hyperforin	Multiple shoot	Karppinen *et al.*, 2007
Lactuca virosa	Sesquiterpene lactones	Hairy root	Kisiel *et al.*, 1995
Leontopodium alpinum	Essential oil	Hairy root	Hook, 1994
Linum flavum	Lignan	Hairy root	Oostdam *et al.*, 1993
Lithospermum erythrorhizon	Shikonin	Hairy root	Fukui *et al.*, 1998
Lobelia cardinalis	Polyacetylene glucosides	Hairy root	Yamanaka *et al.*, 1996
Lactuca virosa	Sesquiterpene lactones	Hairy root	Kisiel *et al.*, 1995
Leontopodium alpinum	Essential oil	Hairy root	Hook, 1994
Linum flavum	Lignan	Hairy root	Oostdam *et al.*, 1993
Lithospermum erythrorhizon	Shikonin	Hairy root	Fukui *et al.*, 1998
Lobelia cardinalis	Polyacetylene glucosides	Hairy root	Yamanaka *et al.*, 1996
Lactuca virosa	Sesquiterpene lactones	Hairy root	Kisiel *et al.*, 1995
Leontopodium alpinum	Essential oil	Hairy root	Hook, 1994
Linum flavum	Lignan	Hairy root	Oostdam *et al.*, 1993
Lithospermum erythrorhizon	Shikonin	Hairy root	Fukui *et al.*, 1998
Lobelia cardinalis	Polyacetylene glucosides	Hairy root	Yamanaka *et al.*, 1996
Lactuca virosa	Sesquiterpene lactones	Hairy root	Kisiel *et al.*, 1995
Myristica fragrans	Myristin	Shoot	Indira Iyer *et al.*, 2009
Nicotiana hesperis	Anatabine	Hairy root	Parr and Hamill., 1987
Ophiorrhiza rugosa var. *decumbens*	Camptothecin	Shoot	Vineesh *et al.*, 2007
Panax ginseng	Glycoside	Hairy root	Jeong and Park, 2007
Papaver somniferum	Codeine	Hairy root	Williams and Ellis, 1992
Perezia cuernavacana	Sesquterpene quinone	Hairy root	Arellano *et al.*, 1996
Pimpinella anisum	Essential oil	Hairy root	Santos *et al.*, 1998
Platycodongrandiflorum	Polyacetylene	Hairy root	Tada *et al.*, 1995
Plumbago zeylanica	Plumbagin	Hairy root	Verma *et al.*, 2002
Podophyllum hexandrum	Podophyllotaxin	Shoot	Li *et al.*, 2009
Psoralea cordifolia	Isoflavones	Multiple shoot	Shinde *et al.*, 2009

Contd...

Table 46.4–*Contd...*

Plant Name	*Active Ingredient*	*Culture Type*	*Reference*
Psoralea corylifolia	Isoflavones	Multiple shoot	Shinde *et al.*, 2009
Rhus javanica	Gallotannins	Root	Tanoguchi *et al.*, 2000
Rubia akane	Anthraquinone	Hairy root	Park and Lee, 2009
Rubia tinctorum	Anthraquinone	Hairy root	Sato *et al.*, 1991
Salvia officinalis	Flavonoid	Multiple shoot	Grzegorczyk and Wysoki-nska, 2008
Scutellaria baicalensis	Flavonoids	Hairy root	Zhou *et al.*, 1997
Serratula tinctoria	Ecdysteroid	Hairy root	Delbeque *et al.*, 1995
Sesamum indicum	Napthaquinone	Hairy root	Ogasawara *et al.*, 1993
Silybium marianum	Silymarin	Hairy root	Rahnama *et al.*, 2008
Silybium mariyanm	Flavonolignan	Root	Alikaridis *et al.*, 2000
Silybum marianum	Silymarin	Hairy root	Rahnama *et al.*, 2008
Solanum aculeatissi	Steroidal saponin	Hairy root	Ikenaga *et al.*, 1995
Swertia japonica	Amarogenetin	Hairy root	Ishimaru *et al.*, 1990
Tagetes patula	Thiophenes	Hairy root	Arroo *et al.*, 1995
Tanacetum parthenium	Sesquiterpene	Hairy root	Kisiel and Stojakowska, 1997
Torreya nucifera	Diterpenoids	Hairy root	Orihara *et al.*, 2002
Vinca major	Vincamine	Hairy root	Tanaka *et al.*, 2004
Withania somnifera	Withaferin	Shoot	Ray and Jha, 2001
Withania somnifera	Withaferin	Hairy root	Banerjee *et al.*, 1994
Withania somnifera	Withanoloid A	Hairy root	Murthy *et al.*, 2008

In vitro organ culture can be given different names depending upon the type of organ used as explant which can be broadly classified in to 3 types – organized organ culture (culture of whole plant organs like embryos, seeds, roots, shoots etc), Non-organized organ cultures (if cells and/or tissues are isolated from organized structures, dedifferentiated in to cells or callus and are then cultured), non-organized/ organized organ culture(cells isolated from organ or tissue are first dedifferentiated to form callus and then redifferentiated to form organs). Such organ cultures do produce similar pattern of SMs as in whole plant. Organ cultures for production of SMs were reported in different plants. For example, production of digoxin in leaf and root culture of *D. lanata*, production of sesquiterpene solavetivone from roots of *Hyoscyamus muticus*, tropaane alkaloids hyoscyamine and scopolamine are produced from root cultures (Verpoorte *et al.*, 1991; Oksman-Caldentey *et al.*, 2000), and essential oils have been produced in shoot cultures (Spencer *et al.*, 1993).The *in vitro* organ culture (shoots from small bulb cuttings) of *Fritillaria unibracteat* was established, chemical composition of the metabolites produced has been confirmed and it is proved that the content of alkaloids and beneficial microelements in the cultured bulbs were higher than found in the wild bulb(Gao *et al.*, 2004). *In vitro* shoot culture or multiplication of *Frangula alnus, F. rupestris* showed increased production of SM

Anthraquinones. Shoot cultures of *Gentianella austriaca* produces more xanthones (Vinterhalter *et al.*, 2008).

Results and Discussion

Improving Productivity

After identifying the proper plant sources and culture methods for SMs production next important measure is to improve its productivity. For this the most common approach is the screening, selection of high yielding cell lines and the optimization of the growth and production media which is a common sort for optimization of microbial production and which can be applicable for plants and oftenly is proved one. This approach was successful in vindoline production in *Catharanthus roseus* whose some production genes are located in chloroplast (De Luca and Cutler, 1987; St-Pierre *et al.*, 1999; Ramani and Jayabaskaran, 2008), berberine production from *Coptis japonica* cell cultures where productivity was increased to 7g/L (Sato *et al.*, 1982; Sato and Yemada, 1984) and for shikonin production upto 3g/L (Fujita, 1988). Even though improved productivities were achieved, some times, the high production levels were not stable and after several subcultures the productivity was much lowered (Ohta and Verporate, 1992), as it depends on various factors like genotype, failure in proper expression of biosynthetic path way due to repeated cell divisions which some times leads to somaclonal variations, some times biosynthesis of SM occurs in different cells just as in vindoline in *C. roseus* (St-Pierre *et al.*, 1999) etc. To overcome all such problems following new biotechnological trends were emerged.

Biotransformation/Addition of Precursor Molecules for SMs Production

Biotransformations are chemical reactions catalyzed by living cells, organs or enzymes produced by them. These enzymes catalyze the transformation of a readily available or inexpensive precursor into a more valuable final product (Figure 46.3). These reactions are different from biosynthesis where complex products are assembled from simple substrates by whole cells, organs or organisms. They are also different from biodegradations in which complex substrates are broken down to simple ones, but, in this only functional groups of organic compounds are modified by living cells. Addition of such appropriate precursors or related compounds to the culture media sometimes stimulates secondary metabolite production. This approach is advantageous if the precursors are inexpensive. This potential of living things (micro organisms) was exploited long long ago (*e.g.* acetification of wine into vinegar) in 6th millennium BC, which can also be achieved by plant cells. Chan *et al.* (1965) have initially examined the production of alkaloids with this approach, later many similar experiments have been carried out. For example, amino acids have been added to cell suspension culture media for production of tropane alkaloids, indole alkaloids, and ephedorin.

Phenylalanine is one of the biosynthetic precursors of rosmarinic acid. Addition of this amino acid to *Salvia officinalis* suspension cultures stimulated the production of rosmarinic acid and shortened the production time as well. It was reported that addition of 500 mM tropic acid to the medium of *Scopolia japonica* increased the

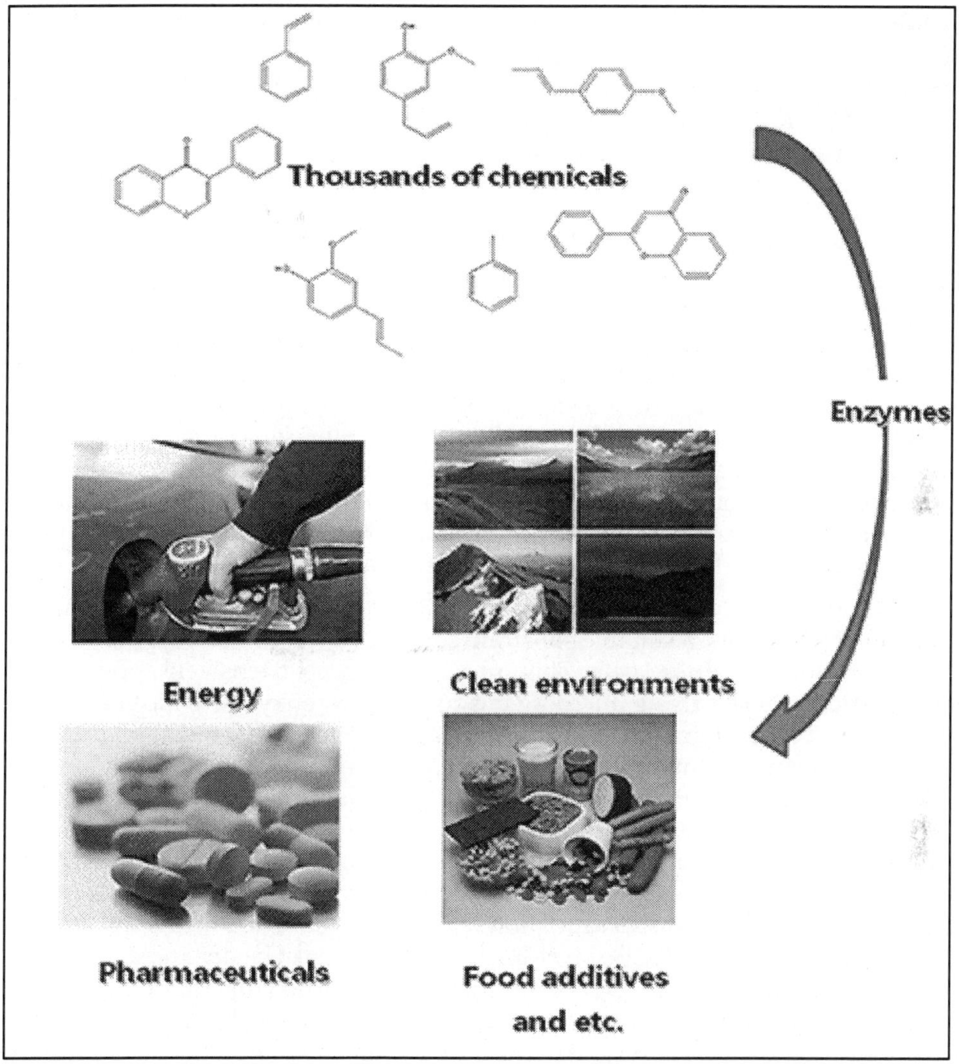

Figure 46.3: Production of Valuable Plant Secondary Metabolites through Biotransformation.

amount of alkaloids by up to 14 times. The level of an anticancer compound, tripdiolide, produced by *T. wilfordii* cultured cells was increased by addition of 100 μg/L farnesol which is dephosphorylated farnesyl pyrophosphate, and an intermediate in the biosynthesis of terpenoids. Addition of phenylalanine into the agar medium of *Taxus cupsidata* cells was found by DiCosmo *et al.* (1992) to stimulate the biosynthesis of an anticancer compound, taxol.

However, biotransformation using intact cells or immobilized cells is an important alternative way, for producing a product by adding precursors into the culture media. It was reported that *Rauwolfia serpentina* formed a new indole alkaloid, 6-hydroxytaumacline, in significant amounts when the cells were cultivated in the presence of ajmalicine. Arbutin, a skin depigmentation agent, is produced by biotransformation of hydroquinone using *Catheranthus roseus* cells. Addition of the precursor into the liquid culture medium of this cell line produced arbutin efficiently. By using *Mallotus japonicus* cells umebelliferone and salicylic acid were converted into o-glucoside by glucosylation which is a prodrug with a yield of 90-95 per cent. The maximum level of the product obtained was 0.9 g/L. The glucoside showed as potent an analgesic activity as salicylic acid, while its effect was more rapid and more long-lived than that of salicylic acid in mice. It is of interest that plant cells are able to form prodrugs having commercial usefulness. Dombrowski and Alfermann, (1992) also reported the glucosylation of salicylic acid and its derivatives by *Salix matsudana*. They found that cultured cells glucosylated salicylic acid to 2-o-salicylic acid-ß-D-glucoside. The cells also converted salicyl alcohol to both salicin (2-o-salicyl alcohol-ß-D-glucoside and isosalicylicin (1-o-salicyl alcohol-ß-D-glucoside). However, salicylic aldehyde was converted to ß-D-glucosides of salicylic acid and salicyl alcohol instead of helicin. The conversion of geraniol and nerol to neral and geranial by *Vitis vinifera* cell suspension cultures is also interesting. Using plant cell culture techniques, radioactive labelled compounds can be formed from appropriate substrates. This is a useful application of the biotechnology because of high value of the product.

Instead of the addition of a particular compound as a precursor into the culture medium of plant cells, a suitable substrate compound may be biotransformed to a desired product using plant cells. This approach has been extensively applied in the fermentation industry using microorganisms and their enzymes which can also be done by plant cells. Biotransformation of ß-methyldigitoxin to ß-methyldigoxin was achieved by using this method with *Digitalis lanata* cells (Alfermann and Reinhard, 1982) which is an important cardiac glycoside with a large market value. About 600-700 mg of ß-methyldigoxin per litre was obtained using a 200 L reactor. This process was studied for commercialization (Boehringer Mannheim Co). Successful processes for production of a very expensive antitumor drug, vinblastine, from catharanthine and vindoline using a biotransformation was established. The process is now being developed for commercialization by Mitsui Petrochemical in Japan (Fujita *et al.*, 1990). Concerning vinblastine-related compounds production, also reported a biotransformation of catharanthine and vindoline to anhydrovinblastine using horseradish peroxidase and glucose oxidase mediated coupling of vindoline and catharanthine.

From all this its become clear that, biotransformation processes including addition of substrates into the cultures are one of the most commercially realistic approaches in plant tissue cultures because of economic reasons. However, the availability of inexpensive precursors is a key issue.

Use of Elicitors

Plants interact with various components of their environment which may be biotic (microbes especially pathogens) and abiotic (stress, certain chemicals, radiations, some heavy metals etc.) and are responsible for producing huge diverse array of secondary metabolites (Figure 46.4) (Zhao *et al.*, 2005).

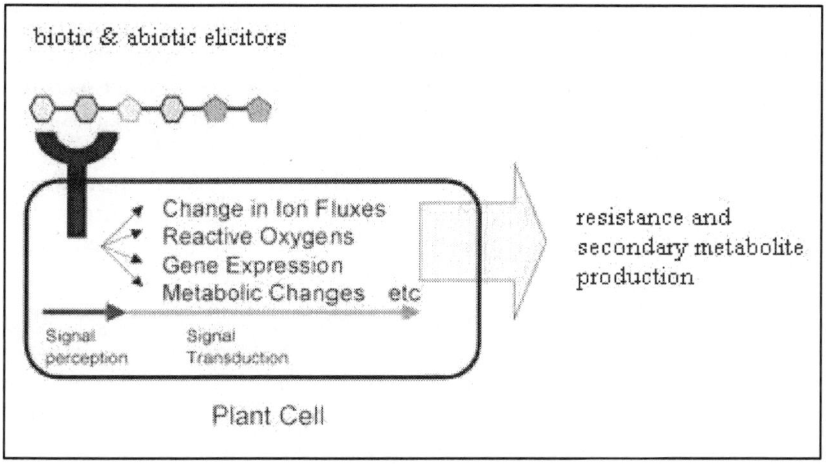

Figure 46.4: Production of Secondary Metabolites in Response to Elicitors.

Such factors/agents that signal triggering the formation of SMs are called elicitors and the process of production of SMs is called elicitation. SMs accumulated in response to microbial attack are referred as phytoalexins which are low molecular weight compounds with antimicrobial activity (Smith, 1996). Formations of these molecules are due to degradation of plant or microbial cell wall and are also used to induce elicitation in plant cultures. Understanding the signal transduction paths underlying elicitor-induced production of SMs is one of the most effective biotechnological strategies for improving the productivity at commercial level (Roberts and Shuler, 1997). Biotic and abiotic elicitors which are classified on the basis of their origin are used to stimulate SMs formation in plant cell cultures, there by reducing the process time to attain high product concentrations (DiCosmo and Tallevi, 1985; Eilert, 1987, Barz *et al.*, 1988).

The best understood elicitor systems are those of fungal pathogens in which case the regulatory molecules have been identified as glucan polymers, glycoproteins and low molecular weight organic acids (DiCosmo and Tallevi, 1985). Examples of microbial elicitor induction include psoralen production in parsley, diosgenin production in the Mexican yam and many others. It was noted that a cell line of *Papaver somniferm* that synthesized and accumulated sanguinarine, a quaternary benzophenanthridine alkaloid when exposed to a homogenate of the fungus *Botrytis* (Tyler *et al.*, 1989). A portion of the sanguinarine was released into the culture medium. Sanguinarine extracted from the intact plants is being used for oral hygienic products.

Effects of elicitors on secondary metabolism have been investigated at the enzymatic levels to determine their mode of action. Addition of autoclaved culture homogenate of yeast, *Rhodotorula rubra* into the suspension culture of *Ruta graveolens* and found that S-adenosyl-L-methionine:anthranilic acid N-methyltransferase was elicited(Eilert and Wolters, 1989). A yeast polysaccharide preparation induced L-tyrosine decarboxylase in suspension cultures of *Thalictrum rugosum* and *Eschscholtzia californica;* the enzyme was induced after 5 hours after addition of the elicitor at 30 to 40 µg/g-cell fresh weight. Recent developments in phytochemical elicitation have shown that simple inorganic and organic molecules can induce product accumulation. Sodium orthovanadate and vanadyl sulphate induced the accumulation of isoflavone glucosides in *Vigna angularis* cultures and indole alkaloid accumulation in *Catharanthus roseus* cultures, respectively. Other substances found to stimulate alkaloid accumulation in *C. roseus* include sodium chloride, potassium chloride and sorbitol as well as abscisic acid. Processes such as these, employing simple and cheap elicitors have much promise in industrial scale plant cell cultures.

Sometimes, addition of abiotic elicitors like organic, inorganic compounds, metals etc... increases elicitation. Addition of oxalate to the medium of *Gossypium hirsutum* suspension culture could reduce the amount of *Verticillium dahliae* elicitor to be employed to stimulate metabolite synthesis. Addition of a fungal elicitor often inhibits the growth of plant cells but a combination of the elicitor and oxalate did not reduce the cell mass of the plant, therefore secondary metabolite synthesis was increased up to ten fold.

Some times combination of biotic and abiotic elicitors are effective in SMs production.Combination of phosphate limitation and fungal elicitation synergistically increased production of secondary metabolites. Either phosphate limitation or elicitation with a mycelial extract of the fungus, *Rhizoctonia solani* alone results in increased production of the sesquiterpene solavetivone by *Agrobacterium rhizogenes*-transformed hairy root cultures of *Hyoscyamus muticus*. However, when phosphate limitation is coupled with fungal elicitation, the productivity increase is considerably greater than that obtained with either method alone.

Although the mechanism by which elicitors increase the productivity of secondary plant metabolites has not been elucidated, their stimulating activity is quite significant if an appropriate elicitor is chosen to stimulate synthesis of a particular product. However, the use of microbial elicitors may not be economical since an elicitor-producing microorganism should be cultivated in a fermentor separately from cultivation of plant cells using another fermentor. The fermentation cost for an elicitor-producing microorganism is not always inexpensive. In this sense, a simple and cheap compound should be employed as an elicitor.

Genetic Engineering/Genetic Transformation

The recent advances and developments in plant genetics and recombinant DNA technology have helped to improve and boost research into PSMs biosynthesis. The major role of research is to identify enzymes of a metabolic pathway and then manipulate these enzymes to provide better control of that pathway. Transformation is currently used for genetic manipulation of more than 120 species of at least 35 families, including the major economic crops, vegetables, ornamental, medicinal, fruit, tree and pasture plants, using *Agrobacterium* mediated or direct transformation methods (Birch, 1997). However, *Agrobacterium*-mediated transformation offers several advantages over direct gene transfer methodologies (particle bombardment, electroporation, etc), such as the possibility to transfer only one or few copies of DNA fragments carrying the genes of interest at higher efficiencies with lower cost and the transfer of very large DNA fragments with minimal rearrangement (Gheysen *et al.*, 1998; Hansen and Wright, 1999; Shibata and Liu, 2000). The gram-negative soil bacteria, *Agrobacterium tumefaciens,* and the related species, *A. rhizogenes,* are causal agents of the plant diseases crown gall tumour and hairy root, respectively. These species, which belong to the Rhizobiaceae, are natural genetic engineers that are able to transform or modify, mainly dicotyledonous plants, although there are reports on the infection of monocotyledonous plants (Tepfer, 1990; Hiei *et al.*, 1994; Ishida *et al.*, 1996). Virulent strains of *A. tumefaciens* and *A. rhizogenes* contain a large megaplasmid (more than 200 kb) which plays a key role in tumor induction and for this reason it was named Ti plasmid, or Ri in the case of *A. rhizogenes*. During infection the T-DNA, a mobile segment of Ti or Ri plasmid, is transferred to the plant cell nucleus and integrated into the plant genome.

A. tumefaciens transfers the T-DNA into the nucleus of infected cells where it is then stably integrated into the host genome and transcribed, causing the crown gall disease (Nester *et al.*, 1984). T-DNA contains two types of genes: the oncogenic genes, encoding for enzymes involved in the synthesis of auxins and cytokinins and responsible for tumor formation; and the genes encoding for the synthesis of opines. *Agrobacterium rhizogenes* has been used regularly for gene transfer in many dicotyledonous plants for hairy root culture (Tepfer, 1990). Plant infection with this bacterium induces the formation of proliferative highly branched adventitious roots at the site of infection; the so-called "hairy roots" (Chilton *et al.*, 1982). This infection is followed by the transfer of a portion of DNA *i.e.* T-DNA, known as the root inducing plasmid (Ri plasmid), to the plant genomic DNA.

The research has been going on for the application of plant transformation and genetic modification using *A. rhizogenes,* in order to boost up the production of these PSMs (Figure 46.5), which are naturally synthesized in the roots of the mother plant. Transformed hairy roots mimic the biochemical machinery present and active in the normal roots, and in many instances transformed hairy roots display higher product yields. Genetic transformation has been reported for various medicinal plants. Successful regeneration of transgenic neem plants (*Azadirachta indica*) using *Agrobacterium tumefaciens* containing a recombinant derivative of the plasmid pTi A6 was reported (Naina *et al.*, 1989). The genetic transformation of *Atropa belladona* has been reported using *A. tumefaciens,* with an improved alkaloid composition (Yun *et al.*, 1992; Cucu *et al.*, 2002). *Agrobacterium* mediated transformation of *Echinacea purpurea* has been demonstrated using leaf explants (Koroch. *et al.*, 2002).

Genetic transformation would be a powerful tool for enhancing the productivity of novel secondary metabolites of limited yield. Hairy roots, transformed with *A. rhizogenes,* have been found to be suitable for the production of secondary metabolites because of their stable and high productivity in hormone-free culture conditions. A number of plant species including many medicinal plants have been successfully transformed with *A. rhizogenes*. The hairy root culture system of the medicinal plant *Artemisia annua* was established by infection with *A. rhizogenes* and the optimum concentration of artimisin was 4.8 mg/L (Cai *et al.*, 1995). Hairy roots development in *Aconitum heterophyllum* using *A. rhizogene* was established (Giri *et al.*, 2003) Pradel *et a,l* (1997) developed a system for producing transformed plants from root explants of

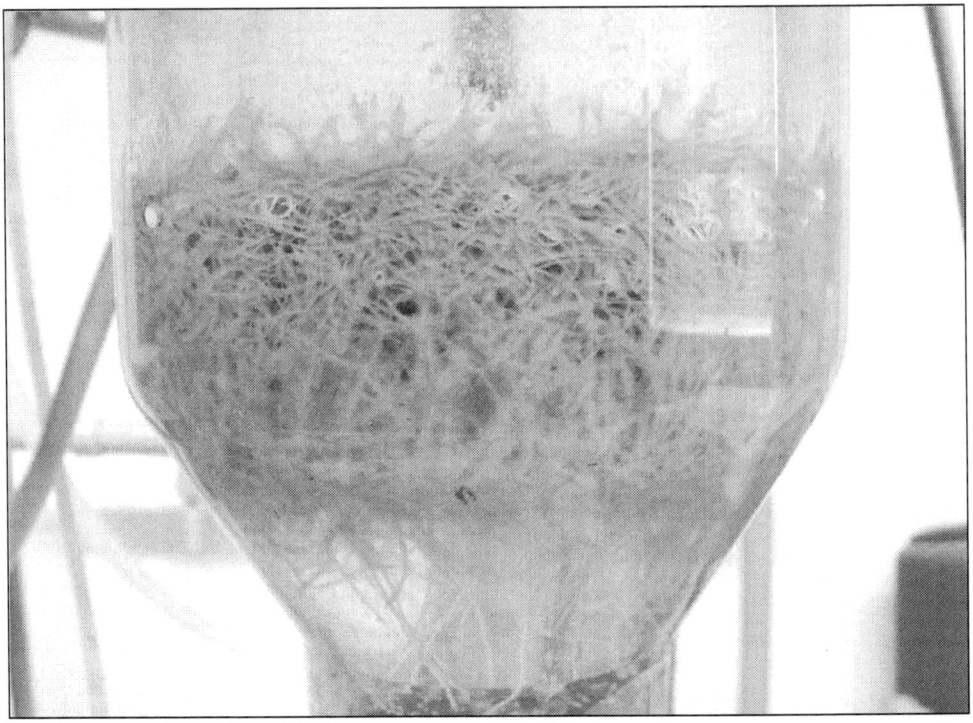

Figure 46.5: Production of PSMs through Hairy Root Cultures.

Digitalis lanata. They evaluated different wild strains of *Agrobacterium rhizogenes* for the production of secondary metabolites obtained from hairy roots and transgenic plants. They reported higher amounts of anthraquinones and flavonoids in the transformed hairy roots than in untransformed roots. An efficient protocol for the development of transgenic opium poppy (*Papaver somniferum*) and California poppy (*Eschscholzia californica*) root cultures using *A. rhizogenes* is reported (Park *et al.*, 2000). Bonhomme *et al.* (2000) has reported the tropane alkaloid production by hairy roots of *Atropa belladonna* obtained after transformation with *A. rhizogenes*. Argolo *et al.* (2000) reported the regulation of solasodine production by *Agrobacterium rhizogenes*-transformed roots of *Solanum aviculare*. Souret *et al.* (2002) have demonstrated that the transformed roots of *A. annua* are superior to whole plants in terms of yield of the sesquiterpene artemisinin. Shi and Kintzios, (2003) have reported the genetic transformation of *Pueraria phaseoloides* with *A. rhizogenes* and puerarin production in hairy roots. The content of puerarin in hairy roots reached a level of 1.2 mg/g dry weight and was 1.067 times the content in the roots of untransformed plants. Thus, these transformed hairy roots have great potential as a commercially viable source of secondary metabolites.

Metabolic Engineering

The main goals of metabolic engineering in industry or agriculture are the stimulation of the production of secondary metabolite end-products, biosynthetic precursors, polymers that have plant origin, and the derivation of new plant organisms with high salt or drought resistance in agriculture. It is not surprising that metabolic engineering applications in plant biotechnology in recent years have had incredible achievements in agriculture, industry, and medicine. This multidisciplinary field draws concepts and methodologies from molecular biology, biochemistry, and genetics, as well as biochemical engineering. In addition, the extension of metabolic engineering to produce desired compounds in plant organisms may answer many fundamental questions applied to plant development, physiology, and biochemistry.

Plant metabolic engineering treats the cell as a factory and adds or optimizes kinds and amounts of metabolites within the cell for some specific design purpose. In other words, metabolic engineering refers to a targeted metabolic pathway being elucidated in plant or bacterial organisms with the purpose of unraveling and utilizing this pathway for future modification of a plant's end-products. It is generally defined as the redirection of enzymatic reactions or manipulation of enzymatic; transport and regulatory functions of the cell with the use of recombinant DNA technology, so as to improve the production of high-value constituents, to produce new compounds in an organism, to mediate the degradation of environmentally toxic compounds, or to create plants that become resistant to environmental stress factors. It involves the targeted and purposeful alteration of metabolic pathways found in an organism to achieve better understanding and use of cellular pathways for chemical transformation, energy transduction, and supramolecular assembly (Lessard, 1996). In addition, metabolic engineering may include not only the manipulation of endogenous metabolic pathways but also the transfer of metabolic pathways into new host organisms (Figure 46.6).

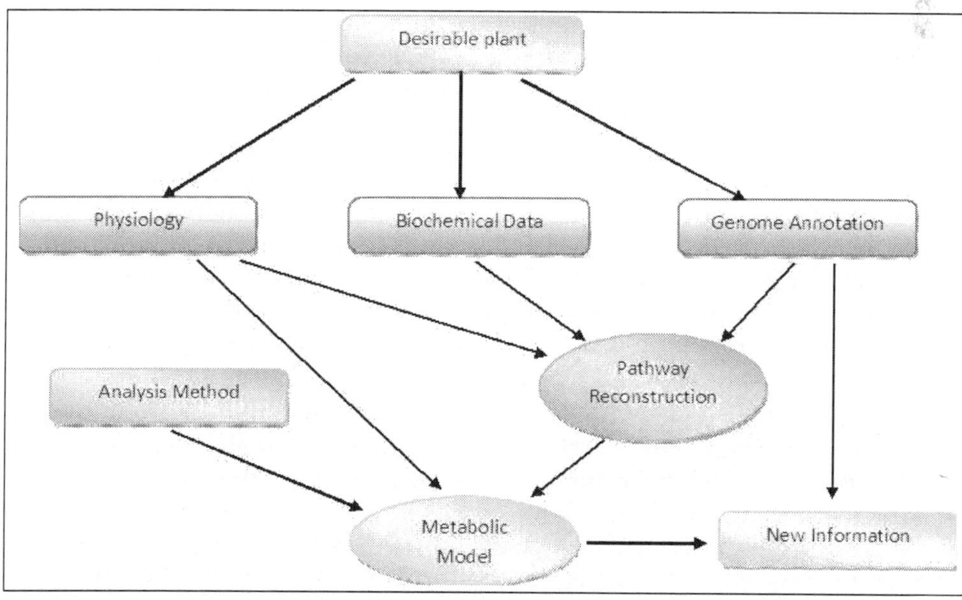

Figure 46.6: Creation of Ability and Production of PSMs through Metabolic Engineering.

This technique applied to plants will permit endogenous biochemical pathways to be manipulated, resulting in the generation of transgenic crops in which the range, scope, or nature of a plant's existing natural products are modified to provide beneficial commercial, agronomic and/or post-harvest processing characteristics (Kinney, 1998). Rational engineering of secondary metabolic pathways in plants requires a thorough knowledge of the whole biosynthetic pathway and a detailed understanding of the regulatory mechanisms controlling the onset and the flux of the pathways.

For example, plant metabolic flux analysis in the primary carbon-based metabolic pathways presents fundamental information on the application of plant metabolic engineering that is based on a thorough knowledge of plant biochemistry and plant physiology. Plant metabolism itself concerns thousands of interacting pathways and processes that are regulated by environmental and genetic stimuli. Therefore, engineering even known metabolic pathways will not always provide the expected results. Despite major advances in metabolic engineering, still only a few secondary metabolic pathways have been enzymatically characterized and the corresponding genes cloned. In this context, the biosynthetic pathways for alkaloids, flavonoids, and terpenoids are presently the best characterized ones at the enzyme and gene levels. More successful cases of gene discovery have also been considered for the lipid biosynthetic pathway, where most genes in plants encoding enzymes for fatty acid biosynthesis have been cloned. This information was applied for eventual manipulation through modification of many fatty acids in transgenic plants by means of metabolic engineering.

As for targeting metabolite manipulation, Della Penna advocated the conversion or chemical modification of an existing compound(s), rather than attempting to increase flux through a metabolic pathway. Another example, he cites, claims that modifications made in the end products or secondary metabolic pathways have been generally more successful than in cases where manipulation of primary and/or intermediary metabolic pathways is attempted (DellaPenna, 2001). Recent achievements have been made in the altering of various pathways by use of specific genes encoding biosynthetic enzymes or genes encoding regulatory proteins (Verpoorte and Memelink, 2002; Maliga and Graham, 2004). Most current metabolic engineering studies have focused on manipulations of enzyme levels and the effect of amplification, addition, or blockage of a particular pathway. A new area is the manipulation of cofactors, which play a major role in plant biochemistry and physiology and in the fermentation process of several end-products. Additionally cofactors are essential for many enzymatic reactions.

Metabolic engineering is becoming a powerful technology for the successful implementation of plant cell biotechnology in the future. This may be possible with the advances we already have mentioned above and some other important issues and key criteria that are cited as follows: (1) Metabolic flux analysis must be applied to well-documented and elucidated metabolic pathways; (2) extension of metabolic cross-talk between the desired metabolite pathway and other pathways for a possible direct impact on plant development and nutritional value must be considered; (3) identification of further elements in the complex regulatory network (such as transcription factors and their binding partners) needs to be examined; and (4) rigorous criteria must be developed for the assessment of the risk and benefit performance of engineered plants. Comprehensive studies in several directions may help to bring metabolic engineering out of the trial-and-error and transform it into industrial applications.

Metabolic engineering approaches can be defined according to several different directions. The first appropriate approach involves increases in the total carbon flux toward the desired secondary metabolite. In addition, decreasing the flux through competitive pathways is an alternative way to increase the biosynthesis of desired metabolite. Other possible directions involve the introduction of an *antisense gene* of the competing enzyme at the branch point, as well as overcoming *rate-limiting steps*, or blocking *competitive pathways*.

Over the last decades, plant cell cultures have been intensively investigated as a possible tool for the production of commercial plant secondary metabolites, including fine chemicals such as pharmaceuticals, agrochemicals, flavors, insecticides, fragrances and cosmetics (Whitmer *et al.*, 2002). In spite of the efforts in the field of *in vitro* production of phytochemicals, few industrial processes have been developed, involving only a limited number of secondary products, such as shikonin, berberine, ginsenosides and paclitaxel (Ramachandra and Ravishankar, 2002). As in many cases production is too low for commercialization, metabolic engineering can provide various strategies to improve productivity, such as: increasing the number of producing cells, increasing the carbon flux through a biosynthetic pathway by overexpression of genes codifying for rate-limiting enzymes or blocking the mechanism of feedback inhibition and competitive pathways and decreasing catabolism.

Many of the isolated pure compounds with biological activity are alkaloids, a diverse group of nitrogen containing chemical ring structure compounds, with alkali-like chemical reactivity and pharmacological activity. Although the pharmacological effects of alkaloids have been studied, the biosynthetic pathways of these compounds are still obscure. Among the most famous are the tropane alkaloids, such as (-)-hyoscyamine, its racemate atropine, and scopolamine (hyoscine), which have an 8-azabicyclo [3.2.1] octane esterified nucleus. These alkaloids are commonly found in plants of different families: Olanaceae, Erythroxylaceae, Convolvulaceae, Proteaceae, Euphorbiaceae, Rhizophoraceae and Cruciferae (Griffing and Lin, 2000). Related to the tropane alkaloids, a new group of nortropane alkaloids, the calystegines, was discovered only 15 years ago. Calystegines bear three to five hydroxyl groups in various positions, making them water-soluble, and they share metabolic steps and enzymes of the formation of tropane alkaloids. Several genes in the biosynthetic pathways for scopolamine, nicotine, and berberine have been cloned, making the metabolic engineering of these alkaloids possible. Expression of two branching-point enzymes was engineered: putrescine *N*-methyltransferase (PMT) in transgenic plants of *Atropa belladonna* and *Nicotiana sylvestris* and (*S*)-scoulerine 9-*O*-methyltransferase (SMT) in cultured cells of *Coptis japonica* and *Eschscholzia californica*. Overexpression of PMT increased the nicotine content in *N. sylvestris*, whereas suppression of endogenous PMT activity severely decreased the nicotine content and induced abnormal morphologies. Ectopic expression of SMT caused the accumulation of enzylisoquinoline alkaloids in *E.californica* (Sato *et al.*, 2001).

Combinatorial Biosynthesis

The basic and novel idea for the production of new and interesting plant secondary metabolites by combining metabolic pathways from different organisms at genetic level by combining genes is called as combinatorial biosynthesis (Figure 46.7) (Michels *et al.*, 1998). The hydroxylations and glycosylations are important bioconversions in pharmaceutical point of view (Knaeblein *et al.*, 2004; Ma *et al.*, 2005). They can yield new drugs and existing drugs can be improved in their increased activity and decreased toxicity.

Recent achievements with the polyketide biosynthesis from microorganisms, especially in Streptomyces, prove the potential of combinatorial biosynthesis (Pfeifer and Khosla, 2001; Weber *et al.*, 2003; Hranueli *et al.*, 2005; Moore *et al.*, 2005). This also showed that this approach can be used to improve the biosynthetic capacity of known producing microorganisms like *Escherichia coli*, *Bacillus subtilis* or *Saccharomyces cerevisiae* or even higher organisms. The heterologous expression of human genes in microorganisms is well known (insulin, Growth factor 8 and 9, Tissue plasminogen activator, α-Anti tripsin) for more than 30 years. Fundamental work on the expression of plant genes from biosynthetic pathways performed since the 1980s and opened a way to research that may even be extended in the future by directed evolution. It is now possible to combine these genes and extend the realm of combinatorial biosynthesis far beyond the polyketide biosynthesis. The diversification of products will increase dramatically when genes of very different origins are used. However there is no need to concentrate on new compounds only; there are many interesting natural products, of which the application (*e.g.* as a drug or fine chemical) is hampered

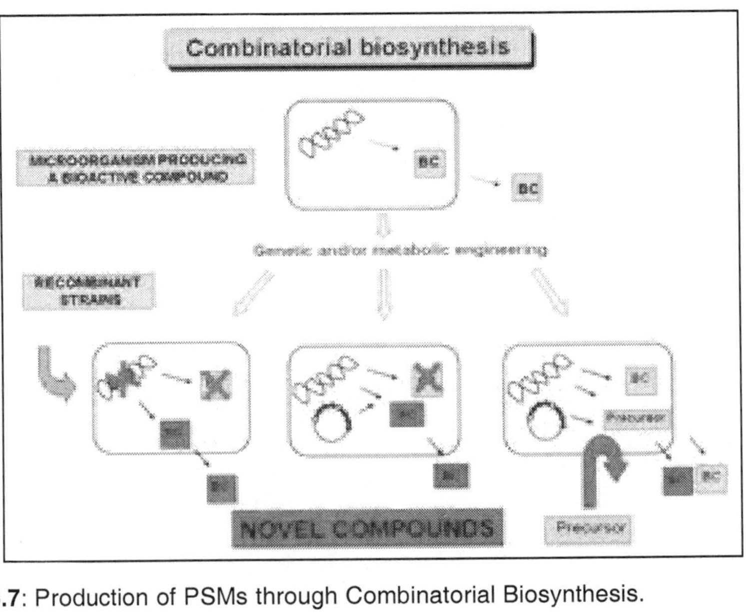

Figure 46.7: Production of PSMs through Combinatorial Biosynthesis.

by its availability. This problem might be solved by using alternative production systems yet to be discovered, that are based on enzymes from other biosynthetic pathways. Podophyllotoxin and paclitaxel are clear examples of pharmaceuticals that can only be produced through the isolation from plants. With regard to the production of Podophyllotoxin it has been shown that cell cultures of *Linum flavum* can be used to convert deoxypodophyllotoxin, a major lignin of *Anthriscus sylvestis* into 6-methoxypodophyllotoxin (Koulman *et al.*, 2003; Van ssssssUden *et al.*, 1997). The combination of the product of one species and the enzymes system of other species to yield a desirable product shows a good example of combinatorial biosynthesis. Combinatorial biosynthesis of alkaloids is known for a few examples like vincristine, vinblastine, ajmaline (Warzecha *et al.*, 2000) and morphine plants rebeccamycin and starosporine from *Streptomyces albus*(Sanchez *et al.*, 2005; Sanchez *et al.*, 2006) which have a long biosynthetic path way involving nearly 30 enzymes.

Conclusion

In conclusion, still it needs further work for isolating and identification of bioactive compounds using biotechnological innovations from medicinal plants, which are essential to formulate life saving drugs.

References

Alfermann, A.W. Reinhard, E. (1982). In *"Proc. 5th Int'l. Cong. Plant Tissue and Cell Culture"* 401-402.

Argolo, A.C., Charlwood, B.V., Pletsch, M. (2000). The regulation of solasodine production by Agrobacterium rhizogenes-transformed roots of Solanum aviculare. *Planta Medica*, **66**: 448-51.

Barz, W., Ellis, B. (1981). Potential of plant cell cultures for pharmaceutical production. In: Beal JL, Reinhard E (eds.) *Natural Products as Medicinal Agents.* Stuttgart, Hippokrates, pp. 471–507.

Barz, W., Daniel, S., Hinderer, W., Jaques, U., Kessmann, H., Koster, J., Tiemann, K. (1988). Elicitation and metabolism of phytoalexins in plant cell cultures. In: Pais, M., Mavituna, F., Novais, J, editors. *"Plant Cell Biotechnology"*. NATO ASI Series. Berlin: Springer-Verlag 211-230.

Berlin, J., Sasse, F. (1985). Selection and screening techniques for plant cell cultures. *Advanced Biochemistry and Engeneering*, **31**: 99-132.

Birch, R.G. (1997). Plant transformation: Problems and strategies for practical application. *Ann Rev Plant Physiol Plant Mol Biol.*, **48**: 297- 326.

Binns, A.N., Thomashow, M.F. (1988). Cell biology of Agrobacterium infection and transformation of plants. *Ann Rev Microbiol.*, **42**: 575- 606.

Bonhomme, V., Laurain-Mattar, D., Lacoux, J., Fliniaux, M., Jacquin-Dubreuil, A. (2000). Tropane alkaloid production by hairy roots of Atropa belladonna obtained after transformation with Agrobacterium rhizogenes 15834 and Agrobacterium tumefaciens containing rol A, B, C genes only. *J Biotechnol* **81**: 151- 8.

Buitelaar, R. M., Tramper, J. (1992). Strategies to improve the production of secondary metabolites with plant cell cultures: a literature review. *Journal of Biotechnology,* **23**: 111-143.

Cai, G., Ye, H., Li, G. (1995). Hairy root culture of Artemisia annua L. by Ri plasmid transformation and biosynthesis of artemisinin. *Chin J Biotechnol* **11**: 227- 35.

Chan, W.N., Staba, E.J. (1965). Alkaloid production by *Datura* callus and suspension tissue culture, *Lloydia* **28**: 55-62.

Chilton, M.D., Tepfer, D., Petit, A., David, C., Casse- Delbart, F.T. (1982). Agrobacterium rhizogenes inserts T-DNA into the genome of the host plant root cells. *Nature,* **295**: 432-4.

Cucu, N., Gabriela, Gavrila, L. (2002).Genetically modified medicinal plants. II. Transfer and expression of a marker kanamycine resistance gene in *Atropa belladonna* plants. *Rom Biotechnol Lett.,* **7**: 869-74.

De Luca, V., Cutler, A, J. (1987). Subcellular localization of enzymes involved in indole alkaloid biosynthesis in *Catharanthus roseus. Plant Physiol.,* **85**: 1099–1102.

DellaPenna, D.(2001). Plant metabolic engineering. *Plant Physiology* **125**: 160–163.

Deus, B., Zenk, M.H.(1982). Exploitation of plant cells for the production of natural compounds. *Biotechnol Bioeng.,* **24**: 1965–74.

Dicosmo, F., Tallevi, S. G. (1985). Plant cell cultures and microbial insult: interactions with biotechnological potential. *Trends in Biotechnology,* **3**: 110-111.

Di Cosmo, F., *et al.* (1992). *2nd National Cancer Inst*. Workshop on Taxol and Taxus, Sep. 23-24, Alexandria, V.A., U.S.

Dixon, R. A. (1999). Plant natural products: the molecular genetic basis of biosynthetic diversity. *Current Opinion in Biotechnology,* **10**: 192-197.

Dombrowski, K., Alfermann, A.W., 1992. Biotransformation of Salicylic Acid Derivatives by Plant Cell Suspension Cultures. *Planta Medica Supplement Issue* **I 58**: A 576-A577.

Eilert, U.(1987). Elicitation: methodology and aspects of application. In: Constabel, F., Vasil, I., editors. *"Cell Culture and Somatic Cell Genetics of Plants"*. 4, San Diego: *Academic Press,* pp.153-196.

Eilert, U., Wolters, B., 1989. Elicitor induction of S-adenosyl-L-methionine: Anthranilic acid N-methyltransferase activity in cell suspension and organ cultures of Ruta graveolens L. *Plant Cell Technol. and Organ Culture,* **18**: 1-18.

Fett-Neto, A. G., Pennington, J. J., DiCosmo, F. (1995). Effect of white light on taxol and baccatin III. Accumulation in cell cultures of *Taxus cuspidata* Sieb and Zucc. *Journal of Plant Physiology,* **146**: 584-590.

Fujita, Y. (1988). Industrial production of shikonin and berberine. In: Applications of plant cell and tissue culture. pp. 228–238. *Ciba Foundation Symposium* 137, Wiley.

Fujita, *et al.* (1990). In "Prog. in Plant Cellular and Molecular Biology", Eds. Nijkamp, H.J.J., 738-743 *Kluwer Acad. Publ. Dordrecht*, Boston, London.

Gao SL, Zhu DN, Cai Z.H., Jiang Y, Xu D.R.(2004). Organ culture of a precious Chinese medicinal plant – *Fritillaria unibracteata*. *Plant Cell Tiss. Org. Cult.*, **59**: 197-201.

Gheysen G., Angenon G., Montague, M.V.(1998). Agrobacterium-mediated plant transformation: a scientifically intriguing story with significant application. In: Lindsey K (ed). *Transgenic Plant Research*. Netherlands: *Harwood Academic Press,* pp1-33.

Giri, A., Banerjee, S., Ahuja, P.S., Giri, C.C.(1997). Production of hairy roots in *Aconitum heterophyllum* Wall. Using *Agrobacterium rhizogenes*. *Invitro Cell Dev Biol- Plant.,* **33**: 293-306.

Goleniowski, M., Trippi, V. S.(1999). Effect of growth medium composition on psilostachyinolides and altamisine production. *Plant Cell Tissue and Organ Culture,* **56**: 215-218.

Gordon, M.C., David, J.N.(2001). Natural product drug discovery in the next millennium. *Pharm. Biol.,* **39**: 8-17.

Griffing, W.J., Lin, G.D.(2000). Chemotaxonomy and geographical distribution of tropane alkaloids. *Phytochemistry,* **53**: 623-637.

Hansen, G., Wright, M.S.(1999).Recent advances in transformation of plants. *Trends Plant Sci.,* **4**: 226-31.

Harborne, J.B.(1996). Recent advances in chemical ecology. *Natural Products Report* **12**: 83-98.

Hiei, Y., Ohte, S., Komari, T., Kumashiro, T.(1994). Efficient transformation of rice mediated by Agrobacterium and sequence analysis of the boundaries of the T-DNA. *Plant J.,* **6**: 271-82.

Hranueli, D., Cullum, J., Basrak, B., Goldstein, P., Long., P. F.(2005). Plasticity of the *streptomyces* genome-evolution and engineering of new antibiotics, *Curr Med Chem.,* **12**: 1697-1704.

Ishida, Y., Saito, H., Ohta, S., Hiei, Y., Komari, T., Kumashiro, T.(1996). High frequency transformation of maize (*Zea mays*) mediated by *Agrobacterium tumefaciens*. *Nature Biotechnol.,* **14**: 745-50.

Kinney, A.J.(1998). Manipulating flux through plant metabolic pathways. *Curr. Opin. Plant Biol.,* **1**: 173-178.

Kliebenstein J.(2004). Biopharmaceuticals expressed in plants, in *Pharmaceutical biotechnology – Drug discoverey and clinical applications,* edited by O Kayser and R H Muller (Wiley VCH, Weinheim), pp.34-36.

Kliebenstein, D.J.(2004). Secondary metabolites and plant/environment interactions: a view through *Arabidopsis thaliana* tinged glasses. *Plant. Cell. Environ.,* **27**: 675–684.

Koroch, A.R., Kapteyn, J., Juliani, H.R., Simon, J.E.(2002). *In vitro* regeneration and Agrobacterium transformation of *Echinecea purpuria* leaf explants. *Trends New Crop New News.*

Koulman, A., Beekman, A. C., Pras, N., Quax, W. J. (2003). The bioconversion process of deoxypodophyllotoxin with *Linum flavum* cell cultures, *Planta Med.*, **69** : 739-744.

Laitinen, M.L., Julkunen-Tiitto, R., Tahvanainen, J., Heinonen, J., Rousi, M.(2005). Variations in birch (*Betula pendula*) shoot secondary chemistry due to genotype, environment, and ontogeny. *J. Chem. Ecol.*, **31**: 697–717.

Lee, C. W. T., Shuler, M. L.(2000). The effect of inoculum density and conditioned medium on the production of ajmalcine and catharanthine from immobilized *Catharanthus roseus* cells. *Biotechnology Bioengineering*, **67**: 61-71.

Lerdau, M.(2002). Benefits of the carbon-nutrient balance hypothesis. *OIKOS.*, **98**: 534–536.

Lessard, P.(1996). Metabolic engineering, the concept coalesces. *Nat. Biotechnol.* **14**: 1654-1655.

Luckner, M., 1986. Secondary metabolism in micro organisms, plants, and animals, *Springer*, Berling, Heidelbeg, New York.

Luckner, M. Nover, L.(1977). Expression of secondary metabolism. An aspect of cell specialization micro organisms, higher plants and animals. In, M., Nover L., bohm H. (eds) Secondary metabolism and cell differentiation. *Mol.Biol.Biochem. Biophys.*, 23: 1.

Ma, J. K., Barros, E., Bock, R., Christou, P., Dale, P. J.(2005). Molecular farming for new drugs and vaccines. Current prospective on the production of pharmaceuticals in transgenic plants, *EMBO rep.*, **6**: 593-599.

Maliga, P., Graham, I.(2004). Molecular farming and metabolic engineering promise a new generation of high-tech crops. *Current Opinion in Plant Biology*, **7**: 149–151.

Mary, Ann Lila.(2006). The nature-versus-nurture debate on bioactive phytochemicals: the genome versus *terroir*. *J. Sci. Food Agric.*, **86**: 2510–2515.

Michelis, P.C., Khmelnitsky, Y. L., Dordick, J. S., Clark, D. S.(1998). Combinatorial biocatalysts: A natural approach to drug discovery, *Trends Biotechnol.*, **16**: 210-215.

Moore, B. S., Kalaitzis, J. A., Xiang, L.(2005). Exploting marine actinomycete biosynthetic pathways for drug discovery, *Antonie Van Leeuwenhoek*, **87**: 49-57.

Naina, N. S., Gupta, P. K., Mascarenhas, A. F.(1989). Genetic transformation and regeneration of transgenic neem (*Azadirachta indica*) plants using *Agrobacterium tumefaciens*. *Curr Sci.*, **58**: 184-7.

Nester, E. W., Gordon, M. P., Amasino, R. M., Yanofsky, M. F.(1984). Crown gall: a molecular and physiological analysis. *Ann Rev Plant Physiol.*, **35**: 387-413.

Ohta, S., Verpoorte, R.(1992). Some accounts of variation (heterogeneity and/or instability) in secondary metabolite production by plant cell cultures. *Ann. Rep. Nat. Sci. Home Econ.*, **32**: 9–23.

Oksman-Caldentey, Arroo R.(2000). Regulation of tropane alkaloid metabolism in plants and plant cell cultures. In: Verpoorte R and Alfermann AW (eds) Metabolic engineering of plant secondary metabolism. *Kluwer Academic Press, Dordrecht.* pp. 253–281.

Park, S.U., Facchini, P.J.(2000). Agrobacterium rhizogenesmediated transformation of opium poppy, Papaver somniferum l., and California poppy, Eschscholzia californica cham, root cultures. *J Exp Botany,* **51**: 1005-16.

Pfeifer, B. A., Khosla, C.(2001). Biosynthesis of polyketides in heterologous hosts *Microbiol Mol Biol Rev.,* **65**: 106-118.

Pradel, H., Dumkelehmann, U., Diettrich, B., Luckner, M.(1997). Hairy root cultures of Digitalis lanata. Secondary metabolism and plant regeneration. *J Plant Physiol.,* **151**: 209–15.

Ramachandra Rao, S.(2000). Biotechnological production of phyto-pharmaceuticals. *Journal of Biochemistry Molecular Biology Biophysics,* **4**: 73-102.

Ramachandra Rao, S. Ravishankar, G. A.(2002). Plant cell cultures: Chemical factories of secondary metabolites. *Biotechnology Advances,* **20**: 101-153.

Ramani, S., Jayabaskaran, C.(2008). Enhanced catharathine and vindoline production in suspension cultures of *Catharanthus roseus* by ultraviolet-B light. *J. Mol. Signal,* **3**: 9-14.

Ravishankar, G. A., Venkataraman, L. V.(1993). Role of plant cell cultures in food biotechnology: commercial prospectus and problems. *New Delhi: Oxford IBH Press*: 255-274.

Roberts, S. C., Shuler, M. L.(1997). Large scale plant cell culture. *Current opinion in Biotechnology* **8**: 154-159.

Robins, R.I.(1994). Secondary products from cultured cells and organs: Molecular and cellular approaches. In: Dixon RA, Gonzales RA (eds). *Plant Cell Culture.* Oxford: IRL Press.

Sanchez, C., Zhu, L., Brana, A.F., Salas, A.P., Rohr, J., Mendez, C., Salas, J.A.(2005). Combinatorial biosynthesis of antitumor indolocarbazole compounds. *Proc. Natl. Acad. Sci.* U.S.A. **102**: 461–466.

Sanchez, C., Mendez, C., Salas, J.A.(2006). Engineering biosynthetic pathways to generate antitumor indolocarbazole derivatives. *J. Ind. Microbiol.Biotechnol.*

Sato, F, Hashimoto, T., Hachiya, A., Tamura, K.I., Choi, K.B., Morishige, T., Fujimoto, H., Yamada, Y.(2001). Metabolic engineering of plant alkaloid biosynthesis. *Proc. Natl. Acad. Sci.* USA **2**: 367-372.

Sato, F., Endo, T., Hashimoto, T., and Yamada, Y.(1982). Production of berberine in cultures of *Coptis japonica* cells. In: Fujiwara A (ed) *Plant Tissue Culture.* Mazuren, Tokyo. pp. 319–323.

Sato, F., Yamada, Y.(1984). High berberine producing cultured *Coptis japonica* cells. *Phytochemistry.,* **23**: 281–285.

Shibata, D., Liu Y.G.(2000). Agrobacterium-mediated plant transformation with large DNA fragments. *Trends Plant Sci.,* **5**: 354-7.

Shi, H.P., Kintzios S.(2003). Genetic transformation of *Pueraria phaseoloides* with Agrobacterium rhizogenes and puerarin production in hairy roots. *Plant Cell Rep* **21**: 1103-7.

Smith, C.J.(1996) Accumulation of phytoalexins: Defence mechanism and stimulus response system. *New Phytol.,* **132**: 1–45.

Souret, F.F., Weathers, P.J., Wobbe, K.K.(2002).The mevalonate-independent pathway is expressed in transformed roots of *Artemisia annua* and regulated by light and culture age. *In-vitro Cell Dev Biol-Plant.,* **38**: 581-8.

Spencer, A., Hamill, J.D., and Rhodes, M.J.C.(1993). In vitro biosynthesis of monoterpenes by *Agrobacterium* transformed shoot cultures of two *Mentha* species. *Phytochemistry,* **32**: 911–919.

St-Pierre B., Vazquez-Flota, F.A., De Luca, V.(1999). Multicellular compartmentation of *Catharanthus roseus* alkaloid biosynthesis predicts intercellular translocation of a pathway intermediate. *Plant Cell.,* **11**: 887–900.

Tepfer, D.(1990). Genetic transformation using Agrobacterium rhizogenes. *Physiol Plant.,* **79**: 140-6.

Tyler, T. *et al.* (1989). In "Primary and Secondary Metabolism of Plant and Cell Cultures" Ed. Kurz W.G.W., *Springer Verlag,* Berlin. pp. 200-207.

Van Uden, W., Bos, J.A., Boeke, G.M., Woerdenbag, H.J., Pras, N.(1997). The large scale isolation of deoxypodophyllotoxin from rhizomes of Anthriscus sylvestris followed by its bioconversion into 5-methoxypodophyllotoxin b- D-glucoside by cell cultures of *Linum flavum*. *J. Nat. Prod.,* **60**: 401–403.

Verpoorte, R.(1999). Secondary metabolism. In: Metabolic engineering of plant secondary metabolism. Verpoorte, R., and Alfermann, A.W. eds. *Klumer Academic Publishers,* the Netherlands. pp. 1-29.

Verpoorte, R., Memelink, J.(2002). Engineering secondary metabolite production in plants. *Current Opinion in Biotechnology,* **13**: 181–187.

Verpoorte, R., van der Heijden, R., van Gulik, W.M., and ten Hoopen, H.J.G.(1991). Plant biotechnology for the production of alkaloids: present status and prospects. *In: Brossi A (ed) The Alkaloids. Academic Press, San Diego.* **40**: 1–187.

Vinterhalter, B., Jankovic, T., Sovikin, L., Nikolic, R., Vinterhalter, D.(2008). Propagation and xanthone content of *Gentianella austiaca* shoot cultures. *Plant Cell Tiss. Org.* Cult., **94**: 329-335.

Wang, H. Q., Yu, J. T. Zhong, J. J.(1999). Significant improvement of taxane production in suspension cultures of Taxus chinensis by sucrose feeding strategy. *Process Biochemistry,* **35**: 479-483.

Warzecha, H., Gerasimenko, I., Kutchan, T.M., Stockigt, J.(2000). Molecular cloning and functional bacterial expression of a plant glucosidase specifically involved in alkaloid biosynthesis. *Phytochemistry,* **54**: 657–666.

Weber, T., Welzel, K., Pelzer, S., Vente, A. Wohlleben, W.(2003). Exploiting the genetic potential of polyketide producing streptomycetes, *J Biotechnol* **106**: 221-232.

Whitmer, S., Van der Heijden, R., Verpoorte, R.(2002). In *Plant Biotechnology and Transgenic Plants* Oksman-Caldentey, K. -M.; Barz W.H., Eds.; Marcel and Dekker: New York-Basel; pp. 373-405.

Wink, M.(1999). Introduction: Functions of plant secondary metabolites. In: Functions of Plant Secondary Metabolites and their Exploitation in Biotechnology. *Wink, M. ed. Sheffield Academic Press Ltd.*, England, pp. 1-16.

Yeoman, M.M., Lindsey, K., Miedzybrodzka, M.B., McLauchalam, W.R.(1982). Accumulation of secondary products as effect of differentiation in plant cell and tissue cultures.In: Yeoman, M.M., Truman, D.E.S. (eds) Differentiation *in vitro* vol.4, British Society for Cell Biology symp. Univ.Press, Cambridge. pp. 65.

Yun, D.J., Hashimoto, T., Yamada, Y.(1992). Metabolic engineering of medicinal plants: transgenic *Atropa belladonna* with an improved alkaloid composition. *Proc Nat Acad Sci* USA **89**: 11799-803.

Zenk, M.H.(1978). The impact of plant cell culture on industry. In: Thorpe TA (ed). *Frontiers of Plant Tissue Culture,* University of Calgary, *Int Ass for Plant Tissue Culture,* pp. 1- 13.

Zhao, J., Davis, L.C., Verpoorte, R.(2005). Elicitor signal transduction leading to production of plant secondary metabolites. *Biotechnol. Adv.,* **23**: 283-333

2015, Major Constrains and Verdict of Crop Productivity *Pages 596–606*
Editor: Dr. U.N. Bhale
Published by: ASTRAL INTERNATIONAL (P) LIMITED, NEW DELHI

Chapter 47

Mutation Breeding to Increase Yield and Productivity in Okra (*Abelmoschus esculentus* L. Moench)

Navnath G. Kashid*

Department of Botany,
Vasant Mahavidyalaya, Kaij, Dist. Beed – 431 123, M.S., India

ABSTRACT

The yield and productivity character in plants is genetically controlled. The expression of yield controlling gene is also influenced by environment and many of the high yielding mutants require different conditions than the one by their parental counterpart for better performance. Okra is a good source of carbohydrates, minerals, vitamin-A and vitamin-C. It is stuffed with meat or cooked into soup. It is also canned and dehydrated for off-season consumption by army at high altitudes and for exports. In Okra variety Parbhani Kranti and Arka Anamika high yielding mutants showed green pod yield higher than control. There was a positive correlation between nodes on main stem and pods per plant. The frequency of viable mutants revealed an increasing trend with gradual increase in dose/concentration of all the mutagens in variety Parbhani Kranti. In variety Arka Anamika the frequency of viable mutants increased with an increasing dose/concentration of EMS and SA treatments while it fluctuated in gamma ray treatment of variety Arka Anamika.

* E-mail: ngkashid@gmail.com

The chemical mutagens showed increasing trend with increasing concentrations in both varieties of okra viz. Parbhani Kranti and Arka Anamika. In both the varieties of the plant, the viable mutants obtained were of the following types. 1) Dwarf mutant 2) Tall mutant 3) Branched 4) Early flowering 5) Long pod 6) Early maturing 7) Late maturing 8) High yielding.

Keywords: *EMS, SA and Gamma rays, Yield, Productivity, Okra.*

Introduction

Okra or 'Lady's Finger', commonly known as "bhindi" in India, comprises one of the most important fruit vegetables grown throughout the tropics and warmer parts of the temperate zone. Bhindi has a vast potential as one of the foreign exchange earning crops and accounts for about 60 per cent of the export of fresh vegetables, excluding potato, onion and garlic. The important countries like Middle East, Western Europe and USA, import bhindi from others (Sharma and Arora, 1993).

The mutation breeding has become an alternative to conventional breeding since last three decades with the sole objective of developing better cultivars of economically important crops. The conventional plant breeder depends solely on the spontaneous mutations. However the frequency of spontaneous mutation is too low to become the basis for systematic breeding, even taking into account the internal and external factors contributing to them. An increase in mutation rate by mutagenic agent increases the chance occurrence of desired mutations and helps in accelerating the breeding process as compared to conventional method of developing novelties.

Several hundreds of crop varieties have been developed through induced mutagenesis and released till to date. In addition to these several varieties were developed in ornamental plants through similar methods. A remarkable change can be seen after going through the practical results monitored by the FAO/IAEA. The number of varieties in crops developed through the use of induced mutation increased from 98 to 335 and in ornamentals increased from 47 to 271 during 1973 to 1984 (IAEA mutation breeding Review 1985) which suggests and demonstrates the enormous potential of mutation breeding in crop improvement.

Induction of mutations is an important method for improving the specific characteristic and rectifying one or few undesirable characters in the existing commercial varieties. The created variability can further be utilized in selection or hybridization to evolve new cultivars.

However several successful plant improvement ventures have been undertaken with cross-fertilizing species exposed to mutagenic agents (Sigurbjornsson and Micke, 1969) and the induction of mutation has been accepted as a useful tool in plant breeding programmes. In case of vegetable crop the utility potential of mutation breeding has been nicely demonstrated by different researchers for increasing the productivity in different systems like Brinjal (Datar and Astapure, 1984), Chilli (Gupta and Yadav, 1984) and in Custer bean (Basha and Rao, 1988).

Materials and Methods

The seed material of two varieties of okra (*Abelmoschus esculentus* L. Moench.) namely Parbhani Kranti and Arka Anamika obtained from the Marathwada Agricultural University, Parbhani was used in the present study.

In the present investigation the above mentioned two varieties of okra were treated by three mutagens. *viz.* physical mutagen, gamma rays and two chemical mutagens namely Ethyl methanesulphonate (EMS) and Sodium azide (SA). Electromagnetic ionizing radiations were applied from a CO^{60} 1000 curie source of the gamma irradiation unit of the Department of Chemistry, Nagpur University, Nagpur. The dose rate was 24.578 rads per hour (Konzak *et al.*, 1965).

The lot of 300 seeds from each treatment was sown in field following randomized block design (RBD) with three replications along with controls for raising the M_1-generation. The seeds were sown at a distance of 30cm between plants and 45cm between rows. Individual plant was bagged to avoid cross-pollination. Seeds of 25 normal looking plants of M_1 selected at random were collected on individual plant basis from all treatments and control. They were used for raising the M_2-generation on plant to a row basis.

From each treatment 25 plants were randomly selected for recording data on different quantitative characters in both M_2 and M_3 generations. Similarly 25 plants were picked up from controls for comparative assessment.

Statistical Analysis

The data were subjected to statistical and biometrical analysis for detecting the micromutations as given below.

The statistical model used for randomised block design comprised.

$$yijk = u + gij + bk + eijk$$

where,

yijk = Phenotypic performance of ij^{th} mutations in k^{th} block,

u = General mean,

gij = The effect of ij^{th} mutations,

bk = The effect of kth block and

eijk = The environmental effect.

The analysis of variance based on this model is given below (Gaul, 1965)

Source of Variation	*d. f.*	*Sum of squares*	*Mean Squares Observed/Expected*	
Replications	r-1	Ssr	Msr	
Treatments	t-1	Sst	MSt	$s^2e + rs^2t$
Error	(r-1)(t-1)	Sse	Mse	s^2e

r: Number of replications; t: Number of treatments.

The Mean squares were tested against error variance by usual 'f' test. The standard error of difference for comparing any two progeny means was estimated by formula. (Mungikar, 1997).

SE(diff) = 2Mse/r

The critical difference was computed by multiplying the standard error of difference with 't' test values for (r-1)(t-1) degrees of freedom at 5 per cent and 1 per cent levels.

Various statistical data were calculated by using the following formulae.

Mean = $\Sigma x/n$

Variance = $(Sx^2/n) - x^2$

Standard deviation (SD) = $\sqrt{\text{Variance}}$

Standard Error (SE) = $SD/\sqrt{N}$

Coefficient of variation (CV) = $(SD/Mean) \times 100$.

Results and Discussion

Viable Mutants

Quite a good number of viable mutants have been observed in M_2 generation of both the varieties of okra. The frequency of viable mutants revealed an increasing trend with gradual increase in dose/concentration of all the mutagens in variety Parbhani Kranti. In variety Arka Anamika the frequency of viable mutants increased with an increasing dose/concentration of EMS and SA treatments while it fluctuated in gamma ray treatment of variety Arka Anamika.

The chemical mutagens EMS and SA showed increasing trend with increasing concentrations in both varieties of okra *viz.* Parbhani Kranti and Arka Anamika. In both the varieties of the plant, the viable mutants obtained were of the following types. 1) Dwarf mutant 2) Tall mutant 3 Branched 4) Early flowering 5) Long pod 6) Early maturing 7) Late maturing 8) High yielding.

Viable Mutants in Variety Parbhani Kranti (Table 47.1)

Dwarf Mutant

These mutants were characterized by an extreme reduction in plant height. It was 88.72 cm as against 108.36 cm for control. All the characters in this mutant demonstrated reduced values as compared with control.

Tall Mutant

The tall mutant showed a height of 136.28 cm as against 108.36 cm in control. These mutants showed positive correlation between heights of the plant with nodes on main stem. The period for maturity in this mutant was comparable with that of control.

Table 47.1: Effect of Mutagens on Frequency and Spectrum of Viable Mutants in M_2 Generation of *Abelmoschus esculentus* (L.) Moench.

Variety: Parbhani Kranti

Mutagens	*Concentration*	*Frequency of Viable Mutants*	*Relative Percentage*							
			Dwarf Mutant	*Tall Mutant*	*Branched Mutant*	*Early Flowering*	*Long Pod Mutant*	*Early Maturing*	*Late Maturing*	*High Yielding*
Control	Control	–	–	–	–	–	–	–	–	–
	5kR	10.15	1.17	1.49	1.97	1.29	1.29	0.78	0.96	1.20
Gamma rays	10kR	12.92	3.27	2.64	2.90	1.45	–	0.65	1.13	1.88
	20kR	13.90	1.84	1.08	–	2.41	1.26	1.58	3.21	2.52
	0.05	14.26	2.16	1.37	2.50	0.97	2.05	2.81	1.14	1.26
EMS (per cent)	0.10	15.01	0.62	2.81	1.43	3.16	1.10	2.98	2.03	0.88
	0.15	15.76	2.08	1.68	–	2.92	2.00	3.28	1.76	2.04
	0.01	14.42	0.78	2.81	1.89	2.19	1.27	0.86	1.98	2.64
SA (per cent)	0.02	16.52	4.16	–	2.53	2.77	1.18	2.52	2.26	1.10
	0.03	21.93	3.76	2.32	2.88	3.11	–	4.89	4.28	2.69

Branched Mutant

These mutants showed 4-5 branches. They exhibited reduction in nodes on branches and pod length (14.8 cm).

Early Flowering Mutant

These mutants attained flowering almost seven days earlier than the control population. They flowered in 36.37 days as against 43.62 days for control. These mutants showed positive correlation between days to flowering and 100 seed weight (6.10 gm).

Long Pod

In these mutants the length of pod increased than control. The length of the pod in these mutants was 20.32 cm as against 17.24 cm in control. They matured at par with control. These mutants showed positive correlation between long pod with green pod yield per plant.

Early Maturing Mutant

These mutants matured almost fifteen days earlier than the control population. They matured in 104.02 days quite earlier to control (120 days). There was no significant difference in pod yield per plant (372.63 gm) in such mutants as against (398.27 gm) in control.

Late Maturing Mutant

These mutants matured quite late than control. They took 130.04 days for maturity as against 120.00 days in control. They exhibited reduction in 100 seed weight (4.2 gm).

High Yielding Mutant

These mutants showed green pod yield higher than control. They showed 448.16 gm of yield per plant against the yield value of 398.27 gm for control. There was a positive correlation between nodes on main stem and pods per plant.

Viable Mutants in Variety Arka Anamika (Table 47.2)

Dwarf Mutant

These mutants were characterized by an extreme reduction in plant height. It was 79.43 cm as against 102.26 cm for control. All the characters in this mutant demonstrated reduced values as compared with control.

Tall Mutant

The tall mutant showed a height of 128.14 cm as against 102.26 cm in control. These mutants showed positive correlation between heights of the plant with nodes on main stem. The period for maturity in this mutant was comparable with that of control.

Table 47.2: Effect of Mutagens on Frequency and Spectrum of Viable Mutants in M_2 Generation of *Abelmoschus esculentus* (L.) Moench.

Variety: Arka Anamika

Mutagens	*Concentration*	*Frequency of Viable Mutants*	*Relative Percentage*							
			Dwarf Mutant	*Tall Mutant*	*Branched Mutant*	*Early Flowering*	*Long Pod Mutant*	*Early Maturing*	*Late Maturing*	*High Yielding*
Control	Control	–	–	–	–	–	–	–	–	–
	5kR	13.25	2.04	–	3.18	1.17	0.89	2.26	1.32	2.39
Gamma rays	10kR	13.74	1.26	1.05	0.99	4.12	–	0.68	2.46	3.18
	20kR	12.16	–	2.25	0.67	3.20	1.28	0.92	1.36	2.48
	0.05	14.71	1.98	2.23	–	3.68	0.98	1.84	2.28	1.72
EMS (per cent)	0.10	14.96	2.12	–	–	4.19	1.27	2.29	1.68	3.41
	0.15	16.04	2.86	1.92	1.58	2.89	2.12	0.78	1.76	2.13
	0.01	13.92	0.87	2.98	3.12	2.87	–	1.44	0.96	1.68
SA (per cent)	0.02	15.83	3.48	1.67	1.89	0.73	2.24	2.02	2.68	1.12
	0.03	18.34	2.18	1.22	2.24	3.81	1.92	1.62	2.28	2.07

Branched Mutant

These mutants showed 5-6 branches. They demonstrated reduction in number of nodes on branches and reduced height.

Early Flowering Mutant

These mutants attained flowering almost eight days earlier than the control population. They flowered in 40.98 days as compared with 49.02 days for control. These mutants showed positive correlation between days to flowering and 100 seed weight (6.5 gm) and days to flowering.

Long Pod

These mutants showed long pod character (14.32 cm) as compared with control (12.18 cm). They matured earlier than that of control. These mutants showed positive correlation between pod lengths with green pod yield per plant.

Early Maturing Mutant

These mutants matured almost ten days earlier than the control population and they matured in 108.28 days and had decreased 100 seed weight (6.1 gm).

Late Maturing Mutant

These mutants matured late than control. They took a maturity period of 139.20 days as against 118.42 days for control. They demonstrated reduced pod per plant.

High Yielding Mutant

These mutants showed higher pod yield (300.13 gm). These mutants showed positive correlation between nodes and pods on plants.

In M_1 generation studies were carried out on the differential effects of different mutagens on biological parameters and morphological abnormalities. In M_2 generation grown from the harvested seeds of M_1, studies were organized pertaining to chlorophyll mutations especially their frequency and spectrum, effectiveness and efficiency of the mutagens used besides screening and analysis of the quantitative characters, and several viable macro mutants which were selected from M_2 population have also been and reported.

In M_3 generation grown from the harvested seeds of M_2, the study included screening and analysis of quantitative characters and field testing of the viable macro-mutants isolated

The mean values for green pod yield showed increased feature in majority of treatments in both the varieties of okra, in M_2 generation. In variety Parbhani Kranti, the mean values exhibited shift in positive direction in all the treatments. The mean values demonstrated a positive shift than control in variety Arka Anamika except 0.02 per cent SA treatment.

In M_3 generation the significant positive shift in mean values was observable at 0.05 per cent, 0.10 per cent and 0.15 per cent EMS treatments in variety Parbhani Kranti and at 0.15 per cent EMS and 0.03 per cent SA treatment in variety Arka

Anamika. At remaining treatments the mean values with positive and negative shifting feature could be evidently seen.

In variety Parbhani Kranti high yielding mutants showed green pod yield higher than control. They showed 448.16 gm of yield per plant against the yield value of 398.27 gm for control. There was a positive correlation between nodes on main stem and pods per plant. In variety Arka Anamika high yields mutants showed higher pod yield (300.13 gm). These mutants showed positive correlation between nodes and pods on plants.

It has been decisively demonstrated by several researches like Gregory (1955), Rawlings *et al.* (1958) and Oka *et al.* (1958) that in progeny population derived from mutagenised parental plants, the variability as regards quantitative characters increases while the mean of characters may or may not change. This increase in variability has been ascribed to mutations of minor genes controlling characters.

Looking at the immense economic importance of the okra crop as a fine source of green vegetable with good medicinal value, the improvement in its yielding potential has been considered necessary. This crop though unstable in its area and productivity in our state is gradually gaining importance chiefly due to its multifaceted utility value.

Some mutants showing high yielding character were also very much prevalent in the M_2 population of okra. Similar reports have been made in the progeny of gamma ray treated okra by Nandpuri *et al.* (1971). The yield of fruits per plant was higher only in gamma radiation treatment than the chemical mutagen treatment. (Jahagirdar and Chandrasekharan, 1978). Tickoo and Jain (1979) with EMS and gamma rays obtained high yielding and disease resistant mutants in *Vigna radiata*, Pawar *et al.* (1981) in mungbean.

A wider variability in the yield parameters of mutagen treated French bean has been reported by Swarup and Gill (1968). Similar features in M2 and M3 generations have been recorded by Hakande (1992) in winged bean, More (1992) in *Medicago sativa*, Satpute (1994) in safflower, Khandelwal (1996) in winged bean, Sawant (2008) in sesame and Bhosale (2009) in cluster bean.

References

Basha S. K. and Rao P. G. (1988). Gamma ray and Sodium azide induced heterophylly of bhindi. *J. Nuclear Agric. Biol.*, **17**: 133-136.

Bhosale, S. S. (2009). Genetic lmprovement of Cluster bean (*Cyamopsis tetrogonoloba* (1.) Taub.) by mutation breeding.Ph. D. Thesis, Dr. Babasaheb Ambedkar Marathwada University, Aurangabad, lndia.

Datar, V. V. and Astapure J. U. (1984). Reaction of brinjal variety, F1-hybrids and solanum species to little leaf diseases. *J. Maharashtra Agric. Uni.*, **9**(3): 325-334.

Gaul, H. (1965). The concept of macro and micro mutations and results on induced mutations in plant breeding (FAO/IAEA meeting Rome) Pergamon Press, p.407.

Gregory, W. C. (1955). X-ray breeding of peanut (*Arachis hypogea* L.), *Agron. J.*, **47:** 396-399.

Gupta, C. R. and Yadhav, R. D. S. (1984). Genetic variability and path analysis in chilli. *Genetic Agraria*, **38**: 425-432.

Hakande, T. P. (1992). Cytogenetical studies in *Psophocarpous tetragonolobous* (L.) DC Ph. D. Thesis, Marathwada University, Aurangabad, India.

Jahagirdar K. S. ans Chandrasekaran P. (1978). Comparative mutagenic effects of gamma rays and diethyl sulphate in bhindi. *Madras Agri. J.*, **65**(4): 211-217.

Khandelwal, A. R. (1996). Cytological Biochemical and Mutational Studies in *Psophocarpus tertragonolobus* L. (DC.). Ph.D Thesis, Dr. B. A. Marathwada University, Aurangabad. MS. India.

Konzak, C. F., Nilan, R. A., Wagner J and Foster R. J. (1965). Efficient chemical mutagenesis. In: "The use of induced mutation in plant breeding", *Rad. Bot. (Supl.)*, **5**: 49-70.

More, A.D. (1992). Cytogenetical studies in *Medicago sativa* L. Ph. D. Thesis, Marathwada University, Aurangabad, India.

Mungikar, A M. (1997). An introduction to Biometry, book published by Saraswati Printing Press. Motikaranja, Aurangabad. M. S. India.

Nandpuri, K. S., Sandhu, K. S. and Randhava, K. S. (1971). Effect of irradiation on variability of okra (*Abelmoschus esculentus* (L.) Moench) *J. Res. P.A. U.*, **8**: 183-188.

Oka, H. I, Hayashi J. and Shiofiri, I. (1958): Induced mutation of polygenes for quantitative characters in rice. *J. Hered.*, **49**: 11-14.

Pawar, S. E., Joshua, D. C., Mitra, R. and Bhatia, C. R. (1981). Variation in some physiological components of yield in induced mutants of mung bean. Proc., *Int. symp. Vienna 9-13 March, IAEA*, pp.213-226.

Rawlings, J.O., Hanway, D.G. and Gardner, G.O. (1958). Variation in quantitative characters of soybean after seed irradiation. *Agron. J.*, **40**: 424-528.

Satpute, R. A. (1994).Mutational studies in safflower (*Carthamus tictorious* L.) Ph. D. Thesis, Dr. B. A. Marathwada University, Aurangabad, India.

Savant, K D. (2008). Genetic improvement of sesame (*Sesamum indicum* L. through induced mutations. Ph. D. Thesis, Dr. Babasaheb Ambedkar Marathwada University, Aurangabad, India.

Sharma, B.R. and Arora, S. K. (1993).Advances in Horticulture vol-5. Vegetable crop: part-1.

Sigurbjornsson, B. and Micke, A. (1969): Progress in mutation breeding. *Proc. "Induced mutation in plants", IAEA, Vienna*, pp.673-698.

Swarup, V. and Gill, H. S. (1968). X-ray induced mutations in French bean and roses. Intern. Symp. "Use of Isotopes, radiations in agricalte and Animal Husbandry Research". New Delhi., pp.13-25.

Tickoo, J. L. and Jain, H. K. (1979). Breeding high yielding varieties of mung (*Vigna radiata* (L) Wilczek) through mutagenesis Proc. Symp. On "The role of induced mutations in crop Improvement." Dept. of Genetics, Osmania University, Hyderabad. September 10-13, 1989, pp.198-203.

2015, Major Constrains and Verdict of Crop Productivity Pages 607–610
Editor: Dr. U.N. Bhale
Published by: ASTRAL INTERNATIONAL (P) LIMITED, NEW DELHI

Chapter 48

Cultivation Practices of *Kalmegh* in Marathwada Region, M.S.

D.A. Dhale*

Post-graduate Department of Botany,
SSVPS's, Late K. Dr. P.R. Ghogrey Science College,
Dhule – 424 005, M.S., India

Andrographis paniculata Nees. belonging to family Acanthaceae known as 'King of the bitters' or 'Kalmegh'. It is one of the medicinal plants that are recommended for cultivation of India, as there is great demand for the plant by pharmaceutical industries mainly for exports. It is source of several diterpenoids of which andrographolide (alkaloid) is important. The drug is used for treating general debility, dyspepsia, chronic malaria, jaundice and dysentery. Some scientists have observed that andrographolide has the potential to be included in the cocktail vaccine against AIDS by virus of its antagonistic property with HIV II virus (Weibo, 1995). It has been already used for treating cancer as it promotes cell differentiation in tumour cells (Mastuda *et al.*, 1994). Though leaves contain maximum Andrographolide (not less than 1.0 per cent), the entire plant is used for extracting the active ingredient. Kalmegh contains bitter principles andrographolide, a bicyclic diterpenoid lactone and kalmeghin (0.85 – 2.5 per cent).

India possesses all types of climatic conditions varying from temperature in Himalayas to tropical in south India, dry in Central India to humid and wet in Assam and Kerala, thus providing different type of conditions in various regions. Such types of condition affects on the growth and commercial cultivation of the medicinal plant.

* *Corresponding Author.* E-mail: datta.dhale@yahoo.com

Figure 48.1: *Andrographis paniculata* Nees.

Cultivation practices for Kalmegh are carried out in India at various Agricultural Universities and Institutes. The data shows great variations in the yield, so it is necessary to take experiment for authenticate the yield value by which cultivators promote toward it.

A microplot experiment was conducted on Herbal garden, department of Pharmacy, Homoeopathic Medical College, MIDC, Latur. The experiment was laid out in a Randomized Complete Block design. The net plot size was 1 x 1 m with proper irrigation channels. Well decomposed farm yard manure is the rate of 2.5 kg per plot was applied and mixed thoroughly. Similar Recommended dose of N (75 kg/ha), K (41 kg/ha) and P (22 kg/ha) were applied in the form of urea, muriate of potash and single super phosphate. Half the doses of N along with full dose of P and K were applied at the time of planting, while the remaining half of N was applied four weeks later. Seedlings of kalmegh were first raised in nursery by sowing seeds n polythene bags containing a mixture of soil: sand: FYM in ratio of 1:1:1. Fifty day old seedlings were transplanted to the field. Plots were irrigated twice a weak for the first 4 weeks and subsequently at weekly intervals to maintain enough moisture and the plants raised for 90 days after transplanting. The observations on growth parameters dry matter of whole plant were measured physically on top loading balance and resulting weight were recorded in gram.

India (8°-30° N and 68-97.5° E) exhibits a wide range in topography and climate, which has been directly affects on total yield. Many of institutions show great variations in net yield of total dry mass of kalmegh. These are varying from minimum of 700-800 kg/ha (Dumbre and Bhoete, 2000; Joshi, 2000). or the moderate 2000-2500 kg/ha (Kattimani *et al.*, 2003; Baliraja, 2001) while moderately highest 3000-4000 kg/ha (Bhosale, 2001) where as the highest amount of yield shown as 3500-4000 kg/ha (Deshmukh, 2002). Beyond these the average result (Table 48.1) shows the highest value of yield (4560-5112 kg/ha) Further studies in order.

Table 48.1: Observations of Yield of Kalmegh

Sl.No.	*Weight of 10 Plants in kg*	*Yield (24000 Plants/ha) in kg/ha.*	*Average Yield in kg/ha*
1.	1.750	4200	
2.	1.830	4392	
3.	1.550	3720	
4.	2.680	6432	
5.	1.490	3576	4560-5112
6.	1.900	4560	
7.	2.200	5280	
8.	1.450	3480	
9.	2.400	5760	
10.	2.130	5112	

Recommended practices for Kalmegh Cultivation

Soil type	Loamy medium black
Transplanting	15th July (Nursery in June)
Spacing	45 x 60 cm.
Seed - Seedling	Seed 200 gram/ha 24000 plants/ha
Manure/ha	Basal FYM 10 ton.and RDF
Irrigation	5 to 10
Harvesting	90 to 100 Days
Yield (Total herbage)	4560 to 5112 kg/ha. (Dry)

References

Baliraja, (2001). Special issue on Medicinal and Aromatic Plants. Nov. 2001, p. 52.

Bhosale Nitin 2001. Special issue on Medicinal and Aromatic Plants. *Baliraja Magazine* Nov. 2001, pp. 52.

Deshmukh P.G. (2002). Information Booklet *Vanayshadhi Lagwad*. PP. 64-66.

Dumbre, A. and Bhoete K.(2000). Medicinal and Aromatic Plants Cultivation, MPKV EXTN PUB No. 358. Published by Mahatma Phule Agricultural University, Rahuri, pp. 21-23.

Joshi, B.P. (2000). Cultivation of Medicinal and Aromatic Plants, Conntinental publication, Pune, pp.84-86

Kattimani, K. N., Hegde L. and Venugopal C.K. (2003). Cultivation practices of Medicinal and Aromatic Plants, Publication of University of Agricultural Sciences, Dharwad, p.32.

Mastuda, T. M., Kuroyanagi, S., Sugiyama, K., Umehara, A. U. and Nishi, K. (1994). cell differentiation inducing diterpenes from *Andrographis paniculata* Nees. *Chem. Pham. Bull.*, **42**: 1216-1225.

Weibo, L. (1995). Prospects for study on treatment of AIDS with traditional Chines medicine. *J.Traditional Chinese Med.*, **15:** 3-9

2015, Major Constrains and Verdict of Crop Productivity Pages 611–623
Editor: Dr. U.N. Bhale
Published by: ASTRAL INTERNATIONAL (P) LIMITED, NEW DELHI

Chapter 49

Survey and Prioritization of Crop Wild Relatives of Balaghat Ranges of Maharashtra, India

R.D. Gore*, K.U. Garad and S.P. Gaikwad

Life Science Research Laboratory,
Walchand College of Arts and Science,
Solapur – 413 006, Maharashtra, India

ABSTRACT

The application of the various criteria for selection of crop wild relative species led to a list of 96 flowering plant species belongs to 43 genera and 20 families in Balaghat Ranges. Of them 10 species are relatives of pulses, 15 species of oil and fiber crops, 26 species are food and fodder crops, 27 species are vegetables, 7 species are fruit crops and 11 species are relatives of other useful crops. In order to establish some estimate of the degree of relatedness between crops and their wild relatives in Balaghat Ranges of Maharashtra adopted gene pool concept of Harlan and de Wet and Maxted et al., classification. Out of 96 CWR species, about 19 CWR species fall into gene pool-1b category, 10 into gene pool-2 category, 4 into gene pool-3 category, 6 into taxon group-1b, 3 into taxon group-2, 8 into taxon group-3 and 46 into taxon group-4. In addition to this, information on distribution, life form, economic types and threat assessment is provided in present paper. The present study will be helpful for future target collections and conservation to serve as reference for management of genetic resources.

Keywords: *Crop wild relatives (CWR), Collection, Prioritization, Balaghat ranges, Maharashtra.*

* *Corresponding Author.* E-mail: ramdgore@gmail.com

Introduction

Crop wild relatives (CWR) can be defined as taxa related to species of direct socio-economic importance, including food and fodder crops, medicinal plants, condiments, ornamental and forestry species, as well as plants used for industrial purposes. It has known from thousands of years that CWR germplasm (genetic material) can be used for improvement of the quality and yield of crops. Plant breeders have been utilizing CWR genes to improve crops for resistance to pests and diseases, and abiotic stresses such as drought and salinity since hundreds of years. With increasing genetic uniformity of modern crops and associated vulnerability to new strains of pests and diseases it is widely recognized that CWR are vital for improving agricultural production, increasing food security, and sustaining the environment.

Crop wild relatives are essential components of natural and semi-natural habitats, as well as agricultural systems, and are critical for maintaining ecosystem health. Their conservation and sustainable use is vital for improving agricultural production and increasing food security, as well as maintaining the environment. Crop improvement, particularly under climate change, depends on the genetic diversity of our plant genetic resources, which are arguably inadequately conserved and poorly used. There is wide recognition that the Convention on Biological Diversity's 2010, targets to reduce the loss of biodiversity has not been met. Biodiversity is at risk from multiple threats, including climate change, and the genetic diversity contained within plant genetic resources, particularly of species that are wild relatives of our crops, faces similar threats but is essential to our ability to respond to the new stresses in the agricultural environment resulting from climate change. CWR are a critical source of genes for resistance to pests and diseases and abiotic stresses. Hence, it is important to consider the genetic value of these crop-wild-relatives and think how they may be conserved, and what new technologies can be implemented to enhance their use.

The native cultivated 166 plant species have been found in Indian gene centers (Vavilov, 1949-1950; Zeven and de Wet, 1982) and over 320 plant species are the wild relatives (Arora, 2000). In India the wild relatives of crop plants largely occur as constituents of disturbed habitats within the major vegetation types with distribution in the warm humid tropical and sub-tropical regions and in the Western Himalaya with low representation in the drier parts of northwestern region (Arora and Pandey, 1996; Maxted *et al.*, 1997; Arora, 2000; Vincent *et al.*, 2012). More than 100 wild relatives and related taxa and endemic, rare, endangered species occur predominantly in the hot-spots of India (Ahmedullah and Nayar, 1986; Nayar, 1996; Pandey and Arora, 2004 and Pandey *et al.*, 2005). But the key matter is to delimitate and which species are to be included in the so-called group of CWR. Taking into consideration the idiosyncrasy of each region it is important to apply a precise definition of the concept of CWR in order to determine a clear group of plant species that can be thought for further action. The case of Balaghat Ranges, one of the regions with greatest CWR diversity in Maharashtra with well 96 species, constitute a challenge for deciding how many and which species to be included in CWR list. Similarly, relatives of promissory species in emerging markets which currently are as important as major crops and relatives of ornamentals should be considered in CWR prioritization actions.

Materials and Methods

Study Area

Balaghat Ranges is eastern spur of Western Ghats of India lies in between 18° 47' 40.26" - 18° 32' 29.28" N longitude and 75° 20' 26.90" - 76° 48' 56.18" E latitude, mainly spread over in Ahmednagar, Beed, Latur, Osmanabad and Solapur districts of the Maharashtra State of India in the basins of river Manjra and its tributaries Terna, Bori, Tiru and Manyad (Figure 49.1). Balaghat occupies an area about 18,111. 34 sq.km. The terrain of Balaghat shows hills and hillocks of varying heights (1520 ft - 2700 ft.) which support unique tropical dry deciduous forests, open scrub vegetation and vast grasslands. The weather in general said to be dry and moderately extreme. The average temperature ranges from 27.7° C to 42° C and the relative humidity is

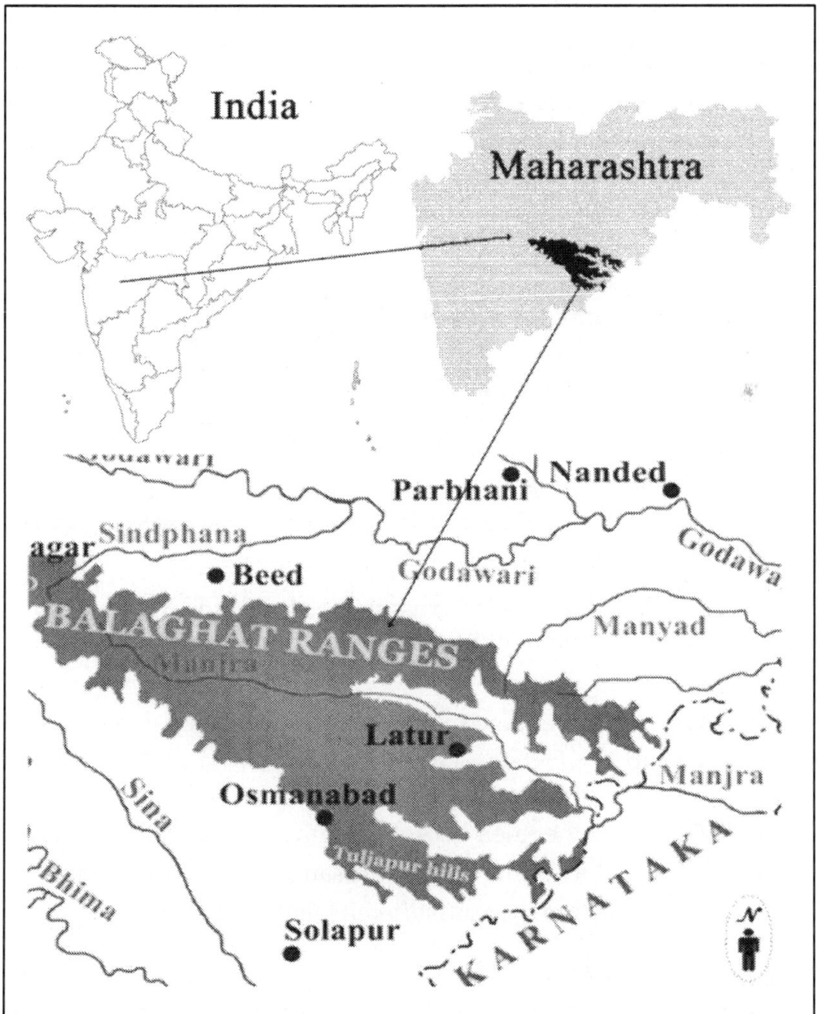

Figure 49.1: Location Map of Balaghat Ranges of Maharashtra.

extremely low (35-50 per cent) for major part of the year while it is highest (75 per cent) during the monsoon season (June - October). The average rainfall is about 729 mm mainly from June to October.

Data Collection

First of all a preliminary list of Crop Wild Relative species for Balaghat ranges of Maharashtra was prepared by referencing to all available literature. After that field trips of 1-4 days were undertaken to different nooks and corners of the Balaghat region for their collection. During the field visits, observations were made on distribution, present status, reproductive behavior and habitat of the CWR species. The taxonomic identifications made with the help of available literature (Cooke, 1958 (Repr. ed.); Naik, 1979, 1998; Lakshminarasimhan, 1996; Singh and Karthikeyan (eds.), 2000, 2001 and Gaikwad *et al.*, 2012a, 2012b). Doubtful identifications were confirmed by comparison of our specimens with authentic specimens deposited in Botanical Survey of India, Pune (BSI), Blatter Herbarium, Mumbai (BLAT) and Dr. BAMU Herbarium, Aurangabad (BAMUA).

To establish some estimate of the degree of relatedness between crop and wild relatives in order to prioritize wild genetic resources for conservation adopted the Gene Pool Concept proposed by Harlan and de Wet (1971), which uses a hierarchy of three gene pools.

> **Gene Pool 1:** within which Gene Pool 1A is the crop itself and Gene Pool 1B the wild forms of the crops including those which grow as agricultural weeds.
>
> **Gene Pool 2:** which includes less closely related species from which gene transfer to the crop is possible but more difficult using conventional breeding techniques.
>
> **Gene Pool 3:** which includes the species from which gene transfer to the crop is impossible, or if possible only using sophisticated techniques such as genetic engineering.

In the absence of knowledge of crossing relationships Maxted *et al.* (2006) classification of crop plant relatives was used. It is as follows

> **Taxon Group 1a**: the crop.
>
> **Taxon Group 1b**: the same species as the crop but not domesticated.
>
> **Taxon Group 2**: members of a group of closely related species forming a series or section of the genus in which the crop occurs.
>
> **Taxon Group 3**: other members of the same subgenus as the crop.
>
> **Taxon Group 4**: other members of the same genus as the crop.
>
> **Taxon group 5**: same tribe but different genus to the crop.

Results and Discussion

The application of the above mentioned criteria for selection of crop wild relative species in Balaghat region led to a list of 96 flowering plant species belongs to 43 genera and 20 families (Table 49.1). Out of 43 genera, 13 are representing category of

Table 49.1: Showing Crop Wild Relatives (CWR) of Balaghat Ranges of Maharashtra

Sl.No.	*Name of the Species*	*Name of the Family*	*Habit*	*Type of Crop*	*Prioritization Category*
1.	*Abelmoschus ficulneus* (L.) Wight and Arn. ex Wight.	Malvaceae	Herb	Vegetable	GP-2
2.	*Abelmoschus moschatus* Medik.	Malvaceae	Herb	Other Uses	TG-4
3.	*Abelmoschus tuberculatus* Pal and Singh	Malvaceae	Herb	Vegetable	GP-2
4.	*Amaranthus caturus* Heyne ex Hook.*f.*	Amaranthaceae	Herb	Food and Fodder	GP-2
5.	*Amaranthus caudatus* L.	Amaranthaceae	Herb	Food and Fodder	GP-2
6.	*Amaranthus roxburghianus* Nevski.	Amaranthaceae	Herb	Vegetable	TG-4
7.	*Amaranthus spinosus* L.	Amaranthaceae	Herb	Vegetable	TG-4
8.	*Amaranthus tricolor* L.	Amaranthaceae	Herb	Vegetable	GP-1b
9.	*Amaranthus viridis* L.	Amaranthaceae	Herb	Vegetable	TG-4
10.	*Asparagus racemosus* Willd. var. *javanicus* (Kunth.) Baker	Liliaceae	Under shrub	Other Uses	TG-4
11.	*Avena ludoviciana* Dur.	Poaceae	Herb	Food and Fodder	TG-4
12.	*Brassica juncea* (L.) Czern. and Coss.	Brassicaceae	Herb	Oil and Fiber	GP-1b
13.	*Cajanus platycarpus* (Benth.) van der Maesen	Fabaceae	Twinner	Pulses	GP-3
14.	*Cajanus scarabaeoides* (L.) Du-Petit-Thours	Fabaceae	Twinner	Pulses	GP-2
15.	*Canavalia cathartica* Thours	Fabaceae	Twinner	Vegetable	TG-4
16.	*Chenopodium album* L.	Chenopodiaceae	Herb	Vegetable	TG-1b
17.	*Chenopodium murale* L.	Chenopodiaceae	Herb	Vegetable	TG-2
18.	*Citrullus colocynthis* (L.) Schrad.	Cucurbitaceae	Climber	Vegetable	TG-4
19.	*Citrus medica* L.	Rutaceae	Shrub	Vegetable	TG-3
20.	*Corchorus aestuans* L.	Tiliaceae	Herb	Oil and Fiber	TG-4
21.	*Corchorus depressus* (L.) Stock	Tiliaceae	Herb	Other Uses	GP-3
22.	*Corchorus olitorius* L.	Tiliaceae	Herb	Oil and Fiber	GP-1b

Contd...

Table 49.1–*Contd...*

Sl.No.	Name of the Species	Name of the Family	Habit	Type of Crop	Prioritization Category
23.	*Crotalaria decasperma* Naik.	Fabaceae	Herb	Oil and fiber	TG-3
24.	*Crotalaria juncea* L.	Fabaceae	Herb	Oil and fiber	TG-1b
25.	*Crotalaria leptostachya* Benth.	Fabaceae	Herb	Oil and Fiber	TG-4
26.	*Crotalaria retusa* L.	Fabaceae	Herb	Oil and Fiber	TG-2
27.	*Crotalaria verrucosa* L.	Fabaceae	Herb	Oil and fiber	TG-2
28.	*Cucumis callosus* (Roettl.) Cogn.	Cucurbitaceae	Climber	Vegetable	TG-3
29.	*Cucumis melo* L. var. *agrestis* Naudin	Cucurbitaceae	Climber	Vegetable	TG-1b
30.	*Cucumis melo* L. var. *utilissimus* Roxb.	Cucurbitaceae	Climber	Vegetable	GP-1b
31.	*Cucumis prophetarum* L.	Cucurbitaceae	Climber	Vegetable	TG-3
32.	*Cucumis sativus* L.	Cucurbitaceae	Climber	Vegetable	TG-4
33.	*Digitaria abludens* (R. and S.) Veldk.	Poaceae	Herb	Food and Fodder	TG-4
34.	*Digitaria bicornis* (Lam.) R. and S. ex Loud.	Poaceae	Herb	Food and Fodder	TG-4
35.	*Digitaria ciliaris* (Retz.) Koeler.	Poaceae	Herb	Food and Fodder	TG-4
36.	*Digitaria longiflora* (Retz.) Pers.	Poaceae	Herb	Food and Fodder	TG-4
37.	*Digitaria stricta* Roth. ex Roem. and Schult	Poaceae	Herb	Food and Fodder	TG-4
38.	*Digitaria ternata* (A. Rich) Stapf	Poaceae	Herb	Food and Fodder	TG-4
39.	*Echinochloa stagnina* (Retz.) P. Beauv.	Poaceae	Herb	Food and Fodder	TG-4
40.	*Eleusine indica* (L.) Gaertn.	Poaceae	Herb	Food and Fodder	TG-4
41.	*Gomphrena serrata* L.	Amaranthaceae	Herb	Other uses	TG-4
42.	*Hibiscus cannabinus* L.	Malvaceae	Herb	Oil and Fiber	GP-1b
43.	*Hibiscus lobatus* (Murr.) O. Ktze.	Malvaceae	Herb	Other uses	TG-4
44.	*Hibiscus panduraeformis* Burm.*f.*	Malvaceae	Herb	Other uses	TG-4

Contd...

Table 49.1–*Contd...*

Sl.No.	*Name of the Species*	*Name of the Family*	*Habit*	*Type of Crop*	*Prioritization Category*
45.	*Imperata cylindrica* (L.) P. Beauv.	Poaceae	Herb	Food and Fodder	GP-3
46.	*Ipomoea hederifolia* L.	Convolvulaceae	Twinner	Other uses	GP-1b
47.	*Ipomoea marginata* (Desr.) Manitz forma *marginata*	Convolvulaceae	Twinner	Other uses	TG-4
48.	*Ipomoea nil* (L.) Roth.	Convolvulaceae	Twinner	Other uses	TG-4
49.	*Ipomoea pes-tigridis* L.	Convolvulaceae	Twinner	Other uses	TG-4
50.	*Jatropha curcas* L.	Euphorbiaceae	Shrub	Oil and fiber	GP-1b
51.	*Jatropha glandulifera* Roxb.	Euphorbiaceae	Shrub	Oil and Fiber	TG-4
52.	*Jatropha gossypifolia* L. var. *elegans* Muell.	Euphorbiaceae	Shrub	Oil and Fiber	TG-4
53.	*Lablab purpureus* (L.)Sweet var. *lignosus* (L.) King	Fabaceae	Twinner	Vegetable	TG-1b
54.	*Lablab purpureus* (L.)Sweet var. *purpureus* Verd.	Fabaceae	Twinner	Vegetable	TG-4
55.	*Lens culinaris* Medik.	Fabaceae	Herb	Pulses	GP-1b
56.	*Linum mysorense* Heyne ex Benth.	Linaceae	Herb	Oil and Fiber	GP-2
57.	*Manilkara hexandra* (Roxb.) Dub.	Sapotaceae	Tree	Fruit	GP-1b
58.	*Momordica cymbalaria* Fenzhl.	Cucurbitaceae	Climber	Vegetable	GP-1b
59.	*Momordica dioica* Roxb. ex Willd.	Cucurbitaceae	Climber	Other uses	GP-2
60.	*Oryza rufipogon* Griff.	Poaceae	Herb	Food and Fodder	GP-1b
61.	*Oryza sativa* L.	Poaceae	Herb	Food and Fodder	GP-1b
62.	*Panicum sumatrense* Roth ex Roem. and Schult.	Poaceae	Herb	Food and Fodder	GP-1b
63.	*Pennisetum hohenackeri* Hochst. ex Steud.	Poaceae	Herb	Food and Fodder	TG-4
64.	*Pennisetum pedicellatum* Trin.	Poaceae	Herb	Food and Fodder	TG-4
65.	*Pennisetum polystachyon* (L.) Schult.	Poaceae	Herb	Food and Fodder	TG-4
66.	*Phoenix sylvestris* Roxb.	Araceae	Tree	Fruit	TG-4

Contd...

Table 49.1–*Contd...*

Sl.No.	Name of the Species	Name of the Family	Habit	Type of Crop	Prioritization Category
67.	*Rorippa indica* (L.) Hiern.	Brassicaceae	Herb	Vegetable	GP-1b
68.	*Rumex dentatus* var. *klotzechious* (Meissn.) Rech.*f.*	Polygonaceae	Herb	Vegetable	TG-4
69.	*Rumex vesicarius* L.	Polygonaceae	Herb	Vegetable	TG-4
70.	*Saccharum spontaneum* L.	Poaceae	Herb	Food and Fodder	Gp-2
71.	*Sesamum laciniatum* Klein ex Willd.	Pedaliaceae	Herb	Oil and Fiber	TG-4
72.	*Sesamum orientale* L.	Pedaliaceae	Herb	Oil and Fiber	TG-1b
73.	*Setaria forbesiana* (Nees ex Steud.) Hook.*f.*	Poaceae	Herb	Food and Fodder	TG-4
74.	*Setaria homonyma* Chiov.	Poaceae	Herb	Food and Fodder	TG-4
75.	*Setaria intermedia* Roem. and Schult.	Poaceae	Herb	Food and Fodder	TG-4
76.	*Setaria verticillata* (L.) P. Beauv.	Poaceae	Herb	Food and Fodder	TG-4
77.	*Solanum incanum* L.	Solanaceae	Under shrub	Vegetable	TG-4
78.	*Solanum torvum* Sw.	Solanaceae	Under shrub	Vegetable	TG-4
79.	*Solanum virginianum* L.	Solanaceae	Herb	Vegetable	GP-3
80.	*Sorghum controversum* (Steud.) Snow.	Poaceae	Herb	Food and Fodder	TG-4
81.	*Sorghum deccanense* Stapf. ex Bor	Poaceae	Herb	Food and Fodder	GP-2
82.	*Stylosanthes fruticosa* (Retz.) Alston.	Fabaceae	Herb	Food and Fodder	GP-1b
83.	*Syzygium cumini* (L.) Skeels	Myrtaceae	Tree	Fruit	TG-4
84.	*Trichosanthes cucumerina* L.	Cucurbitaceae	Climber	Vegetable	TG-4
85.	*Trigonella occulta* Del.	Fabaceae	Herb	Vegetable	TG-4
86.	*Vigna aconitifolia* (Jacq.) Morchal.	Fabaceae	Herb	Pulses	TG-3
87.	*Vigna indica* Dixit *et al.,*	Fabaceae	Herb	Pulses	TG-3
88.	*Vigna mungo* (L.) Hepper.	Fabaceae	Herb	Pulses	TG-3

Contd...

Table 49.1–*Contd...*

Sl.No.	*Name of the Species*	*Name of the Family*	*Habit*	*Type of Crop*	*Prioritization Category*
89.	*Vigna radiata* var. *sublobata* (Roxb.) Verdc.	Fabaceae	Herb	Pulses	GP-1b
90.	*Vigna trilobata* var. *trilobata*	Fabaceae	Herb	Pulses	TG-3
91.	*Vigna umbellata* (Thunb.) Ohwi. and Ohashi	Fabaceae	Herb	Pulses	TG-1b
92.	*Vigna unguiculata* spp. *cylendrica* (L.) Eselt.	Fabaceae	Herb	Pulses	GP-1b
93.	*Ziziphus caracutta* Roxb.	Rhamnaceae	Tree	Fruit	TG-4
94.	*Ziziphus mauritiana* L.	Rhamnaceae	Tree	Fruit	GP-1b
95.	*Ziziphus oenoplia* (L.) Mill.	Rhamnaceae	Shrub	Fruit	GP-2
96.	*Ziziphus rotundifolia* Lamk.	Rhamnaceae	Shrub	Fruit	GP-1b

food and fodder crop relatives, 3 are of category of pulses, 14 are of vegetables, 7 are of oil and fiber crops, 4 are of fruit crops and 7 genera are representing category of socio-economical important (other useful) crops (Table 49.2).

Prioritization of CWR

The Gene Pool Concept proposed by Harlan and de Wet (1971) has been used for prioritization of crop wild relatives of Balaghat ranges of Maharashtra.

1. **Gene Pool 1B:** About 19 CWR species are put into this category as they are wild forms of our crops including those which grow as agricultural weeds.
2. **Gene Pool 2:** About 10 CWR species have been put into this category which includes less closely related species from which gene transfer to the crop is possible but more difficult using conventional breeding techniques.
3. **Gene Pool 3:** About 4 CWR species have been put into this category which includes the species from which gene transfer to the crop is impossible, or if possible only using sophisticated techniques such as genetic engineering.

In the absence of knowledge of crossing relationships followed Maxted *et al.* (2006) classification for prioritization of CWR.

- ✰ **Taxon Group 1b**: About 6 species fall into this category, which are the same as a crops but not domesticated.
- ✰ **Taxon Group 2**: About 3 CWR species fall into this category. They are members of a group of closely related species forming a series or section of the genus in which the crop occurs.
- ✰ **Taxon Group 3**: About 8 CWR species fall into this class which are other members of the same subgenus of a crop.
- ✰ **Taxon Group 4**: About 46 CWR species fall into this class, which are other members of the same genus of a crop.

Rare and Endangered Species

The collected diversity represented large number of endangered species namely *Asparagus racemosus* var. *javanicus, Avena ludoviciana, Lens culinaris, Linum mysorense, Oryza rufipogon, Sorghum controversum, S. deccanense, Trigonella occulta* etc.; they are reported to be endangered mainly due to habitat destruction. *Cucumis prophetarum* was among the rare species collected from dry arid regions in Solapur district.

Wild/Weedy Types of Cultivated Species

The wild and weedy relatives of cultivated species hold characters for hardiness and disease resistance that have been lost during the process of domestication and deserve to conserve as component of crop genepool (Gadgil *et al.*, 1996). Thus, this type of material is of immense value in improvement of cultivated species.

Trait Specific Germplasm

The wild relatives of crops have donated many useful genes for crop improvement/breeding programs. The germplasms with known desirable traits for

Table 49.2: CWR Genera Classified into Six Used Categories

Sl.No.	Food and Fodder Crop Genera	Pulses Crop Genera	Vegetable Crop Genera	Oil and Fiber Crop Genera	Fruit Crop Genera	Other Useful Crop Genera
1.	*Amaranthus* (2)	*Cajanus* (2)	*Abelmoschus* (2)	*Brassica* (1)	*Manilkara* (1)	*Abelomoschus* (1)
2.	*Avena* (1)	*Lens* (1)	*Amaranthus* (4)	*Corchorus* (2)	*Phoenix* (1)	*Asparagus* (1)
3.	*Digitaria* (6)	*Vigna* (7)	*Canavalia* (1)	*Crotalaria* (5)	*Syzygium* (1)	*Corchorus* (1)
4.	*Echinochloa* (1)	*Chenopodium* (2)	*Hibiscus* (1)	*Ziziphus* (4)	*Gomphrena* (1)	
5.	*Eleusine* (1)	*Citrullus* (1)	*Jatropha* (3)	*Hibiscus* (2)		
6.	*Imperata* (1)	*Citrus* (1)	*Linum* (1)	*Ipomoea* (4)		
7.	*Oryza* (2)	*Cucumis* (5)	*Sesamum* (2)	*Momordica* (1)		
8.	*Panicum* (1)	*Lablab* (2)				
9.	*Pennisetum* (3)	*Momordica* (1)				
10.	*Saccharum* (1)	*Rorippa* (1)				
11.	*Setaria* (4)	*Rumex* (2)				
12.	*Sorghum* (2)	*Solanum* (3)				
13.	*Stylosanthes* (1)	*Trichosanthes*(1)				
14.	*Trigonella* (1)					

resistance to diseases and pest and abiotic stress were collected from wider habitat range to make best use in selection for desirable types. These species have either been utilized earlier in crop improvement programs or have known potential for desirable traits for example *Cajanus platycarpus* and *C. scarabaeoides* in cultivated species of *Cajanus cajan; Digitaria abludens, D. bicornis* in *Digitaria exillis; Citrullus colocynthis, Cucumis melo* var. *agrestis, Cucumis melo* var. *utilissimus, Cucumis prophetarum* and *Cucumis sativus* in cucumber; *Hibiscus lobatus* and *H. panduraeformis* in *H. cannabinus; Imperata cylindrica* and *Saccharum spontaneum* in *Saccharum officinarum*.

Acknowledgement

Authors are grateful to the Principal, Walchand College of Arts and Science, Solapur for providing available research facilities; Director, Botanical Survey of India, Western Circle, Pune (BSI); Blatter Herbarium, Mumbai (BLAT) and Dr. Babasaheb Ambedkar Marathwada University Herbarium, Aurangabad (BAMUA) for providing herbarium facilities.

References

Ahmedullah, M. and Nayar, M.P. (1986). *Endemic plants of the Indian region*. Botanical Survey of India, Calcutta, India, Vol 1 and 2.

Arora, R.K. (2000). Wild relatives of cultivated plants. In: *Flora of India: Introductory volume* part II. (Singh, N.P., D.K. Singh, P.K. Hajra and B.D. Sharma eds.), Botanical Survey of India, Calcutta, pp. 218-234.

Arora, R.K. and Pandey, A. (1996). *Wild edible plants of India: diversity, conservation and use*. National Bureau of Plant Genetic Resources, New Delhi, India, p.294.

Cooke, T.C. (1958). *Flora of the Presidency of Bombay Presidency* (Repr. ed.). Botanical Survey of India, Kolkata, Vol. 1, 2 and 3.

Gadgil, M., S.N. Singh, H. Nagendra and Subash Chandran, M.D. (1996). *In situ conservation of wild relatives of cultivated plants: guiding principles and a case study.* Food and Agriculture Organization of the United Nations and Indian Institute of Science, Bangalore, India, pp. 1-21.

Gaikwad, S.P., R.D. Gore and Garad, K.U. (2012a). Additions to the flora of Marathwada region of Maharashtra, India. *Journal of Threatened Taxa*, 4(4): e2515-2523.

Gaikwad, S.P., R.D. Gore and Garad, K.U. (2012b). Additions to the flora of Marathwada region of Maharashtra, India. In: *Plant Science for Sustainable Agriculture and Rural Development.* (Bhale, U.N. and V.S. Savant eds.), Daya Publishing House, New Delhi, pp. 136-144.

Harlan, J.R. and de Wet J.M.J. (1971). Towards a rational classification of cultivated plants. *Taxon,* **20**(4): 509-517.

Lakshminarasimhan, P. (1996). *Monocotyledones*. In: *Flora of Maharashtra State* (B.D. Sharma, Karthikeyan, S. and N.P. Singh ed.), Botanical Survey of India, Calcutta, pp. 1-794.

Maxted, N., B.V. Ford-Lloyd, S.L. Jury, Kell, S.P. and Scholten M.A. (2006). Towards a definition of a crop wild relative. *Biodiversity and Conservation,* **140:** 142-159.

Maxted, N., Ford-Lloyd B.V. and Hawkes J.G. (1997). *Plant genetic conservation: the in situ approach.* London, Chapman and Hall, p. 476.

Naik, V.N. (1979). *Flora of Osmanabad.* Venus publishers, Aurangabad, p. 466.

Naik, V.N. (1998). *Flora of Marathwada.* Amrut Prakashan, Aurangabad, vol. 1 and 2. p. 1118

Nayar, M.P. (1996). *Hot-spots of endemic plants of India, Nepal and Bhutan*. Tropical Botanical Garden and Research Institute, Thiruvananthapuram, Kerala, India, p.252.

Pandey, A. and Arora, R.K. (2004). Potential for domestication of wild species in the Indian gene centre. In: *Plant genetic resource management.* (Dhillon, B. S., R.K. Tyagi, L. Arjun and S. Saxena eds.). Narosa Publishing House, Delhi, pp. 56-78.

Pandey, A., D.C. Bhandari, K.C. Bhatt, S.K. Pareek, A.K. Tomer and Dhillon B.S. (2005). *Wild relatives of crop plants in India: collection and conservation.* National Bureau of Plant Genetic Resources, New Delhi, India, pp. 73.

Singh, N.P and Karthikeyan, S. (ed.). (2000). *Flora of Maharashtra State, Dicotyledons*. Botanical Survey of India, Calcutta, vol. 1, pp. 898.

Singh, N.P and Karthikeyan, S. (ed.). (2001). *Flora of Maharashtra State, Dicotyledons*. vol. II. Botanical Survey of India, Calcutta, vol. 2, pp. 1080.

Vavilov, N.I. (1949-1950). The origin, variation, immunity and breeding of cultivated plants. *Chronica Botanica,* **13:** 366.

Vincent, H., J. Wiersema, S. Dobbie, S.P. Kell, H. Fielder, N. Castenada, R. Eastwood, Guarino, L. and Maxted, N. (2012). *Global checklist of priority crop wild relatives* [online] Available at: http: //www.cwrdiversity.org/checklist/cwr-species-list.php?search-type=location and term=India/Accesed on 19 December 2012.

Zeven, A.C. and de Wet J.M.J. (1982). *Dictionary of cultivated plants and their regions of diversity*. Centre for Agricultural Publishing and Documentation, Wageningen, Netherland, pp. 263.

2015, Major Constrains and Verdict of Crop Productivity *Pages 624–635*
Editor: **Dr. U.N. Bhale**
Published by: **ASTRAL INTERNATIONAL (P) LIMITED, NEW DELHI**

Chapter 50

Impacts of Macrophytes Diversity and Weed Risk Model in Major Freshwater Bodies in Cuddalore District, Tamil Nadu

Saravanakumar Kaliyaperumal*

Botany Wing–DDE,
Annamalai University, Annamalai Nagar – 608 002, Tamil Nadu, India

ABSTRACT

Freshwater ecosystems are important natural resources for the survivability of all the living organisms of the biosphere. Aquatic plants are a relatively small group of species within a wide number of families, but most have a similar range of adaptive characteristics to enable them to survive in the aquatic environment. Impacts of aquatic weeds are well known, with obstruction of waterbodies and displacement of desirable species often the most cited impacts arising from these plants. Within the context of risk management, water plants have a number of notable differences from their terrestrial counterparts. The potential for new species to further impact on indigenous freshwater biodiversity is limited in comparison to that which has already arisen from historical introduction of weed species. Furthermore, the spread of already naturalised weed species represents the most immediate threat to the ecological values and biodiversity of the remaining non-impacted or minimally impacted habitats. Nevertheless, there is still further potential for new species to modify the nature and extent of existing impacts. Conservation of freshwater is an absolute necessity today, hence the present research planned to assess the changes on water quality and ecosystem changes of by the occurrence of aquatic floral diversity in the three major freshwater bodies of Cuddalore District, Tamil Nadu. Point intercept

* *Corresponding Author.* E-mail: drsaravanak2007@gmail.com

method was used to measure the diversity, species richness, frequency etc. and the data were recorded and the weed risk model used for the six species, viz. Limnophyton obtusifolium, Ottelia alsimoides, Ipomoea aquatica, Ceratophyllum demersum, Typha latifolia and Myriophyllum spicatum. From the present work, it is clear, firstly, each potential adverse impact factor, such as those listed above, could be considered in isolation to identify whether there are any new species that might qualify as a risk for that factor. The alternative approach, which is emphasized in this work, is to focus on each new species and to consider what combination of cumulative adverse impacts might be based on all possible factors, so that the species itself can be ranked for risk.

Keywords: *Macrophyte diversity, Cuddalore District, Weed risk model.*

Introduction

The inland freshwater ecosystem, both lentic and lotic contributes a greater fraction towards the available water resource on the planet which, is now being increasingly subjected to greater stress from various human activities. The alarming rate of deterioration of water quality of freshwater resources like lakes, ponds, rivers etc. is now a global problem. Freshwater ecosystems are considered as one of the most important natural resources for the survivability of all the living organisms of the biosphere. Freshwater habitats are of much importance to mankind but they occupy a relatively small portion of the earth's surface as compared to the marine and terrestrial habitats (Santra, 2001). The freshwater ecosystem cover only 0.2 per cent of the earth's surface with a volume of 2.04×105 km3 (Lieth, 1975). It is the availability of water which determines the nature, composition and abundance of terrestrial life. Over-exploitation, misuse and pollution of water are responsible for making it scarce and unfit for consumption (Sharma, 1999).

Aquatic plants are a relatively small group of species within a wide number of families, but most have a similar range of adaptive characteristics to enable them to survive in the aquatic environment. These include an ability to rapidly spread through a water body, often by asexual means, and a general lack of lignified structural tissue, with support being provided by the surrounding water. The spread of introduced aquatic species is often constrained by a lack of natural dispersal vectors from one water body to another. They are reliant on human activities for distribution, either accidental or deliberate. Impacts of aquatic weeds are well known, with obstruction of water bodies and displacement of desirable species often the most cited impacts arising from these plants. Within the context of risk management, water plants have a number of notable differences from their terrestrial counterparts, and in this respect they need to be addressed separately (Champion, 1994).

Aquatic macrophytes play important roles in the structures and functions of aquatic ecosystems (Wetzel, 2001). They act physically on the environment by hampering wave action and water flow and stabilizing sediment (Brix, 1997; Madsen *et al.*, 2001) in addition to strongly affecting nutrient cycling and the physico-chemical characteristics of both water and sediment (Wigand *et al.*, 1997; Havens, *et al.*, 2004). Aquatic vegetation also interacts closely with other members of aquatic communities, supplying food, shelter, and refuge for a diversity of organisms such as fishes, invertebrates, and waterfowl (Pelicice *et al.*, 2005; Rybicki and Landwehr, 2007).

Impact of aquatic flora on water quality has been studied periodically by number of researchers in India (Sharma and Solomon *et al.*, 2005; Das, 2008; and Sushilkumar, 2011). Biota of a water body is basically responsible for its natural renewal of purity. The capacity of a system to purify waters is an important issue to be investigated in the context of water resources management. The study of biota holds the basic clue to this. Conservation of freshwater, therefore, is an absolute necessity today (WHO, 1992). Hence the present research proposal planned to assess the changes on water quality and ecosystem changes of by the occurrence of aquatic floral diversity in the three major water reservoir of Cuddalore District, Tamil Nadu.

Materials and Methods

Study Area

Perumal Lake

The Perumal Lake is oldest and largest Lakes of Tamil Nadu. This lake is situated (Lat. 11°35′N; Long. 79°40′E) in the Cuddalore District in the state. The lake is being used mainly for irrigation and fish catching. The length of the lake is 17 km North-South and width of 3 km East-West. The depth is nearly5.44 m and water holding capacity is 574 M. cft.

Veeranam Lake

Veeranam tank is one of the biggest freshwater reservoir. The Veeranam lake (Lat. 11° 17′N); Long. 79° 32′E) is located 14 km SSW of Chidambaram in Cuddalore District of Tamil Nadu and is located 235km from Chennai. The length of tank in North to South is 16 km, and in East to West it is 5 km. The full tank level is 45.50 feet and water catchment area is 165 sq. miles.

Wellington Lake

Wellington Reservoir is situated in Tittagudi Taluk of Cuddalore District at a distance of 240 Km in the South West of Madras. The Reservoir is located in Vellar Basin across a tributary stream Periya Odai of Vellar River. It receives Regulated Supply diverted from Vellar River at Tholdur Regulator and an additional catchment area of 129 Sq. Km., of its own during North East Mansoon. The Reservoir was constructed during 1913-1923 and irrigates an Ayacut of 11, 200 Hectare.

In the present study, a survey was conducted during January 2012 to May 2013 using point intercept survey.

Assessment of macrophytes in Cuddalore district using MAF weed assessment model (Phelong, 1996). This model is based on adaptations of the systems used in Esler *et al.* (1993) and Champion (1995). Attributes of the plants ecology, biology and weediness are assessed based on observations of their behaviour in Cuddalore district and/or information from other states of India. The attributes of greatest importance are ranked on a scale of 0-10, of intermediate importance 0-5 and 0-3 and of minor importance 0-1. These attributes are briefly discussed below and are fully outlined in Table 50.1.

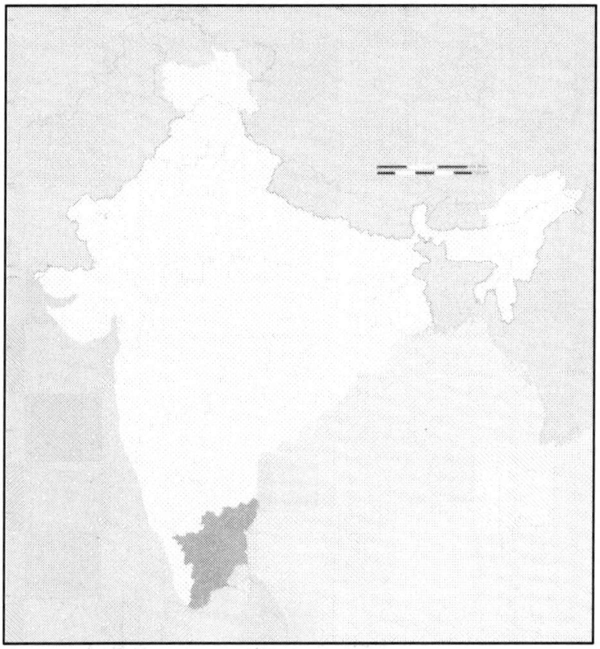

India

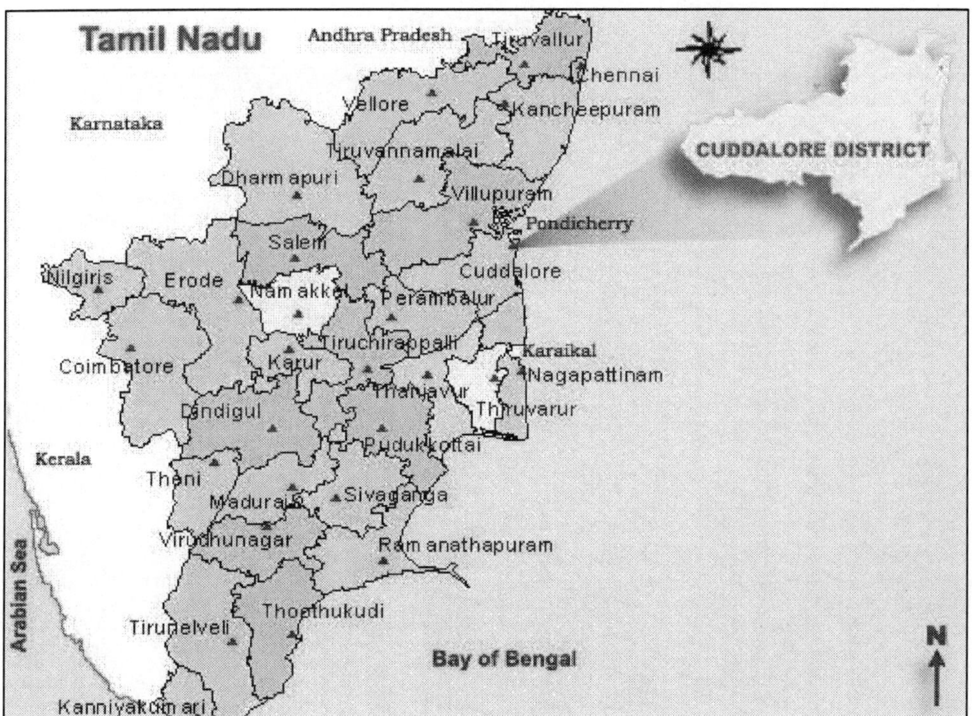

Tamil Nadu

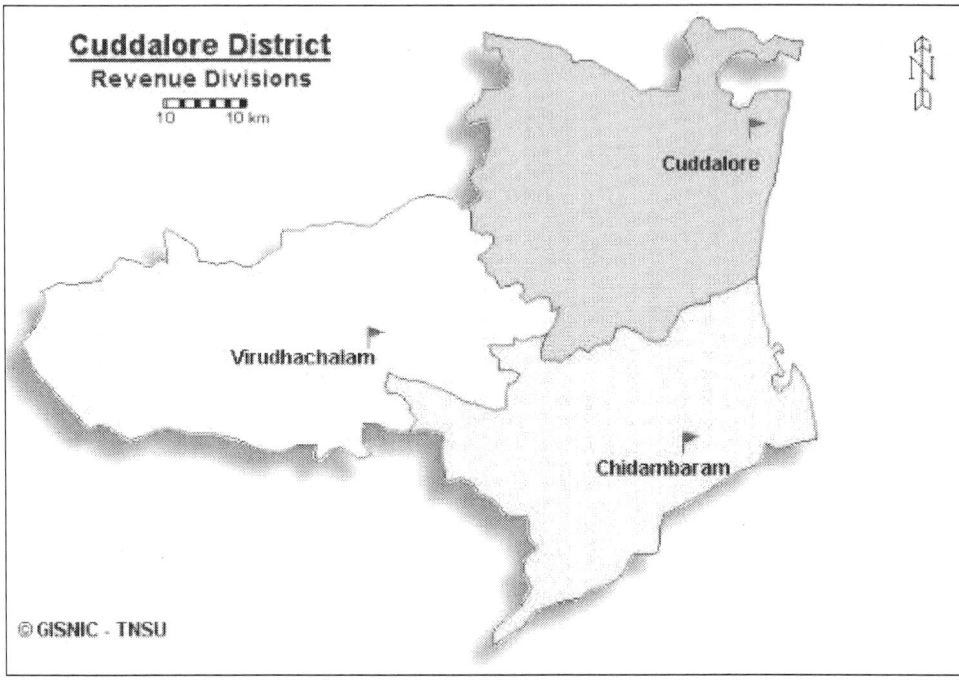

Cuddalore District

Wellington Lake

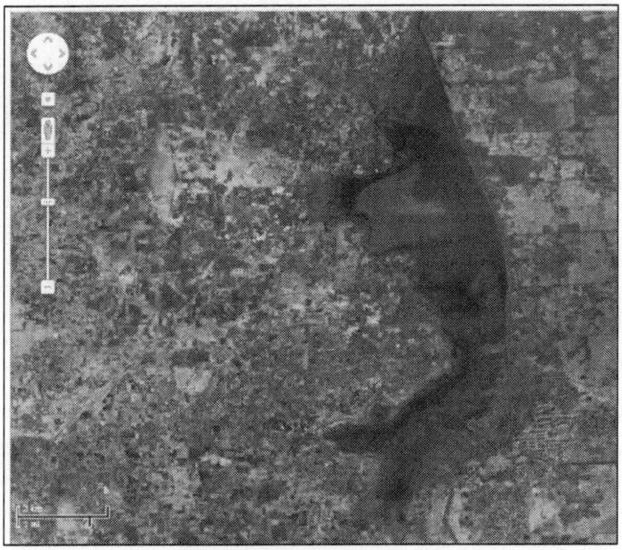

Veeranam Lake

Perumal Lake

Courtesy: Google map

Versatility (2-10)

This relates to the tolerance of plant species to a range of environmental variables such as low temperature, salinity, water depth/exposure, trophic status and water clarity.

Competitive Ability (0-10)

This compares the competitive ability of the plant to displace other species within the same life-class (*e.g.* submerged, floating, emergent) and between life classes. This competitiveness is determined from field observations and/or inter-species competitive trials, *e.g.* Hofstra *et al.* (1999).

Propagule Dispersal (0-10)

This relates to the range and effectiveness of dispersal mechanisms into new catchments including natural agents (birds, or wind), human activity (accidental or deliberate), and the ability to spread within a catchment via seed or plant fragments.

Degree of Obstruction (0-10)

This relates to potential obstruction problems caused by the plant affecting recreational water use, access to waterbodies, hydro-electric power generation, irrigation, flood control and aesthetic qualities (visual and olfactory).

Damage to Natural Ecosystems (0-10)

This relates to ecosystem values such as reduction in biodiversity (Adair and Groves 1998), reduction in water quality (especially deoxygenation) and negative impacts on physical processes (*e.g.* substrate stability, increased/decreased flooding)

Extent of Suitable Habitat not Occupied within Cuddalore District (0-10)

This evaluates the current distribution of the plant in New Zealand and its potential distribution. The score relates to available habitat not presently occupied by the plant.

Resistance to Management (0-10)

This combines various aspects of weed control, including ease of recognizing a weed problem, accessibility, scope of control methods, suitability of control methods, and the effectiveness and duration of control.

Weed History in different Habitats (1-9)

This combines the potential degree of weediness of the species in lentic (flowing), lotic (static) waters, and wetland habitats.

Seeding Ability (0-5)

This evaluates the potential maximum seed (or other perennating structures) production, its viability and persistence.

Cloning Ability (0-5)

This relates to the ability of the plant to spread by fragmentation, rhizome, or stolon extension.

Table 50.1: Weed Risk Model of the Members Studied

	Limnophyton obtusifolium	*Ottelia alismoides*	*Ipomoea aquatica*	*Ceratophyllum demersum*	*Typha latifolia*	*Myriophyllum spicatum*
Versatility	**7**	**5**	**7**	**4**	**7**	**7**
Temperature	2	1	1	1	1	3
Salinity	0	0	1	0	1	0
Habitat	2	2	3	1	2	2
Water substrate	2	1	1	1	2	2
Clarity	1	1	1	1	1	0
Habitat	**7**	**5**	**4**	**4**	**4**	**5**
Lentic	2	2	1	2	0	2
Lotic	3	3	1	2	3	3
Wetland	2	0	3	0	1	0
Competition	**6**	**10**	**10**	**4**	**5**	**7**
Within	6	8	8	4	5	7
Between	0	2	2	0	0	0
Dispersal	4	4	4	4	6	4
Bird/wind	0	0	1	0	5	0
Accidental	2	2	2	2	1	2
Deliberate	1	1	0	1	0	1
Within	1	1	1	1	0	1
Maturation	**3**	**3**	**3**	**1**	**1**	**3**
Seeding	**3**	**5**	**3**	**2**	**4**	**6**
Quantity	1	3	2	1	3	0
Viability	2	2	1	1	1	0
Cloning	4	4	4	3	3	5
BIOLOGICAL SUCCESS	34	36	35	22	30	37
Obstruction	**8**	**5**	**8**	**2**	**4**	**9**
Water use	2	1	1	1	1	2
Access	0	1	2	0	1	1
Flow	2	0	2	0	0	2
Irrigation	2	2	2	0	1	2
Aesthetic	2	1	1	1	1	2
Natural Areas	**5**	**5**	**5**	**2**	**6**	**9**
Biodiversity	4	4	4	1	5	5
Water quality	0	0	1	0	1	3
Physical	1	1	0	1	0	1

Contd...

Table 50.1–*Contd...*

	Limnophyton obtusifolium	*Ottelia alismoides*	*Ipomoea aquatica*	*Ceratophyllum demersum*	*Typha latifolia*	*Myriophyllum spicatum*
Other	**0**	**0**	**0**	**0**	**0**	**1**
Health	0	0	0	0	0	1
Weed	0	0	1	0	0	0
Habitat	**8/8**	**5/5**	**4/4**	**1/1**	**10/10**	**10/10**
Resistant to management	**6**	**3**	**3**	**0**	**1**	**5**
Implementation	1	1	1	0	1	1
Recognition	1	0	0	0	0	1
Scope	1	0	0	0	0	0
Suitability	1	1	0	0	0	1
Effectiveness	1	0	1	0	0	1
Duration	1	1	1	0	1	1
Other countries	5	5	5	0	1	1
WEEDINESS	32	23	25	5	22	35
Total Score	**66**	**59**	**60**	**27**	**52**	**72**

Behaviour in other Countries (0-5)

This evaluates the weediness of the plant in other temperate or tropical countries and its history of naturalisation into other countries.

Maturation Rate (1-3)

This evaluates the time taken to produce dispersive propagules.

Other Undesirable Traits (0-3)

This relates to other features not accounted for under the other attributes and includes aspects of health impairment (drowning risk, toxicity, wounding and mosquito breeding habitat) and weediness in terrestrial systems.

Results and Discussion

From the present observation, it is clear that there are two approaches to the undertaking of a risk assessment. Firstly, each potential adverse impact factor, such as those listed above, could be considered in isolation to identify whether there are any new species that might qualify as a risk for that factor. The alternative approach, which is emphasized in this report, is to focus on each new species and to consider what combination of cumulative adverse impacts might be based on all possible factors, so that the species itself can be ranked for risk. Such adverse impacts need to be assessed for their potential scale or extent. They also need to be placed in the context of the range of positive and negative potential impacts posed by each species, so that their final risk ranking will be based on a cumulative index of all factors.

Comparing to three freshwater lakes, the maximum score observed in *Myriophyllum spicatum* followed by *Limnophyton obtusifolium.* The minimum score recorded in *Ceratophyllum demersum* followed by *Typha angustata* and both occupied near the edges of the lake. Among the weeds in individual lake concern *Limnophyton obtusifolium* is dominant in Perumal lake, *Myriophyllum spicatum* is dominant in Wellington lake and *Ceratophyllum demersum* is dominant in Veeranam lake.

It is correlated with earlier report in various parts of India *viz. Ottelia alismoides* and *Ceratophyllum demersum* are the most problematic weeds in Ansupha lake of Orissa. *Typha* is a big problem in reservoirs and pond of Punjab (Sharma and Chandi, 1996). *Ipomoea aquatica* is in the first order among water weeds causing menace in Tamil Nadu (Sushilkumar, 2011). Aquatic weeds blocked the drainage canals of the colonies and irrigations canals badly due to its profuse growth (Sushilkumar *et al.*, 2009). On the other hand, the flow of water in canals is reduced drastically, 40 to 90 per cent by submersed weeds such as pond weeds (*Potamogeton* spp.). Vast swampy areas, ditch banks, drainage channels and flood-control channels are infested with cattails (*Typha* spp.) throughout India which are often designated as India's worst weed (Gupta, 2001, Varshney *et al.*, 2008).

Negative effects can range from a shift in the aquatic plants toward predominantly oating and emergent species (*e.g.*, Egertson *et al.*, 2004) to a complete collapse of the macrophyte community(*e.g.*, Philips *et al.*, 1978; Rasmussen and Anderson, 2005). Aquatic macrophytes in these communities have unique ways of occupying the different environments, with diversied life-and growth-forms and distinct distribution patterns in the pool space (Pott *et al.*, 1989) over time. Since a plant life-form is considered a survival strategy (Grime *et al.*, 1977), the proportion of each life-form present in a community – *i.e.*, its biological spectrum indicates not only how adapted the species are to the prevailing climatic conditions but also how strong the possible stressors are and how they impact the community (Nikolic *et al.*, 2011).

In summary, the two major difficulties in developing a meaningful Weed Risk Assessment for aquatic plants have been inadequate tools (*i.e.* models for determining potential weediness) and unreliable data (*i.e.* identity of potential ecological weeds already in India and volume of traffic entering the border by illegal means). This stage of the report (Weed Risk Model) has focused on resolving the first problem by developing a specific risk assessment model for aquatic plants, while Stage two of the Border Control Programme (Weed Risk Assessment) will correct the data on aquatic species in this country and volume of traffic entering by the various pathways.

References

Adair, R.J. and Groves, R.H. (1998). Impacts of environmental weeds on biodiversity; a review and development of a methodology. National Parks and Wildlife, Canberra, Australia.

Brix, H. (1997). Do macrophytes play a role in constructed treatment wetlands?. *Water Science and Technology*, **35**: 11–17.

Champion, P.D. (1994). The special case of aquatic weeds. Pp. 47-48 in A.I. Popay and M.J. Hartley (eds). Potential Problem Weeds. NZ Plant Protection Society Inc.

Champion, P.D. (1995). Assessment of plant pests proposed for inclusion in Auckland Regional Council's Pest Management Strategies. NIWA Consultancy Report ARC315.

Das, C.R. (2008). Integrated sustainable environmental conservation and large freshwater Ansupa lake, a famous wetland of Orissa. *Proceedings of Taal 2007: The world 12th lake conference*. 1571-1577.

Egertson, C.J., Kopaska, J.A., and Downing, J.A. (2004). A century of change in macrophyte abundance and composition in response to agricultural eutrophication. *Hydrobiologia*, **524**: 145–156.

Esler, A.E., Liefting, L.W., and Champion, P.D. (1993). Biological success and weediness of the noxious plants of New Zealand. MAF Quality Management, Lynfield, Auckland.

Grime, J.P. (1977). Evidence for existence of three primary strategies in plants and its relevance to ecological and evolutionary theory. *American Naturalist*. **111:** 1169–1194.

Gupta, O.P. (2001). Weedy aquatic plants: their utility menace and management. *Agrobios* (India), p. 273.

Havens, K.E, Sharfstein, B., Brady, M.A., East, T.L., Harwell, M.C., Maki, R.P., and Rodusky, A.J. (2004). Recovery of submerged plants from high water stress in a large subtropical lake in Florida, USA. *Aquat. Bot.* **78**: 67–82.

Hofstra, D.E., Clayton, J.S., Green, J.D. Auger, M. (1999). Competitive performance of *Hydrilla verticillata* in New Zealand. *Aquatic Botany*, **63**: 305-324.

Lieth, H. (1975). Primary Productivity in ecosystems. Comparative analysis of global pattern. pp. 67-88). In: W.H. Van Dobben and R.H. Lowe Mc. Connel (Eds) Unifying concepts of Ecology. Dr. W. Junk Pub. The Haque.

Madson, J., Chambers, P., James, W., Koch, E., and Westlake, D. (2001). The interaction between water movement, sediment dynamics and submersed macrophytes. *Hydrobiologia*, **444**: 71–84.

Nikoli ´c, L., Cobanovic, K., and Nicin, S. (2011). Relationships between plant life forms and ecological indices in a lacustrine ecosystem. *Central European Journal of Biology*, **6**: 275–282.

Pelicice, F. M., Agostinho, A. A. and Thomaz, S. M. (2005). Fish assemblages associated with *Egeria* in a tropical reservoir: investigating the effects of plant biomass and dial period. *Acta Oecologica*, **27**: 9–16.

Pheloung, P.C. (1996). Predicting the weed potential of plant introductions. *Proceedings of Eleventh Australian Weeds Conference*, Weed Science Society of Victoria Inc., Australia.

Philips, G.L., Eminson, D., and Moss, B. (1978). A mechanism to account for macrophyte decline in progressively eutrophicated freshwaters. *Aquat.Bot.*, **4:** 103–126.

Pott, V.J., Bueno, N.C., Pereira, R.A.C., Salis, S.M., Vieira, N.L. (1989). Distribuic ̧ ão de macrótas aquáticas numa lagoa na fazendo Nhumirim, Nhecolândia, Pantanal, MS. *Acta Botanica Brasilica,* **3**: 153–168.

Rasmussen, P., and Anderson, N. (2005). Natural and anthropogenic forcing of aquatic macrophyte development in a shallow Danish lake during the last 7000 years. *J. Biogeogr.,* **32**: 1993–2005.

Rybicki, N. B. and Landwehr, J. M. (2007). Long-term changes in abundance and diversity of macrophyte and waterfowl populations in an estuary with exotic macrophytes and improving water quality. *Limnology and Oceanography,* **52**: 1195–1207.

Santra, S. C. (2001). Environmental Science. New Central Book Agency (P) Ltd. Kolkata.

Sharma, A., Jarousha, A. M. Kh. and Srivastava, P. (1999). Chlorococcales from Jal Mahal Lake, Jaipur- their relevance to composition of water. *J. Envnt.* and *Poll.,* **6**(4): pp. 231-235.

Sharma, L. L., and Solomon II J. (2005). Weed menace in Udaipur lakes and options for control: a limnological perspective. 1-13. In: *Aquatic weeds: problems, control and management.* (Eds. Mathur, S.M, Mathur, A. N, Trivedy, R.K, Bhatt, Y.C and Mohnot P). Himanshu Publication, Udaipur.

Sharma, M.K and Chandi, J.S. (1996). Eradication of water hyacinth (*Eichornia crassipes*) in irrigation drains and ponds of Punjab. 65-70. In: Proceedings of Workshop on "Aquatic Weed- Problem and Management" (Ed. Varma, CVJ) held at Bangalore (Karnataka), 5-7, June 1996.

Sushilkumar, Shobha Sondhia, and Kamlesh Vishwakarma, (2009). Occurrence of alien alligator weed in India with special reference to its infestation in some districts of Madhya Pradesh. *Indian J. Weed Sci.,* **41**(3 and 4): 185-187.

Sushilkumar, (2011). Biological based chemical integration for early control of water hyacinth. *Indian J. Weed Sci.,* **43**(3&4): 211- 214.

Varshney Jay G, Sushilkumar and Mishra J.S. (2008). Current status of aquatic weeds and their management in India. Proceedings of the Tal: (Eds. Sengupta M. and Dalwani R.) The 12 World Lake Conference, Jaipur, 28 October - 2 November 2007, 1039-1045.

Wetzel, R. G. (2001). Limnology: Lake and River Ecosystems, 3rd ed. Academic Press, San Diego.

WHO, (1992). Our planet our health- report of the WHO commission on health and Environment, Published by WHO, Geneva, pp. 130-131.

Wigand, C., Stevenson, J. C. and Cornwell, J. C.(1997). Effects of different submersed macrophytes on sediment biogeochemistry. *Aquatic Botany,* **56**: 233–244.

2015, Major Constrains and Verdict of Crop Productivity *Pages 636–644*
Editor: Dr. U.N. Bhale
Published by: ASTRAL INTERNATIONAL (P) LIMITED, NEW DELHI

Chapter 51

Productivity of Finger Millet: A Review

Swarupa Kulkarni[1]* and Vijaya Nikam[2]

[1]Shivaji University, Kolhapur – 416 004, M.S., India
[2]K.B.P. Mahavidyalaya, Pandharpur – 413 304, M.S., India

ABSTRACT

Finger millet is an ancient crop. In spite of its importance in nutrition, it has been ignored by scientific community. There are many aspects which define the productivity of finger millet. Productive response of different cultivars and local land races of finger millet were compared by various research workers. Very few attempts have been made on national and international level to study productivity of finger millet. Productivity of finger millet was calculated by using RGR, leaf area, per cent of flowering, finger length, number of grain per ear, harvest index etc. In present article different growth and productivity related studies are enlisted. Both internal and external factors affect the productivity of finger millet. Environmental conditions such as light, temperature, moisture also influence productivity of finger millet.

Introduction

Finger millet (*Eleusine coracana* Gaertn.) is a cereal crop which is highly nutritious. It is rich in protein, fat and minerals than rice, corn or *Sorghum* (Reed, 1976). This crop is a source of World's food. The Uganda region is regarded as the center of its origin. Finger millet was domesticated about 5000 years ago. In India, finger millet is commonly known as ragi, which is widely cultivated in hilly, lateritic soils in the 500-1000 mm rain fall belt of the tropics and subtropics (Rachie, 1975). In certain parts of India, two crops are taken per year; the first between May-August, and second

* *Corresponding Author.* E-mail: vkn_vijaya@yahoo.co.in

between July-November. Although its high importance in nutrition, it is ignored by scientific community. Unfortunately it is known as 'Poor person's crop', 'Famine crop' or 'Bird seed'. In many areas of world, number of this crop starts declining. In India and South Africa, many tribal people are living in the hilly areas and their staple food is finger millet. These tribal populations are practicing old cultivation methods. So, in present review an account of various investigations related to productivity has been enlisted; which may be beneficial for farmers to cultivate finger millet at different environmental conditions.

Methodological Approach

Various research workers have studied different yield parameters using many genotypes. In some cases local landraces have also been used as a source of plant material (Amgain *et al.*, 2005). Productivity of finger millet was measured using different parameters such as length of flag leaf, flag leaf area, finger width, number of grains per spikelet, dry weight of plant parts,DM yield, LAI, CGR, grain yield, harvest index etc. (Agalodia *et al.*1979, Perumal, 1980; Amgain *et al.*, 2005). Analysis of productivity under different environmental conditions was done using growth chamber, stoichiometric crop weather model (Koyama *et al.*1983; Muralidhara and Rajegowda, 2002). Productivity responses of finger millet were measured using organic, inorganic and different forms of chemical fertilizers (Muthuvel and Krishnamoorthy, 1981; Bhoite and Nimbalkar, 1995; Vaiyapuri and Sriramchandrasekharan, 2004).

Factors Influencing Yield of Finger Millet

Internal Factors

There are various internal factors which influence yield of finger millet crop. Many research workers studied effect of these internal factors on productivity of the finger millet. Chaudhari and Acharya (1969) and Agalodia *et al.*(1979) studied 48 genotypes and they reported that grain yield is positively associated with grain weight of main ear, number of productive tillers and plant height, width of the main ear, number of finger/ear and grain weight of the main ear. On the other hand it is negatively related with days required for heading. Study of forty two genotypes by Perumal (1980) revealed that, days to panicle initiation, leaf dry weight at 50 per cent flowering, number of grains per ear and biological yield are positively correlated with productivity of finger millet.After comparing 90 genotypes of finger millet, Thakur and Saini in 1995 reported that yield of finger millet decreases as germination of seeds delayed or heading and maturing of ear heads were affected.Five landraces from mid-hill (Begnas, Kaski) and two landraces from central Terai (Kachorwa, Bara) of Nepal, with one improved Nepalese cultivar *Okhal-1* were studied by Amgain *et al.* (2005). They found thatgrain yield per plant, tillers per plant,leaf number, blade length of flag leaf, blade breadth of flag leaf, finger width, grain per spikelet and finger number were responsible for maximum grain yield. On the other hand characters like grain yield per plant ($r=-0.397$) and days to maturity ($r=-0.351$) determine loss of yield.After analyzing thirty two cultivars of finger millet John (2007) reported that days to maturity, number of productive tillers per plant, number of finger per ear, main ear length and harvest indexall these characters were responsible for yield.

External Factors

Different internal as well as external factors affect productivity of the crop. There are various external aspects such as light, temperature, altitude which constrain yield of finger millet.

Temperature

Patil, (2007) and Sharath *et al.* (2008) noted that various factors which govern finger millet yield are flexible under diverse environmental conditions. Nature of response of *Eleusinecoracana* and other seventeen tropical grasses to various day/ night temperatureswere studied by Koyama *et al.* (1983). In a growth chamber these grasses were planted and studied for 35 days after sowing at 31/25, 27/21 or 23/ 27°C. They reported that at increasing temperature, dry matter, leaf area index and yield was increased. Rupa Kumar (1986) reported that changes in weather condition causes reduction in yield of finger millet. At minimum temperature all treatments gave similar results. The rate of growth and yield of finger millet was calculated by Muralidhara and Rajegowda (2002) using a stoichiometric crop weather model. This model predicts that at tillering, ear emergence and 50 per cent flowering phases, higher Growing Degree Days (GDD)are effective.On the other hand, evapotranspiration is effective during tillering and grain formation phases. They further noted that late sown crops gave maximum yield. At Central Soil and Water Conservation Research and Training Institute, Research Center, Sunbeda, Orissa, Sudhishri and Dass (2006) raised three improved cultivars (*Bhairabi, VL-149* and *Subhra*) of finger millet for studying their accumulated thermal time and heat use efficiency (HUE). They concluded that for meeting physiological maturity finger millet crops required 1840 growing degree- days with HUE of 1.05 kg ha^{-1} °C day. They observed an exponential relationship and linear relationship between HUE and grain yield and accumulated growing degree days. In the summer and winter season of the year 1984-85, field trials of finger millet varieties *AKP-2, VZM-1, Sharada* and a *local* were taken for comparing DM yield, LAI, CGR, grain yield, harvest index and 1000-grain weight by Padma and Shivraj (1990). They noticed that as compared to summer season grown plants, maximum grain yield and LAI were noticed in winter season grown plants. Bandyopadhyay (2004) studied effect of diurnal temperature on growth, development and physiological basis of yield in finger millet cultivars *TNAV874* and *GPU 45* in the year 2000-2001(June 12th and 11th resp.). He noticed similar response of various flowering groups such as early, intermediate and late to physiological basis for yield.

Altitude

At Ranchauri (at an altitude of 2100 m) 27 ragi varieties were cultivated by Bandyopadhyay (1998). He noted that at high altitude duration of grain filling were shorted and high harvest index would enhance grain production. Further in (1999) Bandyopadhyay compared 22 genotypes at an altitude of 210m above sea level for 8 yield characters namely plant height, tiller number, total biological yield, grain yield and harvest index. He opined that for enhancing grain yield, harvest index is the most effective character at high altitude. Experiments of Kumar *et al.* (2002) on twelve genotypes of finger millet on hilly regions under rainfed condition revealed fluctuation in rates of photosynthesis and related parameters at vegetative and reproductive stages of crop. They further stated that at high altitude early maturing varieties are beneficial under hill environment because grain filling stage matches with low temperature phase which ultimately decreases yield. John (2007) stated that plant height is not a sturdy criterion to measure crop yield under low temperature and at high altitude. Impact of various environmental factors causes implication of different cultural practices which ultimately influence yield of finger millet.

Soil

Rajgopal (1969) cultivated finger millet in calcareous sandy loam soil at Adutharai (India). His results revealed that at 80 per cent field capacity 21.2g yield and13.8g grain yield at 25 per cent field capacity.

Abiotic Stress

In two moisture regimes (drought and irrigation) finger millet landraces *TZA-01* and *TZM-01* were cultivated by Muhammad and Ali (2007). Treatment of drought was started at 28 DAS (date after sowing) and growth and development were observed between 21DAS and 105 DAS. They recorded reduction in accumulation of dry matter, seed weight, radiation use efficiency and yield of finger millet. Treatment of drought affects number of leaves per plant, leaf area per plant from 42 DAS to 105 DAS. Lower grain yield 1.92 t/ha was recorded due to drought in *TZM-01*. Reduction in grain yield was observed in cultivars of finger millet *CO.7, ECW.840* and *CO.8* due to enhancement in salinity treatment (Morchan *et al.*, 1971).

Agricultural Practices

Agricultural practices also influence the yield of finger millet. Ashoka and Halikatti (1997) studied effect of sowing dates on yield of finger millet cv. *GPU 26, VL-149, PR-202* and *Indaf-8*. Maximum yield (3.21 t/ha) was observed at early sowing on the other hand reduction in grain yield noticed during late sowing. In (2000) these workers further noted that in all studied genotypes, the highest mean number of effective shoots (33), grain yield (32.08) and harvest index (0.37) were observed with earlier sown seeds and these values declined due to late sowing. To evaluate the favorable timing for sowing of finger millet at Karaikal region, Pandiselvi *et al.* (2010) performed experiment inPandit Jawaharlal Nehru College of Agriculture and Research Institute, Karaikal (TN) during June to September 2007. For this study they evaluated three varieties *TRY 1, CO13* and *CO14* and six transplanting dates at weekly intervals (May 17th, May 24th, May 31st, June 14thand June 21st).They recorded highest grain yield of 1827 kg/ha in earlier crop planted (variety *TRY 1)i.e.* on May 17th. Under different sowing times Nagaraju *et al.* (2009) performed varietal trials among six finger millet varieties (*Indaf 8, L-5, PR-202, GPU28, Indaf -9* and *GPU26*). They stated that delay in sowing causes reduction in grain yield.Their study concludes that transplantation is best practice to obtain maximum yield after late sowing.Three Japanese and 6 Nepalese cultivars of finger millet was studied by Kasajima *et al.* (2008) for observing influence of direct sowing and transplanting on the yielding potential and agronomic traits. They found that transplanting did not cause any effect onJapanese cultivars while overall transplanting cause production of lower yield, tillering and ear number.

Application of Fertilizers

Fertilizer is any organic or inorganic compound of natural or synthetic origin which is added to soil to supply one or more plant nutrients essential to the growth of crops and to increase productivity. In single as well as in mixed forms these fertilizers are added.

There are number of reports which highlighted positive yield response of finger millet to nitrogen fertilizer application. According to Muthuvel and Krishnamoorthy (1981), highest concentration of nitrogen (135 kg/ha) gave maximum finger millet yield. They also stated that lower concentration of nitrogen is effective under conditions of soil moisture stress to raise the crop yield. Agrawal *et al.* (1999) applied N in 0,40 and 80 kg/ha to finger millet. They observed enhancement in N content in leaves, stem and grainwith increasing N application. Application of 0-51 kg K/ha as muriate of potash on rice and ragi caused increase in grain yield as well as K uptake in both crops (Roy and Kumar, 1995). Influence of application of potassium to drought stressed finger millet crops was studied by Yadav *et al.* (1999). They recorded that under control condition application of 30 ppm K and 60-70 per cent (field capacity), 60 ppm under 45-50 per cent and 25-30 per cent field capacity levels nullified the negative effects of water stress and grain yield increased. Bhoite and Nimbalkar (1995) recorded maximum grain yield in variety Godawari of finger millet with application of 30 kg N and 20 kg P_2O_5. Pilane *et al.* (1997) reported highest grain yield and net return of finger millet after application of 100 kg nitrogen with 50 kg P_2O_5. Reddy *et al.*(2004) performed an experiment of application of varying levels of N (40, 60 and 80Kg/ha), P (20, 60 and 80 kg/ha) NP (20:15, 40:30 and 80:60 kg/ha) NPK of (40:20:20, 80:40:20, 100:50:50, 120:60:60 and 160:80:80) on finger millet crop. They noticed enhancement in biomass and grain yield with increasing levels of fertilizer up to 60 kg P/ha and 40:30 kg N: P/ha in field experiment and in the pot trail NPK up to 160:80:80 kg/ha was effective. They also recorded that fertilizer application did not cause any noticeable change in harvest index. Vaiyapuri and Sriramchandra-sekharan (2004) studied influence of sulphur (15, 30 and 45 kg/ha) on finger millet growth using ammonium sulphate, potassium sulphate, gypsum and iron pyrite as sulphur source. They concluded that as S level increasesgrowth and yield of ragi enhanced irrespective of S sources. Sahu and Pal (1984) treated acidic soilof Orissawith Boron, Molybdenum, Copper and Zinc in combination like B+Mo+Zn, B+Mo+Cu and Cu+B+Zn. Enhancement in yield was noticed by application of these combination of fertilizers. Along with chemical fertilizers, organic fertilizers also proved helpful for finger millet growth and yield. Ashwini and Sridhar (2006) reported highest growth of finger millet in terms of shoot and root lengths was observed in compost treated plants, while highest accumulation of dry matter was noticed in FYM+ mixed litter compost treated plants. According to Ganapathi *et al.*(2008) careful application of chemical fertilizers with farm yard manure at 10 t/ha concentration in every year caused maximum sustainability yield index (SYI) of 0.80 to 0.89 as compared to 0.66 and 0.41 for use of only FYM and sole NPK resp.

Conclusion

Productivity of finger millet depends on number of internal factors such as leaf area; grain per spikelet, leaf dry weight at 50 per cent flowering.As germination of seeds delayed there was decrease in grain yield. Various external factors such as light, temperature, high and low altitude, moisture conditions, salinity, drought etc. are also affects the grain yield. Agricultural practices such as seed sowing, application of fertilizers, transplantation dates also causes considerable effect on finger millet productivity. But finger millet is a sturdy crop grows in various climatic conditions. This crop has potential to grow efficiently and give grain yield under stressful conditions also.

Acknowledgment

One of the authors (S.S. Kulkarni) is thankful to the Shivaji University, Kolhapur for providing financial support. She is also grateful to the Department of Botany, Shivaji University, Kolhapur for providing necessary laboratory facilities.

References

Agalodia, A.V., Desai, K. B. and Tikka, S. B. S. (1979). Analysis of the parameters of variability and interrelationship of yield attributes in finger millet. *Indian J of Agric Sci.*, **49**(12): 924-928.

Agrawal, V. K., Dwivedi, S. K., Patel, R. S., Tiwari, A. B. (1999). Correlation and regression analysis of phenological parameters with grain yield of ragi (*Eleusine coracana* (L.) Gaertn.) under different levels of plant density and nitrogen. *Crop Research* (Hisar), **17**(3): 376-380.

Amgain, R. B., Joshi, B. K., Shrestha, P., Chaudhary, B., Adhikari, N. P., Baniya, B. K., Sthapit, B. R., Upadhyay, M. P., Shrestha, P. K. and Jarvis, D. I. (2005). Intra- and interpopulation variation in finger millet (*Eleusine coracana* (L.) Gaertn) landraces grown in Kachorwa, Bara, Nepal. International Plant Genetic Resources Institute (IPGRI), Rome, Italy, In: *On-farm conservation of agricultural biodiversity in Nepal*. Volume 1: assessing the amount and distribution of genetic diversity on-farm. *Proc. of the Second National Workshop*, Nagarkot, Nepal, 25-27 August, 2004, pp. 84-95.

Ashoka, M. B. and Halikatti, S. I. (1997).Response of finger millet (*Eleusine coracana* Gaertn.) genotypes to time of sowing in northern transitional tract of Karnataka. *Karnataka J. Agric Sci.*, **10** (2): 292-297.

Ashoka, M. B. and Halikatti, S. I. (2000). Performance of finger millet genotypes to sowing dates. *Advances in Agricultural Research.*, **10**: 9-11.

Ashwini, K. M. and Sridhar, K. R. (2006).Evaluation of pill millipede compost on plant growth and dry matter yield. *Electronic J. Envirn. Agric. and Food Chem.*, **5**(2): 10.

Bandyopadhyay, B. B. (1998). Character association and path coefficient analysis of grain yield components in finger millet at high altitude. *Annals of Agric Res.*, **19** (4): 415-417.

Bandyopadhyay, B. B. (1999). Selection criteria for grain yield improvement in finger millet at high altitude. *Indian Journal of Genetics and Plant Breeding*, **59** (4): 535-538.

Bandyopadhyay, B. B. (2004). Physiological basis of yield and influence of diurnal temperature on growth and development in finger millet, Sorghum Improvement Conference of North America, Lubbock, USA. *Int. Sorghum and Millets Newsletter*, **44**: 159-163.

Bhoite, S. V. and Nimbalkar, V. S. (1995). Response of finger millet cultivars to nitrogen and phosphorus under rainfed condition. *Journal of Maharashtra Agricultural Universities*, **20** (2): 189-190.

Chaudhari, L. B. and Acharya, R. C. (1969).Genetic variability and path coefficient analysis of components of ragi (*Eleusine coracana* Gaertn.). *Exp Agric.*, **5**(4): 295-306.

Ganapathi, Shankar, M. A. and Gajanan, G. N. (2008). Effect of long term use of chemical fertilizers and organic manure on soil fertility and sustainable productivity of fingermillet in alfisols of dryland: *Environment and Ecology*, **26** (2A): 872-876.

John, K. (2007). Estimates of genetic parameters and character association in finger millet (*Eleusine coracana* Gaertin). *Agricultural Science Digest*, **27**(2): 95-98.

Kasajima, S., Kurauchi, N. and Inoue, N. (2008). Influence of direct sowing and transplanting on the yielding ability and agronomic traits of finger millet. *Tropical Agriculture and Development*, **52**(2): 58-62.

Koyama, N., Ibaraki, K. and Tokunaga, H. (1983).Effect of temperature on dry matter production of tropical grasses. *Bulletin of the Kyushu National Agricultural Experiment Station*, **23**: 63-75.

Kumar, A., Gangwar, J. S., Prasad, S. C. and Harris, D. (2002). On farm seed priming increases yield of direct sown finger millet in India. *International Sorghum and Millets Newsletter*, **43**: 90-92.

Morchan, Y. B., Rajagopal, A. and Sankaran, N. (1971). Effect of salinity on the South Indian field crops, yield in ragi (*Eleusine coracana* Gaertn.). *Madras Agric J.*, **58**(8): 694-698.

Muhammad, M. and Ali, S. N. A. (2007). Effects of environmental stress on growth, radiation use efficiency and yield of finger millet (*Eleucine coracana*). *Pak J. Bot.*, **39**(2): 463-474.

Muralidhara, K. S. and Rajegowda, M. B. (2002). Stoichiometric crop weather model to predict the finger millet grain yield. *Journal of Agrometeorology*, **4** (1): 53-57

Muthuvel, P. and Krishnamoorthy, K. K (1981). Influence of soil moisture regimes and nitrogen levels on ragi yields. *Mysore J. Agric Sci.*, **15**(2): 251-255.

Nagaraju, A. P., Kumar, H. K. M. and Nataraju, M. S. (2009). Performance of finger millet [*Eleusine coracana* (L.) Gaertn.] varieties on sowing dates and methods of establishment. *Mysore J. Agric Sci.*, **43** (2): 208-212

Padma, V. and Shivraj, A. (1990).Growth and development of ragi (*Eleusine coracana* Gaertn) in winter and summer season. *J. Res APAU.*, **18**(2): 196-200

Pandiselvi, T., Narayanan, A.L. and Karthikeyan, R. (2010). Evaluation of optimum time of sowing of finger millet (*Eleusine coracana* G.) varieties in Karaikal region. *International J. Agric. Sci.*, **6**(1): 94-96.

Patil, H.E. (2007). Stability analysis for grain yield in finger millet (*Eleusine coracana* G.). *Intn. J. Agric. Sci.*, **3** (1): 84-86.

Perumal, K.R. (1980). Genotypic variability in photosynthetic efficiency and translocation and its relation to leaf characters, growth and productivity in finger millet *Eleusine coracana* Gaertn. *Mysore J. Agric Sci.*, **15**: 339.

Pilane, M.S., Salve, R. B., Pawar, V. S. and Bhoi, P. G. (1997).Response of fingermillet (*Eleusine coracana*) varieties to nitrogen and phosphorus. *Indian Journal of Agronomy*, **42** (4): 637-640.

Rachie, K.O. (1975). The millets: Importance, utilization and outlook. International Crop Research Institute for the Semi-arid Tropics, 1-11-256, Begumpet, Hyderabad (AP), India

Rajgopal, C.K. (1969). Influence of soil moisture stress on nutrient uptake of ragi. *Madras Agric J.*, **56**: 642-652.

Reddy, D. V. V., Udayakumar, M., Prasad, T. G., Seethram, A. and Nanjareddy, Y. A. (2004). Influence of NPK or relative stability of harvest index in finger millet. *Karnataka J. Agric Sci.*, **17** (4): 691

Reed, C.F. (1976). Information summaries on 1000 economic plants. Type scripts submitted to the USDA

Roy, H. K. and Kumar, A. (1995). Response of rice and ragi to K fertilizer use. *Journal of Research, Birsa Agricultural University* **7** (2): 113-116.

Rupa Kumar, K. (1986). Yield-weather relationships of finger millet under different manorial treatments. *Int. J. Tropical Agric.*, **4**(2): 146-153.

Sahu, S. K. and Pal, S. S(1984). Effect of soil applied boron, molybdenum, copper and zinc in combinations, vis-à-vis, limiting and F.Y.M. on the yield of wheat (*Triticum aestivum* CL) followed by ragi (*Eleusine coracana* Gaertn.). *Indian J of Agric Chem.*, **17**(2): 189-192.

Sharath, B., Sonad, K., Shanthakumar, G. and Salimath, P.M. (2008).Genotype environment interaction effect on seed yield and its component characters in white ragi (*Eleusine coracana* Gaertn). *J. Agri Sci.*, **21** (2): 190-193.

Sudhishri, S. and Dass, (2006).Thermal indices for major crops and prediction of growth and yield of finger millet in hilly southeastern Orissa state. *Journal of Agrometeorology*, **8** (2): 186-19.

Thakur, S. R. and Saini, J. P. (1995).Variation, association and path analysis in finger millet (*Eleusine coracana*) under aerial moisture-stress condition. *Indian J. Agric. Sci.*, **65** (1): 54-57.

Vaiyapuri, V. and Sriramachandrasekharan, M. V. (2004).Response of ragi to sulphur fertilization. *Journal of Ecobiology*, **16**(2): 99-104.

Yadav, D. S., Goyal, A. K. and Vats, B. K. (1999).Effect of potassium in *Eleusine coracana* (L.) Gaertn. under moisture stress conditions. *Journal of Potassium Research,* **15 (¼):** 131-134.

2015, Major Constrains and Verdict of Crop Productivity Pages 645–668
Editor: Dr. U.N. Bhale
Published by: ASTRAL INTERNATIONAL (P) LIMITED, NEW DELHI

Chapter 52

Crop Plant Diseases at the Verge of Changing Climate

A.P. Suryawanshi* and Utpal Dey

Department of Plant Pathology,
Vasantrao Naik Marathwada Krishi Vidyapeeth,
Parbhani – 431 402 (M.S.), India

Introduction

The agricultural sector represents 35 per cent of India's Gross National Product (GNP) and as such plays a crucial role in the country's development. In India, agriculture and allied activities constitute the single largest component of Grass Domestic Product (GDP) contributing nearly 25 per cent of the total, and it provides employment to two-thirds of the total workforce. The share of agricultural products in exports is also substantial, with 15 per cent of export earnings. Agriculture is one of the most climate-sensitive industries, with outdoor production processes that depends on particular levels of temperature and precipitation. Water -limiting environments, pests and diseases, declining soil fertility, availability and degradation of the soil resources, urbanization etc. are among key constraints to production of quality foods (Woods *et al.*, 2005; Admassu *et al.*, 2008; Elphinstone and Toth, 2008).

The earth's climate has always been changing in response to changes in the cryosphere, hydrosphere, biosphere and other interacting atmospheric factors. Among the most serious challenges that agricultural production facing is climate change. Since, climate change effects are challenging to study, but are of potentially of great importance. The climate change is mainly associated with increase in temperature

* *Corresponding Author.* E-mail: apsmkv@rediffmail.com

and the effects of green house gases, especially CO_2 and Ozone and it adds on an extra larger of complexity to an already complex agro-ecological system. It is widely accepted that human activities are now increasingly influencing changes in global climate and has significantly elevated the CO_2 emission and mean temperature globally, since 1750. It has been projected that at the end of 21 st century atmospheric CO_2 will increase by 1250 ppm and temperature by 3.4° C. (Pachauri and Resinger, 2007; Gautam *et al.*, 2013).

Climate change is predicted to have a progressively negative effect on food crop yields, particularly in absence of efforts to mitigate global CO_2 emissions. For example, CO_2 increase to 550 ppm increases yields of rice, wheat, legumes and oilseeds by 10-20 per cent. A single factor of climate change like temperature can have a catastrophic effect on crop yield, for example, over and above a 1°C increase in temperature may reduce yields of wheat, soybean, mustard, groundnut, and potato by 3-7 per cent. Temperature increases of 1°C, 2°C and 3°C in Punjab, would reduce the grain yield of rice by 5.4 per cent, 7.4 per cent and 25.1 per cent, respectively (Aggarwal *et al.*, 2009). Slow-onset climate changes are expected to increasingly hit potentially catastrophic impacts on food production of the developing world, which would have significant impacts on food security during the future. Projected changes in temperature, precipitation emissions of the gases *viz.* CO_2, CH_4, N_2O and O_3 and related outbreaks of pests and diseases are likely to cause fluctuations in crop production and productivity, which ultimately would magnify the challenge of increasing agricultural production to feed ever escalating population.

As a discipline, Plant Pathology is dedicated to ensure sustainable production in agricultural systems through the management and improvement of plant health. The discipline plays a major role in assessment of how plant diseases will impact on agricultural systems under climate change in order to minimize losses to production and quality. Despite significant improvements in crop protection and crop production technologies, plant diseases continued to impose significant constraints of crop production. Under prevailing climate, losses due to diseases range from 10 to 16 per cent in the major crops *viz.*, rice, wheat, barley, maize, potato, soybean, cotton etc. and overall losses of atleast 10 per cent of global food production; representing a threat to food security (Strange and Scott, 2005). Agrios (2005) estimated that an annual loss by diseases costs US $ 220 billion. Besides direct losses, the cost of disease management, environmental contamination by the chemicals applied, chemical residues in food and socio-economic problems etc. are the additional concerns.

According to the classic disease triangle for a disease to occur and/or develop, the interaction of susceptible host, a virulent pathogen and a favorable environment are required. The close relationship between the environment and plant diseases suggest that the impacts of climate change can be positive, negative or natural; since there can be decrease, an increase or no effect on the various pathosystems from region to region and current phytosanitory scenario/practices. As there exists a very high complexity of interactions between a changing climate and plant diseases, there is inevitable uncertainty in understanding plant disease development, unpredictable alterations in the pathosystem and exacerbated losses to crop production due to plant diseases. However, based on limited literature available, the probable impacts of climate change on plant diseases and related events are being reviewed in this chapter.

What has Already Happened

Continued world population growth resulted in increased emission of gasses from agriculture, combustion of fossils-fuels and industrial processes, which caused changes in chemical composition of the atmosphere. The atmospheric concentration of Carbon dioxide (CO_2) has reached the levels significantly higher than in the last thousands of years. Since, 2000, CO_2 concentration is increasing more rapidly, compared to the previous decades. Some of the continents/countries on the verge of rising CO_2 emission are: USA (5.9 billion tons), China (4.7 billion tons), Russia (1.7 billion tons), Japan (1.3 billion tons) and India (1.1 billion tons). Similar trend have also been observed for methane (CH_4), Nitrous oxide (N_2O) and other greenhouse gases (Spahni *et al.*, 2005; Canadell *et al.*, 2007). The average global surface temperature has been increased by 0.2°C per decade in the past 30 years and alterations in water cycle have also been observed (Hansen *et al.*, 2006). Atmospheric CO_2, a major greenhouse gas, has increased by nearly 30 per cent and temperature has risen by 0.3° to 0.6°C. NO_2 and CH_4 have increased by 45 per cent and 15 per cent, respectively. The Inter- government Panel on Climate Change predicts that with the current emission scenario, global mean temperature would rise between 0.9° and 3.5°C by the year 2100. India is the most affected country due to climate change and temperature increase by 0.5°C has lead to a decrease of 0.45 tonns of wheat per hectare and decrease of 25 to 30 per cent sugarcane yield per hectare.

In India, during last decade the disease scenario of chickpea and pigeonpea has changed drastically; dry root rot (*Rhizoctonia bataticola*) of chickpea and *Phytophthora* blight (*Phytophthora drechsleri* f. sp. *cajani*) of pigeonpea have emerged as a potential threat to the production of these pulses (Pande and Shanna, 2010). Chakarborty *et al.* (1998) observed that of the 10 biotrophic pathogens studied, disease severity was enhanced in six (60 per cent) and reduced in four (40 per cent) at high CO_2; and of 15 necrotrophic pathogens, disease severity increased in nine (60 per cent), reduced in four (26 per cent) and remained unchanged in rest two (16 per cent).

In soybean, elevated concentration of CO_2 and O_2 altered the expression of three soybean diseases, *viz.* downy mildew (*Peronospora manshurica*), brown spot (*Septoria glycines*) and sudden death syndrome (*Fusarium virguliforme*) and plant response to these diseases varied considerably (Eastburn *et al.*, 2010). High CO_2 concentration is likely to reduce stomatal density, production of papillae, accumulation of silicon at the sites of appressorial penetration and changed leaf chemistry increased the host resistance to powdery mildew (*Blumeria graminis*) in barley (Hibberd *et al.*, 1996 a). The severity of downy mildew was significantly reduced at high levels of CO_2 In contrast, high levels of CO_2 alone or in combination with high concentration of O_3, increased the severity of *Septoria* leaf spots (Chakaraborty and Newton, 2011).

Elevated concentration of CO_2 increased the susceptibility in rice to the diseases *viz.*, blast (*Pyricularia oryzae*) and sheath blight (*Rhizoctonia solani*) (Kobayashi *et al.*, 2006). Also on barly, high levels of CO_2 has enhanced the disease severity, incidence and rate of pathogens such as powdery mildew and anthracnose (Chakraborty *et al.*, 2000).

Some of the major weather events happened and their effects on crop pests/ diseases are summarised in the Table 52.1 (Rosenzweig *et al.*, 2000).

Anticipated Climate Change Effects

Global climate models (GMC's) have projected the effects of increasing atmospheric concentrations of greenhouse gases. Recent simulations predict a mean global warming of between 1.5 to 5.8°C and an increase in mean global precipitation of 5 to 15 per cent, by the end of 21st century. Hydrological cycle is likely to further intensify, bringing more floods and more droughts. If the buildup of greenhouse gases in the atmosphere continuous without limit, it is bound sooner or later to warm the earth's surface. Such a warming trend can be expected to affect the biophysical processes of photosynthesis and respiration. Infestations of weeds, insects and diseases and indeed the entire thermal and hydrological regimes governing our agricultural systems would be intensified (Roenzweig *et al.*, 2001).

Climate change has the potential to modify host physiology, host resistance and to alter the stages and rate of development of pathogens. Projected temperature increases can be expected to induce earlier and faster development of the crops, which would cause increased crop damages by pests and diseases at the sensitive earlier stages of crop development. Increasing temperatures will limit the life cycle of some pathogens (*e.g. Puccinia striformis* f.sp. *tritici*), whilst, increasing atmospheric CO_2 may provide more favourable conditions for pathogens such as *Fusarium pseudograminearum*. Projected climate change would potentially threat the agro-ecosystem through possible introduction of new pathogen species, insect vectors etc; changes in land use patterns which will alter plant pathogens populations. The most likely impacts would be shift in the geographical distribution of the host and pathogen; change in physiology of host-pathogen interactions and increase in crop losses (Manning and Tiedemann, 1995; Luck *et al.*, 2012).

As environment and plant diseases are closely related, climate change will probably alter the geographical and temporal distribution of plant diseases as well as phytosanitory problems. New disease complexes may arise and source diseases may cease to be economically important. Plants growing in marginal climate could experience chronic stress that would predispose them to pest and disease outbreak. Warming and other changes could also make plants more vulnerable to damage from pathogens, that are currently not important because of unfavourable climate. With changing climate, pathogens of shorter generation time would evolve at faster rate and pathogens would follow the migrating hosts.

Climate change will result in melting of portions of the Himalayan glaciers, disturb the monsoon pattern, increase flooding/droughts in Asia, as this will affect 25 per cent of the world's cereal production through increased uncertainty of water availability for irrigation and more frequent floods affecting lives and livelihoods. Extreme weather events *e.g.* spells of very high temperature, torrential rains, droughts and hurricanes (storms, cyclones), causing the spread of pathogens to new areas are

Table 52.1: Effect of Extreme Weather Events on Pests/Diseases in Past

Sl.No.	*Extreme Weather Events*	*Conditions Prevailed*	*Consequences on Pests/Diseases*
1.	Flood and heavy rains	Increased soil moisture and relative humidity favoured the epidemics and prevalence of fungal pathogens	1. Rice leaf blight caused Bengal famine (1942) 2. Wheat stripe rust outbreak in China (1960) 3. Outbreak of mycotoxin producing fungi (*Fusarium* spp.) in US (1990) 4. Fungal epidemics in corn, soybean and wheat in US Midwest (1993) 5. Outbreaks of soybean sudden death syndrome in North-Central US (1993) 6. Epidemics of maize gray leaf spot disease in Iowa and Illinois (1996) 7. Increased dissemination of soil borne pathogens to non-infected areas
2.	Droughts	Water stress, loss of plant vigour, altered C/N ratio	1. Increased incidence of *Aspergillus flavus* (Aflatoxin producer) in Southern US (1977 and 1983) 2. Outbreak of soybean cyst nematode in North-Central US (1990) 3. Favoured the attack by fungal pathogens of stems and roots 4. Dry and warm conditions enhanced the growth and population of insect vectors, which increased viral epidemics 5. Plant resistance to nematodes reduced
3.	Warm winters	Enhanced over wintering of pests/ insect vectors	1. Overwintering of wheat rust and scab fungi, potato leaf hopper and corn borer 2. Increased the population of aphids, the vector of soybean mosaic virus
4.	Air current	Large scale/distance transportation of disease agents/insects	1. Spread of corn leaf blight from Mississippi to Midwest Mexico (1970) 2. Spread of wheat stem rust in Mexico and Texas

common, and are expected to increase with the projected increase in the frequency of extreme weather events under climate change.

Though not the total rainfall, but its distribution patterns both regionally and seasonally will be affected drastically due to climate change. Moreover, all climate models predict a gradual rise in CO_2 concentration and temperature all over the world and there will be more extreme weather conditions, with more droughts, heavy rainfall and storms in agricultural production regions. Such extreme weather events will influence where and when diseases will occur, and therefore impose severe risks of crop failure.

Under changing climate, some pests/diseases will be able to invade new areas and become increasingly problematic for maintenance of the biodiversity, functioning of the ecosystems and profitability of crop production. Some pests/diseases existing already to a limited area at low densities may be able to exploit the changing climatic conditions by spreading more widely and reaching damaging population densities.

A temperature increase of 2-5° C is predicted over coming decades (IPCC, 2007) and increasingly variable rainfall is anticipated, with the region becoming generally drier, especially in the east (Scholes and Biggs, 2004). An increase in extreme events (both droughts and floods) is also anticipated (Tyson *et al.*, 2002). The consequence of these combined changes is expected to be reduced yields of several staple food crops (Jones and Thornton, 2003; Lobell *et al.*, 2008; Cooper *et al.*, 2009).

Overall, climate change is predicted to have a progressively negative effect on the yield of food crops; however, some of the probable effects of climate change on agriculture are summarized in the Table 52.2.

Potential Impacts of Climate Change on Plant Diseases

Due to uncertainties in climate change predictions, it seems difficult to predict or forecast exactly the impacts of changing climate on plant diseases. However, after referring the scanty literature available, we have put our sincere efforts in compiling the information on impacts of climate change on various events of the plant diseases/ pathogens, which are being narrated under the following sub-heads.

Effects on Disease Occurrence, Distribution and Yield Losses

Each year an estimated 10-16 per cent of global harvest is lost due to plant diseases (Strange and Scott 2005; Oerke, 2006). In financial terms, disease losses cost US $ 220 billion, and additional post-harvest losses of 6-12 per cent (Agrios, 2005). Plant pests and diseases could potentially deprive humanity of upto 82 per cent of the attainable cotton yield and 50 per cent for other crops (Oerke, 2006) and, combined with post harvest spoilage and deterioration in quality.

Weather is the most significant factor influencing all stages of plant hosts and pathogens life cycles, incidence, severity, host-pathogen interactions, developments and relative importance of the plant diseases. Changes in both physical climate and atmospheric composition influence disease severity. Under changing climate more aggressive strains of the pathogens with broad host range (*Rhizoctonia, Sclerotinia, Sclerotium, Fusarium*) and other necrotrophic pathogens as well as specialized

Table 52.2: Projected Effects of Climate Change on Agriculture over the Next 50 Years

Sl.No.	*Climatic Elements*	*Expected Change by 2050*	*Confidence in Prediction*	*Effect on Agriculture*
1.	CO_2	Increase from 360 ppmto 450-600 ppm	Very high	Good for crops, increased photosynthesis and reduced water use efficiency.
2.	Temperature	Rise by 1-2ºC, increased frequency of heat waves	High	Growing seasons faster, shorter and earlier, heat stress risk andincreased evapo-transpiration
3.	Precipitation	Seasonal changes by ± 10 per cent	Low	Impacts on drought, risks of soil workability and water logging
4.	Storminess	Increased wind speed, especially in north. More intense rainfall events	Very low	Lodging, soil erosion and reduced infiltration of rain water
5.	Sea level rise	Rise by 10-15 cm	Very high	Loss of land, coastal erosion, flooding and salinization of groundwater

biotrophic pathogens may migrate from agricultural crops to natural plant communities and *vice-versa*. Unspecialized necrotrophic pathogens may extend their host range to cause new disease problems in introduced crops. Some workers suggest that elevated CO_2 concentration and climate change may accelerate plant pathogens evolution, affecting their virulence. Enhanced atmospheric CO_2 concentration may increase incidence of the diseases like damaging off, root rots etc and also severity of most foliage diseases. Elevated CO_2 and O_3 (Ozone) may affect leaf surface properties and increase/or decrease the incidence and severity of the diseases like rust.

An increase in CO_2 levels may encourage the production of plant biomass, which modify the microclimate, leaf surface properties and consequently favours pathogenesis related events. Consequently, an increased carbohydrates content in the host tissues promotes the development of sugar loving fungi such as rust, powdery mildews (Chakraborty *et al.*, 2002). In general, increased plant density will tend to increase leaf surface wetness duration and regulation of the temperature, thus facilitating the infection by foliar pathogens (Yáñez-López *et al.*, 2012). Under elevated CO2, reduced stomatal opening and altered leaf chemistry may result in reduced disease incidence and severity in many plant pathosystems (McElrone *et al.*, 2005). Under higher concentration of CO_2, the risk of potato late blight (*P. infestans*) has been predicted to be significantly higher in all potato growing regions (Carter *et al.*, 1996; Luck *et al.*, 2012). There would be increased dissemination/spread of the diseases like rice blast (*Magnaporthe grisea*), wheat scab (*Fusarium* spp.), stripe rust (*P. striiformis*) and powdery mildew (*Blumeria graminis*) (Gareett *et al.*, 2009).

Foliage diseases are favoured by high temperature and humidity, hence increased temperature and precipitation under climate change may result in increased severity of the foliage diseases, which consequently increase the crop losses. Temperature is one of the most important factors affecting the occurrence of bacterial diseases such as *Ralstonia solanacearum*, *Acidovorax avenae* and *Burkholderia glumea*. Thus, bacteria could proliferate in the areas where temperature-dependent diseases have not been previously observed (Kudela, 2009). Some modelling studies predict that with rising temperature and other weather variables there would be increased incidence and severity of the disease such as banana black sigatoka (*Mycosphaerella fijiensis*), grapevine downy mildew and phoma stem canker (*Leptosphaeria maculans*) on oilseed rape (Evan *et al.*, 2008; Ghini *et al.*, 2008). Change in temperature regimes will provide wider opportunities for overwintering of sexual stages, thereby accelerating gene recombination and opportunities for the development of more aggressive pathogen strains (McElrone *et al.*, 2005). Excess soil moisture under heavy precipitation would favour some dreaded soil-borne diseases caused by *Phytophthora*, *Pythium*, *R. solani* and *Sclerotium rolfsii*, especially in pulses (Sharma *et al.*, 2010).

Effects of Altered Crop Physiology/Crop Canopy/Anatomy

Atmospheric CO_2 enrichment promotes increase in plant biomass; as a result maximum avenues will be available for infection by plant pathogens. Also metabolic and physiological processes are accelerated leading to increased root exudation and stimulation of both pathogens and antagonistic microbiota in the rhizosphere. There will be significant increase in photosynthetic rate, increase in water and nitrogen use efficiency and decrease in transpiration rate per unit leaf area. Dense/enlarged crop canopy and plant size may create favourable micro-climate, which may facilitate spore germination, penetration, colonization, growth, population and spread of leaf infecting pathogens, requiring high air humidity (rust, powder mildews, leaf spots, downy mildews, blights).

High CO_2 concentration tend to modify plant structure such as increased size of plant organs, increased leaf area, leaf thickness, higher no. of leaves, stems, branches etc. These changes in plant architecture may reduce leaf stromatal density, delays growth of pathogens germ tube and aspersorium, reduce conidial germination on leaves and may extend incubation period of the pathogens such as rust, downy mildews, bacteria etc. Changed leaf chemistry may reduce stomatal opening, so that infection by pathogens entering through stomata may be reduced (Hibberd *et al.*, 1996 a,b).

Under projected climate change, soil properties *viz.*, nutrient availability, soil temperature, soil water content, nitrogen content and crop residues etc may have significant impacts on the soil microbiota and pathogens survival, especially soil fungal and bacterial, diversity, density and activity (Nosengo, 2003; Garrett *et al.*, 2006; Walters and Bingham, 2007).

Effects on Host-Pathogen Interactions

Host-pathogen interaction is a component of the bigger game plan, struggle for survival, operating in nature. The survival of either host or pathogen is at stake, the pathogens survive by exploiting the plant resources and the resulting damage may threaten the existence of plants. The co-existence and co-evolution of host and pathogen initiated an unending 'race', wherein each tries to have the upper edge. However, weather factors plays crucial role in determining success or failure of host-pathogen interactions. Each component of the host pathogen/microbe interaction and the pathogenic complex will have a different interaction with the environment, and thus will change under different climates (Stonard *et al.*, 2008; Xu and Nicholson, 2009). However, such associations of organisms in a disease complex are very variable due to their opportunistic nature and therefore represent another area of uncertainty in predicting the effects of climate change.

Increased CO_2 concentration alters host-pathogen interactions *via* alterations in plant physiology and enhances inducible defensive responses and genetic resistance in plants against the pathogens (Broga *et al.*, 2006). At elevated CO_2, modified host physiology, especially increased rate of photosynthesis will increase mobilization of resources involved in host resistance. Other changes *viz*, production of papillae, accumulation of silicon at the sites of appressorial penetration, greater accumulation of carbohydrates in leaves, more waxes, extra layers of epidermal cells, increased fibre content, greater number of mesophyll etc could all influence the host resistance. Climate changes could alter stages and rates of development of the pathogens, modify host resistance, and result in changes in the physiology of host-pathogen interactions (Hibberd *et al.*, 1996 a; Chakarborty *et al.*, 1999). At high CO_2, stomatal density and stomotal conductance would be reduced resulting in reduced pathogen penetration.

Elevated temperature will also affect crop physiology and host resistance, wherein high temperature increases plant hosts lignifications, thereby potentially enhancing the level of host resistance and may cause the breakdown of temperature- sensitive resistance. Changes in temperature and precipitation regimes under changing climate may alter the growth stage, development rate and pathogenicity of infectious agents, through altered physiology and resistance of the host plants (Chakraborty and Datta, 2003). Climate change is also putting stem rust resistance due to Sr31 under threat of Ug99 race of stem rust (*Puccinia graminis* f. sp. *tritici*). Elevated temperature and CO_2 concentration are also posing higher threat perception of potato late blight (*Phytophthora infestans*) disease, rice blast (*Pyricularia oryzae*) and rice sheath blight (*Rhizoctonia solani*).

Abiotic stresses (*e.g.* heat, droughts) may contribute to host susceptibility to pathogen or may induce defense pathways that increase the host resistance. Extreme drought conditions will reduce host resistance to the diseases apparently and can exacerbate the diseases caused by the species of *Armillaria, Macrophomina, Fusarium* etc. Precipitation, whether optimal, excessive or insufficient are probably the most important weather variables affecting host-pathogen interactions. Many pests and diseases are favored by warm and humid conditions.

In addition, other changes like concentration of CH_4, other greenhouse gases, UV light and sunshine hours will also have different impacts on pathogens and host–pathogen interactions, resulting in varied incidence and severity of the diseases. Evolution of pathogen populations may accelerate under enhanced UV-B radiation and/or increased reproduction in elevated CO_2 (Chakraborty *et al.*, 2000). Evidence suggests that sunlight affects pathogens due to the accumulation of phytoalexins or protective pigments in host tissues (Chakraborty and Datta, 2003).

Effects on Crop Hosts Resistance/Susceptible

Under elevated CO_2, hosts defense mechanisms such as formation of papillae and accumulation of silicon at sites of appressorial penetration, reduced stomatal density would increase hosts resistance against pathogens (*e.g.* mildews, rusts, leaf spots). At elevated CO_2, there will be increased accumulation of assimilates in the roots of crops such as carrot, sugar beet, radish, and consequently damage to these crops due to soilborne pathogens would be decreased.

Due to stress or stress relief, there may be loss of resistance expression, *e.g.*, in barley a sudden relief of drought stress compromised the effectiveness of the *mlo* resistance gene against powdery mildew caused by *Blumeria graminis* f. sp. *hordei* (Newton and Young, 1996). Stress-related and defence related gene expressions were found to be compromised in stress-relieved barley plants compared to non-stressed plants (Barker, 1998), with the differences in accumulation of defence gene transcripts correlated with the levels of resistance breakdown in different barley genotypes (Stewart, 2002). Under drought-stressed conditions, enhanced level of free radicals reduced greatly the resistance in barly (Goodman and Newton, 2005). Therefore, one cannot assume that resistance genes currently in use will continue to be equally effective under environments with changed abiotic stress such as water stress. Therefore, actions need to be taken to counter the effects of drought stress to improve drought stress tolerance of crops generally, and the effects on resistance mechanism expression in particular.

In rice, for example, some resistance genes, such as *Xa7* effective against some *Xanthomonas oryzae* pv. *oryzae* races, are more effective at high temperatures than low, whereas other resistance genes are less effective (Webb *et al.*, 2010). Diseases such as common bunt (*Tilletia caries*) and Karnal bunt (*Tilletia indica*) in wheat can be of importance under changing climatic conditions in regions with low productivity if proper seed treatment is not followed in these crop (Oerke, 2006). Some climate change models predict that higher atmospheric water vapour concentrations with increased temperature would also favour pathogen and disease development. While physiological changes in host plants may result in higher disease resistance under climate change scenarios. Host resistance to disease may be overcome quickly by more rapid disease cycles, resulting in a greater chance of pathogens evolving to overcome host-plant resistance.

Effects on Pathogens/Diseases

The environment, includes temperature, CO_2 concentration, light and moisture, that connotes all external conditions, affecting the life and development of an organism/pathogen. The weather can affect pathogen/disease development by the way of perpetuation or over-wintering of the pathogen, build-up of both primary and secondary inoculums, dissemination of inoculums, spore germination and host penetration. Plant pathogens will have varying responses (positive, negative or neutral) to climate change. However, increasing temperature will influence life cycle of some pathogens and elevated CO_2 may create more favourable conditions for pathogens.

Favourable microclimate with enlarged crop canopy under elevated CO_2, will influence the pathogens various activities *viz.*, mode of reproduction (asexual, sexual) as well as their rate of reproduction, accelerate the rate of evolution and host colonization, increase pathogens aggressiveness and fecundity etc, thereby providing more opportunities for infection (Lupton *et al.*, 1995; Chakraborty and Datta, 2003; Chakarborty *et al.*, 2006). In sexually reproducing pathogens with broad genetic diversity (rust, downy mildews), increased population size and the number of generations in favorable microclimate would increase the probability of evolving more damaging pathotypes. Minor pests and diseases may suddenly turn serious problems under a changing climate (Sutherst *et al.*, 1996). High CO_2 concentration in soil will decrease the soil water pH which may activate the germination of resting spores of fungi in soil.

Change in temperature will directly influence infection, reproduction, dispersal, survival between seasons and other critical stages in the life cycle of a pathogen, as well as disease development. Temperature can also affect disease resistance in plants, thus affecting the incidence and severity of the diseases. Temperature sensitivity to resistance has been reported in wheat leaf rust (*Puccinia recondita*), broomrape (*Orobanche cumana*) in sunflower, black shank (*Phytophthora nicotianae*) in tobacco and bacterial blight (*Xanthomonas oryzae* pv. *oryzae*) in rice (Gregory *et al.*, 2009).

Under changing climate,genetic changes in the virus through mutation and recombination, changes in the vector populations and long-distance transportation of plant materials or insects-vectors due to trade of vegetables and ornamental plants have resulted in the emergence of the diseases such as tomato yellow leaf curl disease, African cassava mosaic disease, diseases caused by bipartite begomoviruses in Latin America, *Ipomovirus* diseases of cucurbits, tomato chlorosis (criniviruses), and the torrado-like diseases of tomato (Navas-Castillo, 2011).

Effects on Pathogens Vectors

Most plant viruses are transmitted by insect vector aphids, which are more sensitive to temperature changes and therefore, elevated temperature would affect aphids population and activity as virus vector. Increases in frequency and intensity of summer storms with high winds, rain and hail will increase wounding of plants and results increased transmission of viruses and bacteria by mechanical means. Warmer climate will result in severe outbreaks of plant disease vectors like aphids, whiteflies, thrips, beetles, and also introduction and establishment of new vector species. Since winter temperatures are widely predicted to increase (IPCC, 2007), it seems inevitable that some aphid vectors will arrive, or become active, earlier (Harrington *et al.*, 2007), necessitating early pest management practices. CO_2 enriched environment may enhance the virus- vector (insects) reservoir/population in the agro-ecosystem, by increasing the size and persistence of the plant viral diseases.

Effects on Pathogens Adaptations

Plant pathogens are generally highly adaptable and likely to exploit any compromise in plant defence brought about directly or indirectly by climate change (Garrett *et al.*, 2006). However, some changes, such as increased drought, may result in short periods with good infection conditions due to presence of free water, high humidity or open stomata required by many pathogens for initial infection. Under stressed conditions, pathogens ability to mutate may be enhanced due to stimulation of retrotransposon activity in pathogens such as *Fusarium oxysporum, Ophiostoma ulmi* and *Ophiostoma nono-ulmi* (Anaya and Roncero, 1996; Bouvet *et al.*, 2008). Other mechanisms found in fungal pathogens might include alternative intron processing (Costanzo *et al.*, 2007; Ho *et al.*, 2007), alternate splicing of mRNA (Halterman *et al.*, 2003) and other post-translational regulatory mechanisms (Newton, 1988), which may lead to enhanced genome plasticity enabling organisms to respond rapidly to environmental changes, which can later become fixed through conventional mutation and recombination processes.

Effects on Plant Disease Epidemiology

Weather variables *viz.*, temperature, precipitation, humidity, dew, radiation, wind speed and their circulation patterns influence the growth, spread and survival of crop pathogens. Epidemics can be both fast and severe where epidemiologically successful races of pathogens have adapted to the prevailing host resistance mechanisms deployed. These are likely to have the most damaging economic impact, particularly rust on cereals and are likely to have a large impact in terms of food security under climate change scenarios.

Increased temperature and humidity, results spread of diseases as wet vegetation promotes spore germination and proliferation of fungi and bacteria. Temperature, the dominant climate factor may have direct effect on spread and overwintering or oversummering of pathogens and elevated temperature may lead to the geographic expansion of the pathogens and vector distributions, bringing pathogens into contact with more potential hosts and providing new opportunities for pathogen hybridization. A trend of increasing global temperature could impact plant disease epidemics in many ways might lead to appearance of different races of the pathogens hitherto not active and may cause sudden epidemics. Warmer winters could increase disease severity through greater survival of the pathogens and their vectors. Dry and warm conditions promote growth of insect vector populations, increasing epidemics of viral diseases and powdery mildews (Chakraborty *et al.*, 2000).

Under elevated $CO_{2,}$ increased plant biomass, longer crop growth seasons, slower rate of the decomposition of plant litters (due to high C : N ratio), and higher winter temperature could increase pathogens survival on overwintering crop residues and increase the amount of initial inoculums available for subsequent infection, *e.g.* $CO_{2,}$ *Septoria* brown spot of soybean could be increased, but downy mildew decreased.

Floods and heavy rains increases moisture and humidity which favors prevalence and epidemics of foliage fungal pathogens. Increased precipitations saturate the soils with water and such soils facilitates dissemination of soilborne pathogens to non-infected areas. Also, continuous soil saturation causes long term problems of root rot diseases and increase the damage by pathogens. High levels of moisture promote infection and sporulation and facilitate spore release and germination in many fungi (Agrios, 2005). Prolonged or repeated high moisture conditions lead to epidemics.

Most fungi are active over a range of temperatures, but activity is highest at moderate temperatures (18–24°C). With adequate moisture and optimum temperatures, polycyclic fungi completes their disease cycles quickly, often within several days. Overall, with a trend toward warmer, drier summers, plant diseases caused by fungi are expected to decrease. There might be little change in disease levels in the spring and fall, when moisture is more abundant, but disease progress is expected to slow in the warmer and drier summer months. A reduction in the number of disease cycles for polycyclic diseases and a slowing in the progress of some monocyclic diseases will result in reduced primary inoculum for the next season. On the other hand, there are some fungal plant pathogens, such as *Podosphaera*, *Sphaerotheca*, *Uncinula*, and *Ustilago* species, that thrive in warmer, drier conditions (Agrios, 2005; Trigiano *et al.*, 2004), and these are expected to increase significantly with expected climate changes.

Monocyclic fungal diseases, caused by *Ustilago* and *Sclerotinia* fungal species, are not expected to be affected by climate change, as much as polycyclic diseases. In polycyclic diseases, such as those caused by *Colletotrichum*, *Peronospora*, *Phytophthora*, and *Puccinia* species, each additional disease cycle multiplies inoculum manyfold; therefore, an increased duration of the growing season is expected to increase number of disease cycles and inoculum potential. On the other hand, the effect of a longer growing season on polycyclic diseases might be counteracted by drier summers. In contrast, monocyclic vascular wilts and root rots will result in more severe disease symptoms as a result of increased drought stress. Bacterial pathogens require humid and wet conditions for infection and water and rain splash for spread. Therefore, warmer and drier climates expected with climate change would limit bacterial diseases. Survival (overwintering, over summering) of plant pathogenic bacteria in or on host debris, and in vectors, would be increased. Survival of the bacterial pathogens on/in infected host tissues (*e.g. Erwinia amylovora*), host debris in soil or on the soil surface (*e.g. Pseudomonas syringae* pv. *Phaseolicola*) and on infected seeds (*e.g. Pseudomonas syringae* pv. *tomato* and *Xanthomonas campestris* pv. *Vesicatoria*) is expected to increase, under climate change. Most bacterial diseases are considered polycyclic and the spread of bacteria is mainly by water (rain splash) and insects. Moisture is the most important factor in the development of bacterial diseases, because abundant moisture increases multiplication, oozing, and spread of bacteria. In humid, wet conditions the spread of bacterial pathogens will be increased; whereas the warmer, drier summers expected with climate change could limit bacterial diseases. As bacteria often enter their plant hosts through wounds (Trigiano *et al.*, 2004; Agrios 2005), and the expected increase in frequency and intensity of summer storms with high winds, rain, and hail will increase wounding of plants and provide moisture for spread of the bacteria.

Effects on Bio-control Agents

There is practically no information available on the impacts of climate change on bio control agents plant biological control of plant diseases. Very few information available focuses the climate change impacts on the composition and dynamics of the microbial community of the phyllosphere and rhizosphere (Ayres *et al.*, 1996). Biocontrol agents may be more vulnerable to climate change, but general climate change will be beneficial to them, both natural and introduced condiitions (Ghini and BeHoil, 2008); as a consequence, the biological equilibrium of agricultural systems will be benefited. In rhizosphere, elevated CO_2 would interact with nitrogen and other soil factors to modify the number and type of rhizosphere microflora. Mycorrhizal fungi will be stimulated to influence root colonization, root health, nutrient uptake and faster plant growth, resulting to improved plant health.

Effects on Production of Mycotoxins

Plant diseases are major impediments to the production and quality of important food crops. In addition of reducing yield, they are of particular concern because of their direct impacts on human and animal health. As a direct effect of increasing precipitation, temperature and $CO_{2,}$ grain quality may be deteriorated that reduces protein and micronutrient content in grains, which can influence moulds growth and mycotoxin production and consequently affecting quality during storage and transport.

Mycotoxins and pesticide residues in food are two major food safety concerns under changing climate (Miraglia *et al.*, 2009). Mycotoxins are secondary fungal metabolities that cause pathological and/or physiological ill-effects in human and animals. Whereas, mycotoxicoses are the diseases caused due to ingestion of foods/ food stuffs or feeds contaminated by mycotoxins. Grain moisture more than 12 per cent at harvest as well during storage predisposes to the fungal genera *viz., Aspergillus, Fusarium, Penicillium* and *Alternaria,* which are amongst the most common mycotoxin – producing fungal genera (Desjardins, 2006; Paterson and Lim, 2010; Shephard, 2008). The Food and Agriculture Organization (FAO) has estimated that about 25 per cent of the worlds food crops are affected by mycotoxins production during crop growth stages and storage. Currently over 300 mycotoxins are reported, of which the major ones are Aflatoxin (*Aspergillus flavus*), Ochratoxin A (species of *Aspergillus* and *Penicillium*), Trichothecens (*Fusarium, Myrothecium, Phomosis*), Ergot alkaloids (*Claviceps purpurea*), Zeralenone (*Fusarium roseum*), Deoxnivalenol/DON (*F. culmorum, F. graminearum*) etc. Deoxynivalenol (DON) mycotoxin production is a common problem in the grains with high moisture content (>12 per cent) stored at poor ventilation conditions. Under projected climate changes of drought and elevated CO_2 and temperature, mycotoxins (Aflatoxin, DON) concentration in peanut, maize, cottonseed, wheat, barley, oats etc would be increased (Miller, 2009; Chakarborty and Newton, 2011).

Effects on Disease Management Practices

Plant pathogens are difficult to control because their populations are variable in time, space and genotype. Climate change may add extra uncertainty in management strategies for plant disease caused by various pathogens. There is limited empirical research on plant diseases under field conditions that realistically mimic climate change, which restricts the development of disease management options under climate change.

Climate change will influence crop physiology and changed crop physiology interferes with the penetration, uptake and translocation of pesticides in general and mode of action of systemic fungicides particularly, which inturn urges for frequent applications of the fungicides/pesticides (Coakley *et al.,* 1999). For example, for effective *P infestans* (late blight) disease management strategy in potato, pesticide usage may increase if changing crop physiology interferes with the uptake and translocation of pesticides. Changes in other climatic factors (*e.g.* more frequent rainfall, washing away residues of contact pesticides) may also necessitates the need for more frequent applications of pesticides. Faster crop development at increased temperature could also increase the need for application of pesticides. Modified host-pathogen interaction under changing climate would determine the outcome of many plant diseases; which will further interfere with the disease management strategies (Garrett *et al.,* 2011; Chakraborty and Pangga, 2004).

Effectiveness of most plant protection chemicals depends on prevailing climate conditions. Changed duration, intensity and frequency of rainfall events and precipitation during the post application period would impact on the effectiveness of chemical control measures (Schepers, 1996). Precipitation, following fungicide application may improve its distribution, but an increase in rainfall intensity may deplete fungicide deposition on the foliage. Therefore, disease management strategies will be influenced due to altered efficacy of chemical and biological control options.

Further, faster crop development at increased temperature could also increase the need for application of pesticides.

Fungicide and bactericide efficacy may change with increased CO_2, moisture and temperature (Schepers, 1996). The more frequent rainfall events predicted by climate change models could result in farmers finding it difficult to keep residues of contact fungicides on plants, triggering more frequent applications. Systemic fungicides could be affected negatively by physiological changes that slow down uptake rates, such as smaller stomatal opening or thicker epicuticular waxes in crop plants grown under higher temperatures. The same fungicides could be affected positively by increased plant metabolic rates that could increase fungicide uptake.

Effect of Ozone and UV-B Light on Pathogens/Diseases

Current climate change scenarios predict a further increase of tropospheric ozone, which is known to inhibit photosynthesis and growth of plants. Elevated Ozone predisposes plants to enhanced attack by necrotrohs, and root rot fungi. Ozone (O_3) and UV-B radiation are individual climate change factors that have direct biological effects on plants (Manning and Tiedmann, 1995). Such effects may directly or indirectly affect the incidence and severity of plant diseases caused by biotic agents. Ozone (O_3) is likely to have adverse effects on plant growth and unlikely to have direct adverse effect on fungal pathogens. Necrotrophic pathogens may colonize plants weakened by Ozone at an accelerated rate, while obligate biotrophs infections may be reduced.

Ozone and Nitrous oxide indirectly influence disease through their effect on host. Elevated levels of pollutants affect both pathogen and host. Exposure to 5-10 ppm Ozone for few hours can cause visible injury to sensitive crops like barely, tomato, potato, onion, soybean, wheat and tobacco. Higher concentration of nitrous oxide ($N_{2}O$) can cause water soaked lesions, which soon turn brown. Ozone and N_2O injury on plants may add new problem to pathologists in disease diagnosis. Elevated Ozone concentration can change leaf structure, leaf topography, leaf chemical composition, epicuticular waxes, etc. The changes may alter leaf surface properties which will affect pathogens attachment to leaf surface and infection, enhance attacks on plants by necrotrophic fungi and root rot fungi. Elevated ozone (2-5 fold) can increase rust infection on poplar, but this response is reduced by elevated CO2 (Karnosky *et al.*, 2002). However, wheat stem rust was found to be strongly inhibited by ozone but unaffected by CO_2 (Tiedemann and Firsching, 2000) have a similar range of effects (Plazek *et al.*, 2001; Plessl *et al.*, 2005).

UV-B component of solar radiation influences growth of crops and vegetation, life cycle of pathogenic fungi and plant diseases. Reduced net photosynthesis due to increased UV-B could decrease diseases caused by biotrophs and increase diseases caused by necrotrophs. It can influence phylloplane organisms, amount of antifungal compounds in foliage and accumulation of phytoalexins or protective pigments in plant tissues and thereby affect plant diseases directly *via* the pathogen or indirectly *via* the host. Changes in the intensity and quality of UV radiations as a result of thinning of the ozone layer is another factor that had been received least in relation to plant diseases. Ultraviolet light is known to stimulate sporogenesis in fungi and reduced survival of the spores during dispersal and early stages of infection (Paul *et al.*, 1998). An increase in solar UV-B radiations would likely promote sporulation, but daylight already containing sufficient UV radiation also stimulates sporulation of light-dependent fungi. However expected increase in UV radiation is not sufficient to significantly affect the life cycles of plantpathogenic fungi (Manning and Tiedemann, 1995).

Future Prospects

1. A comprehensive analysis of climate change impacts on plant diseases is not possible with present knowledge/literature, which is inadequate and fragmented. To develop process based approaches to mitigate possible diseases, experimental approaches to generate epidemiological data on individual plant diseases need to be addressed critically.
2. A large number of interacting factors of climate change would influence plant diseases directly or indirectly. Therefore, field-based research evaluating the influence of the interacting factors would needs to be acquired, synthesized and generalized to provide a more realistic appraisal of climate change impacts.
3. Present research on impacts of climate change on plant diseases has been limited to the effect of a single weather variable on host, pathogen or the interaction of these two under controlled conditions. Therefore, logistic epidemiological studies considering various dependent (host) and independent (weather) variables need to give more impetuous.
4. Realistic and coordinated surveillance system complement with robust diagnostic networks and widely accessible information systems needs to be revalidated and reestablished to provide immediate diagnoses and advice for more problematic pests and diseases.
5. In monocultures, the use of major genes for resistance to pathogens is likely to lead to strong selection on pathogen population, to overcome them. Therefore, to adapt to pathogen threats following climate change, the crops should have the traits such as durable resistance and resilience in the crops with more genetic variability, both within and between cultivars, thereby mimicking the broader genetic basis of resistance to both abiotic and biotic stresses.
6. Rust and other biotrophic pathogens follow a gene-for-gene model of host-pathogen specificity and for such diseases/pathogens, more sustainable disease management strategies needs to be evolved employing resistance genes, marker assisted selection of desirable genes and transgenic approaches. Whereas, many necrotrophic pathogens that have broad host range, do not follow gene-for-gene specificity. However, for such pathogens the host resistance generally relies upon multiple defense mechanisms, which would offer a partial reduction in disease severity, but not complete protection.
7. Fundamental understandings of pathogen biology, pathogen population, epidemiology, host resistance mechanisms needs be improved significantly,

considering changes in geographical distribution of crops, to better manage the diseases under climate change.

8. Systematic qualitative analysis of the climate change effects will be necessary in developing future disease management strategies, such as plant breeding, development of resistant cultivars, altered planting schedules, chemical and biological methods and monitoring for new disease threats. Further, disease management strategies should be reoriented in changing conditions with amalgamation of new strategies for sustainable food production
9. Changing disease scenario due to climate change has intensified the need for further studies on empirical research on climate change-related issues to develop more realistic disease severity prediction models for important pathogens/diseases of agronomical and horticultural crops and to develop practically feasible disease management strategies.
10. Monitoring and early warning systems for forecasting disease epidemics should be developed for important host–pathogens which have a direct bearing on the earnings of the farmers and food security at large.
11. International phytosanitation measures, plant passports, quarantine measures, new disease resistance and deployment of resistance heterogeneously etc needs to be exploited at local, national, regional and international levels, through policy and extension service bodies, all of which have roles to play in building resilience to changes likely to accelerate due to climate change.
12. Use of botanical pesticides and plant-derived soil amendments (neem oil, neem cake and karanj seed extract) helps in mitigation of climate changes, as these helps to reduce nitrous oxide emission by nitrification inhibitors such as nitrapyrin and dicyandiamide.
13. To provide the rational to climate change impacts on plant diseases, an improved understanding of a disease cycle will enhance our capacity to predict and manage plant diseases under the current and future changing climatic conditions. An improved understanding of the causes, impact and consequences of climate change will be required for amelioration and mitigation of plant diseases.

References

Admassu, B., Lind, V., Freidt, W. and Ordon, F. (2008). Virulence analysis of *Puccinia graminis* f. sp. *tritici* populations in Ethiopia with special consideration of Ug99. *Plant Pathol.*, **58**: 362-369.

Aggarwal, P. K., Singh, A. K., Samra, J. S., Singh, G., Gogoi, A. K., Rao, G. G. S. N. and Ramakrishna, Y. S. (2009). Introduction. In *Global Climate Change and Indian Agriculture* (ed. Aggarwal, P. K.), ICAR, New Delhi, pp. 1–5.

Agrios, G.N. (2005). Plant pathology. 4th ed. Academic Press, San Diego, Calif.

Anaya, N. and Roncero, M.I.G. (1996). Stress-induced rearrangement of *Fusarium* retrotransposon sequences. *Mol. Gen. Genet.*, **253**: 89-94.

Ayres, P.G., Gunasekera, T.S., Rasanayagam, M.S. and Paul, N.D. (1996). Effects of UV-B radiation (280–320 nm) on foliar saprotrophs and pathogens. In: Frankland J, Magan N, Gadd G, eds. Fungi and environmental change. Cambridge, UK: Cambridge University Press, pp. 32–50.

Barker, S.J. (1998). The temporary breakdown of *mlo*-resistance in barley to powdery mildew. D.Phil Thesis, University of Oxford, U.K.

Bouvet, G.F., Jacobi, V., Plourde, K.V. and Bernier, L. (2008). Stress-induced mobility of OPHIO1 and OPHIO2, DNA transposons of the dutch elm disease fungi. *Fungi Gen. Biol.*, **45**: 565-578.

Broga, M.R., Aidar, M.P.M., Marabesi, M.A. and de Godoy, J.R.L. (2006). Effects of elevated CO_2 on the phytoalexin roduction of two soyabean cultivars differing in the resistance to stem canker disease. *Env. Exp. Bot.*, **58**: 85-92.

Canadell, J.G., Le Quere, C., Raupach, M. R., Field, C. V., Buitenhuis, E. T., Ciais, P., Conway, T J., Gillett, N. P., Houghton, R. A. and Marland, G. (2007). Contributions to accelerating atmospheric CO_2 growth from economic activity, carbon intensity and efficiency of natural sinks. *Proceeding of the National Academy of Sciences*, **104**: 1866-1887.

Carter, T. R., Saarikko, R. A. and Niemi, K. J. (1996). Assessing the risks and uncertainties of regional crop potential under a changing climate in Finland. Agric. Food Sci. Finland, **5**: 329–350.

Chakraborty, S., Murray, G.M., Magarey, P.A., Yonow, T., O'Brien, R., Croft, B.J., Barbetti, M.J., Sivasithamparam, K., Old, K.M., Dudzinski, M.J., Sutherst, R.W., Penrose, L.J., Archer, C., Emmett, R.W. (1999). Potential impact of climate change on plant diseases of economic significance to Australia. *Australasian Plant Pathology*, **27**: 15-35.

Chakraborty, S., Tiedemann, A. V. and Teng, P. S. (2000). Climate change: potential impact on plant diseases. *Environ. Pollut.*, **108**: 317–326.

Chakraborty, S., Murray, G. and White, N. (2002). Potential impact of climate change on plant diseases of economic significance to Australia. *Australas. Plant Pathol.*, **27**: 15–35.

Chakraborty, S. and Datta, S. (2003). How will plant pathogens adapt to host plant resistance at elevated CO_2 under a changing climate? *New Phytol.*, **159**: 733–742.

Chakraborty, S. and Pangga, I.B. (2004). Plant disease and climate change. In: Gillings, M. Holmes, A. (Eds.) *Plant Microbiology*. London: Bios Scientific, pp. 163-180.

Chakraborty, S., Liu, C.J., Mitter, V. *et al.* (2006). Pathogen population structure and epidemiology are keys to wheat crown rot and fusarium head blight management. *Australasian Plant Pathology*, **35**: 643–55.

Chakraborty, S. and Newton, A. C. (2011). Climate change, plant diseases and food security: an overview. *Plant Pathol.,* **60**: 2–14.

Coakley, S. M., Scherm, H. and Chakraborty, S. (1999). Climate change and plant disease. *Annu. Rev. Phytopathol.,* **37**: 399–426.

Cooper, P., Rao, K.P.C., Singh, P., Dimes, J., Traore, P.S., Rao, K., Dixit, P. and Twomlow, S.J. (2009). Farming with current and future climate risk: advancing a "Hypothesis of Hope" for rainfed agriculture in the semi-arid tropics. *J. Sat. Agric. Res.,* **7:** 1-19.

Costanzo, S., Ospina-Giraldo, M.D., Deahl, K.L., Baker, C.J. and Jones, R.W. (2007). Alternate intron processing of family 5 endoglucanase transcripts from the genus *Phytophthora*. *Curr. Gen.,* **52**: 115-123.

Eastburn, D. M., Degennaro, M. M., Delucia, E. H., Dermody, O. and Mcelrone, A. J. (2010). Elevated atmospheric carbon dioxide and ozone alter soybean diseases at SoyFACE. *Global Change Biol.,* **16:** 320-330.

Elphinstone, J. and Toth, I.K. (2008). *Erwinia chrysanthemi* (*Dikeya* spp.) - the facts. Potato Council, Oxford, U.K.

Evans, N., Baierl, A., Semenov, M. A., Gladders, P. and Fitt, B. D. L. (2008). Range and severity of a plant disease increased by global warming. *J. Roy. Soc. Interface*, **5**: 525-531.

Garrett, K.A., Dendy, S.P., Frank, E.E., Rouse, M.N. and Travers, S.E. (2006). Climate change effects on plant disease: Genomes to ecosystems. *Ann. Rev. Phytopathology*, **44**: 489-509.

Garrett, K. A., Nita, M., Dewolf, E. D., Gomez, L. and Sparks, A. H. (2009). Plant pathogens as indicators of climate change. In *Climate and Global Change: Observed Impacts on Planet Earth* (ed Letcher, T. M.), Elsevier, Oxford, pp. 425–437.

Garrett, K.A., Forbes, G.A., Savary, S. *et al.,* (2011) Complexity in climate-change impacts: An analytical framework for effects mediated by plant disease. *Plant Pathology*, **60**: 15-30.

Gautam, H. R. Bhardwaj, M. L. and Kumar, R. (2013). Climate change and its impact on plant diseases. *Current Science,* **105**(12): 1685-1691.

Ghini, R., Hamada, E., Gonçalves, R. R. V., Gasparotto, L. And Pereira, J. C. R. (2008). Risk analysis of climate change on black sigatoka in Brazil. *J. Plant Pathol.,* **90:** 105.

Goodman, B.A. and Newton, A.C. (2005). Effects of drought stress and its sudden relief on free radical processes in barley. *J. Sci. Food. Agric.,* **85**: 47-53

Gregory, P. J., Johnson, S. N., Newton, A. C. and Ingram, J. S. I. (2009). Integrating pests and pathogens into the climate change/food security debate. *J. Exp. Bot.,* **60**: 2827–2838.

Halterman, D.A., Wei, F.S. and Wise, R.P. (2003). Powdery mildew-induced Mla mRNAs are alternatively spliced and contain multiple upstream open reading frames. *Plant Physiol.,* **131**: 558-567.

Hansen, J., Sato, M., Ruedy, R., Lo, K., Lea, D. W. and Medina-Elizade, M. (2006). Global temperature change. *Proceeding of the National Academy of Sciences*, **103**: 14288-14293.

Harrington, R., Clark, S.J., Welham, S.J., Verrier, P.J., Denholm, C.H., Hulle, M., Maurice, D., Rounsevell, M.D., Cocu, N. and European, U. E. C. (2007). Environmental change and the phenology of European aphids. *Glob. Ch. Biol.*, **13**: 1550-1564.

Hibberd, J.M., Whitbread, R. and Farrar, J.F. (1996 a). Effect of elevated concentrations of CO_2 on infection of barley by *Erysiphe graminis*. *Physiol. Molec. Plant Pathol.*, **48**: 37-53.

Hibberd, J. M., Whitbread, R. and Farrar, J. F. (1996 b). Effect of 700 µmol mol–1 CO_2 and infection with powdery mildew on the growth and carbon partitioning of barley. *New Phytol.*, **134**: 309–315.

Ho, E.C., Cahill, M.J. and Saville, B.J. (2007). Gene discovery and transcript analyses in the corn smut pathogen *Ustilago maydis*: expressed sequence tag and genome sequence comparison. *BMC Genomics*, **8**: 334 [http: / /www.biomedcentral.com/ 1471-2164/8/334]

IPCC (2007). Intergovernmental Panel on Climate Change, Climate Change, *The Fourth IPCC Assessment Report*, Cambridge University Press, Cambridge, UK.

Jones, P.G. and Thornton, P.K. (2003). The potential impacts of climate change on maize production in Africa and Latin America in 2055. *Gl. Env. Ch.*, **13**: 51-59

Karnosky, D.F., Percy, K.E., Xiang, B.X., Callan, B., Noormets, A., Mankovska, B., Hopkin, A., Sober, J., Jones, W., Dickson, R.E. and Isebrands, J.G. (2002). Interacting elevated CO_2 and tropospheric O_3 predisposes aspen (*Populus tremuloides* Michx.) to infection by rust (*Melampsora medusae* f. Sp. *tremuloidae*). *Glob. Ch. Biol.*, **8**: 329-338.

Kobayashi, T., Ishiguro, K., Nakajima, T., Kim, H.Y., Okada, M. and Kobayashi, K. (2006). Effects of elevated atmospheric CO_2 concentration on the infection of rice blast and sheath blight. *Phytopathology.*, **96**: 425-431

Kudela, V. (2009). Potential impact of climate change on geographic distribution of plant pathogenic bacteria in Central Europe. *Plant Prot. Sci.*, **45**: S27–S32.

Lobell, D.B., Burke, M.B., Tebaldi, C., Mastrandrea, M.D., Falcon, W.P. and Naylor, R.L. (2008). Prioritizing climate change adaptation needs for food security in 2030. *Science*. **319**: 607-610.

Luck, J., Spackman, M., Freeman, A *et al.* (2011). Climate change and diseases of food crops. *Plant Pathology*, **60**: 113-121.

Lupton, J., Chakraborty, S., Dale, M., Sutherst, R. (1995). Assessment of the enhanced greenhouse effects on plant diseases – a case study of *Stylosanthes anthracnose*. In: *Proceedings of the 10th Biennial Australian Plant Pathology Conference*. Lincoln University, New Zealand, 28-30 August.

Manning, W. J. and Tiedemann, A. V. (1995). Climate change: potential effects of increased atmospheric carbon dioxide (CO_2), ozone (O_3), and ultraviolet-B (UVB) radiation on plant diseases. *Environ. Pollut.*, **88**: 219–245.

McElrone, A. J., Reid, C. D., Hoye, K. A., Hart, E. and Jackson, R. B. (2005). Elevated CO_2 reduces disease incidence and severity of a red maple fungal pathogen via changes in host physiology and leaf chemistry. *Global Change Biol.*, **11**: 1828–1836.

Miller, S.A, Beed, F.D. and Harmon, C.L. (2009). Plant disease diagnostic capabilities and networks. *Annu. Rev. Phytopathol.*, **47:** 15-38.

Miraglia, M., Marvin, H. J. P., Kleter, G. A., Battilani, P., Brera, C., Coni, E., *et al.* (2009). Climate change and food safety: An emerging issue with special focus on Europe. *Food and Chemical Toxicology*, **47**(5): 1009-1021. Retrieved from <Go to ISI>: // 000265970600011. doi: 10.1016/j.fct.2009.02.005

Navas-Castillo, J., Fiallo-Olive, E. and Sanchez-Campos, S. (2011). Emerging virus diseases transmitted by whiteflies. *Annu. Rev. Phytopathol*., **49**: 219–248

Newton, A.C. (1988). Mutant instability in *Septoria nodorum*. *Tr. Br. Mycol. Soc.*, **91**: 607-610.

Newton, A.C. and Young, I.M. (1996). Temporary partial breakdown of Mlo-resistance in spring barley by the sudden relief of soil water stress. *Plant Pathol*., **45**: 970-974.

Nosengo, N. (2003). Fertilized to death. *Nature*, **425:** 894-895.

Oerke, E. C. (2006). Crop losses to pests. *J. Agric. Sci.*, **144**: 31–43.

Pachauri, R.K and Reisinger, A. (2007). "Climate Change 2007: Synthesis Report. Contribution of Working Groups I, II and III to the Fourth Assessment Report of the Intergovernmental Panel on Climate Change" [Core Writing Team, (eds.)]. IPCC, Geneva, Switzerland, pp. 104.

Pande, S. and Shanna, M. (2010). Climate change: potential impact on chickpea and pigeonpea diseases in the rainfed semi-arid tropics (SAT). In *Proceedings of the 5th International Food Legumes Research Conference (IFLRC V) and 7th European Conference on Grain Legumes (AEP VII)*, Antalya, Turkey.

Paul, N.D., Ayres, P.G., Rasanayagam, S., and Royle, D.J. (1998). Stratospheric ozone depletion, UVB radiation and *Septoria tritici* infection of wheat. [online]. *In* 7th International Congress of Plant Pathology. 9–16 August 1998, Edinburgh. Vol. 1. *Compiled by* British Society for Plant Pathology, Birmingham. Abstr. 4.2.6S. Available from http: //www.bspp.org.uk/icpp98 [accessed 16 August 2004].

Plazek, A., Hura, K., Rapacz, H. and Zur, I. (2001). The influence of ozone fumigation on metabolic efficiency and plant resistance to fungal pathogens. *J. Appl. Bot.*, **75**: 8-13.

Plessl, M., Heller, W., Payer, H.D., Elstner, E.F., Habermeyer, J. and Heiser, I. (2005). Growth parameters and resistance against *Drechslera teres* of spring barley

(*Hordeum vulgare* L. cv. Scarlett) grown at elevated ozone and carbon dioxide concentrations. *Pant Biol.,* **7**: 694-705.

Rosenzweig, C., Iglesias, A., Yang, X. B., Epstein, P. R. and Chivian, E. (2000). Implications of climate change for US Agriculture: Extreme Weather Events, Plant Diseases and Pests. Cambridge, Massachusetts: Centre for Health the Global and Environment, Harvard Medical School Cambridge, M. A. pp.56.

Rosenzweig, C., Iglesias, A., Yang, X. B., Epstein, P. R. and Chivian, E. (2001). Climate change and extreme weather events. *Global Change and Human Health,* **2:** 90-104.

Schepers, H. T. A. M. (1996). Effect of rain on efficacy of fungicides on potato against *Phytophthora infestans*. *Potato Res.,* **39:** 541–550.

Scholes, R.J. and Biggs, R. (2004). Ecosystem services in Southern Africa: A regional assessment. Council for Scientific and Industrial Research, Pretoria, South Africa.

Sharma, M., Mangala, U. N., Krishnamurthy, M., Vadez, V. And Pande, S. (2010). Drought and dry root of chickpea. In *Proceedings of the 5th International Food Legumes Research Conference (IFLRC V) and 7th European Conference on Grain Legumes (AEP VII),* Antalya, Turkey.

Spahni, R., Chappellaz, J., Stocker, T. J., Loulergue, L., Hausammann, G., Kawamura, K., Fluckiger, J., Schwaander, J., Raynaud, D., Mansson-Delmotte, V. and Jouzel, J. (2005). Atmospheric methane and Nitrous Oxide of the late Pleistocene from Antarctic ice cores. *Science,* **310**: 1317-1321.

Stewart, K.J. (2002). Abiotic stress and *mlo*-resistance breakdown to barley powdery mildew. D.Phil Thesis, University of Oxford, U.K.

Stonard, J.F., Marchant, B., Latunde-Dada, A.O., Liu, Z., Evans, N., Gladders, P., Eckert, M. and Fitt, B.D.L. (2008). Effect of climate on the proportions of *Leptosphaeria maculans* and *L. biglobosa* in phoma stem cankers on oilseed rape in England. *Asp. Appl. Biol.,* **88**: 161-164.

Strange, R. N. and Scott, P. R. (2005). Plant disease: a threat to global food security. *Ann. Rev. Phytopathol.,* **43**: 83-116.

Sutherst, R.W., Yonow, T., Chakraborty, S., O'Donnel, C. and White, N. (1996). A genetic approaches to defining impacts of climate change on pests, weeds and diseases in Australia. In Bouma, W.J., Pearman, G.I., Manning, M.R. (Eds.), Greenhouse, Coping with climate change. CSIRO, Australia, pp. 281-307.

Tiedemann, A.V. and Firsching, K.H. (2000). Interactive effects of elevated ozone and carbon dioxide on growth and yield of leaf rust-infected versus non-infected wheat. Env. Poll. **108**: 357-363.

Trigiano, R.N., Windham, M.T., and Windham, A.S. (*Editors*). (2004). Plant pathology: concepts and laboratory exercises. CRC Press, New York.

Tyson, P., Odada, E., Schulze, R. and Vogel, C.I. (2002). Regional-global change linkages: Southern Africa. In: Tyson P, Fuchs R, Fu C, Lebel L, Mitra AP, Odada E, Perry J, Steffen W, Virji H (eds), Global-regional linkages in the earth system. START/ IHDP/IGBP/WCRP. Springer, Berlin and London.

Walters, D. R. and Bingham, I. J. (2007). Influence of nutrition on disease development caused by fungal pathogens: implications for plant disease control. *Annals of Applied Biology,* **151**: 307-324.

Woods, A., Coates, K.D. and Hamann, A. (2005). Is an unprecedented Dothistroma needle blight epidemic related to climate change? *Bioscience,* **55**: 61-69.

Xu, X.M. and Nicholson, P. (2009). Community ecology of fungal pathogens causing wheat head blight. *Ann. Rev. Phytopathol.,* **47**: 83-103.

Yáñez-López, R., Torres-Pacheco, I., Guevara-González, R. G., Hernández-Zul, M. I., Quijano-Carranza, J. A. and Rico-García, E. (2012). The effect of climate change on plant diseases. *Afr. J. Biotechnol.,* **11:** 2417–2428.

2015, Major Constrains and Verdict of Crop Productivity *Pages 669–677*
Editor: **Dr. U.N. Bhale**
Published by: **ASTRAL INTERNATIONAL (P) LIMITED, NEW DELHI**

Chapter 53

Effect of Root Exudates on Rhizosphere Fungi of different Varieties of Pigeonpea [*Cajanus cajan* (L.) Mill Sp.]

V. Jalander[1]* and B.D. Gachande[2]

[1]Department of Botany, Telangana University, Nizamabad, A.P., India
[2]Department of Botany, N.E.S. Science College, Nanded, M.S., India

ABSTRACT

Rhizosphere soil fungi were isolated from different varieties of pigeonpea at different stages of plant growth and the effect of root exudates on the growth of rhizosphere soil fungi was studied. The root exudates obtained from resistant varieties of pigeonpea was stimulated the growth rhizosphere fungi except F. oxysporun and Fusarium sp., The growth of these two species was inhibited root exudates obtained from wilt resistant varieties but stimulated by the root exudates of susceptible variety ICP-2376. The stimulation and inhibition of fungal growth was may be due to variation in composition of root exudates from resistant and susceptible varieties.

Keywords: *Cajanus cajan, Root exudates, Rhizospheric fungi and Growth.*

Introduction

As plant roots grow through soil they release water soluble compounds such as amino acids, sugars and organic acids that supply food for the microorganisms. The

* *Corresponding Author.* E-mail: jalandervaghmare @gmail.com

food supply means microbiological activity in the rhizosphere is much greater than in soil away from plant roots. In return, the microorganisms provide nutrients for the plants. All these activities make the rhizosphere the most dynamic environment in the soil. Plant root exudates are a complex mixture of chemicals and organic compounds secreted into the soil by the roots (Bais, 2004). Root exudates include amino acids, organic acids, carbohydrates, sugars, vitamins, mucilage and proteins. The exudates act as messengers that stimulate biological and physical interactions between roots and soil organisms. They modify the biochemical and physical properties of the rhizosphere and contribute to root growth and plant survival.

The region of contact between root and soil where the soil is affected by roots was designated as the rhizosphere. In fact, in 1904, Hiltner first defined the rhizosphere as that zone of soil in which the microflora is influenced by plant roots. The rhizosphere is a centre of intense biological activity due to the food supply provided by the root exudates. Bacteria, actinomycetes, fungi, protozoa, slime moulds, algae, nematodes, enchytraeid worms, earthworms, millipedes, centipedes, insects, mites, snails, small animals and soil viruses compete constantly for water, food and space. Soil chemistry and pH can influence the species mix and functions of microbes in the rhizosphere.

In general, it has been discovered that, plant families will employ similar secondary metabolites for similar purposes. For instance, most of the plants from family Leguminosae use isoflavonoids as antimicrobials, while Soalnaceae members typically use sesquiterpenes. For example, the legume plant *Glycine max* produces glyceollin, *Vigna unguiculata* produces medicarpin, *Mundulea seica* produces munduserone, *Phaseolus vulgaris* produces coumestrol, and *Cajanus cajan* produces cajanin; all of which are isoflavonoids that function as phytoalexins (defensive antimicrobial proteins) in the rhizosphere (Dakora, 2003). Many of the different compounds that make up root exudates can have both antibacterial as well as antifungal properties, but most often, the compounds work synergistically to achieve general antimicrobial results (Bais, 2004).

Bacterial and fungal pathogen interactions have also been well documented with a significant increase of interest in the area of biocontrol. Many of the anti-fungal interactions documented have involved *Pseudomonas* species (Whipps, 2001). However, other species of bacteria that are typically referred to as plant growth promoting rhizobia that live in the rhizosphere dependent on plant root exudates, specifically *Bradyrhizobium* sp., and forty nine different strains of *Sinorhizobium meliloti*, have also been prooved to reduce a wide range of plant pathogens. In some cases, these rhizobia can decrease sporulation in fungal pathogen up to 75 per cent and inhibit growth up to 50 per cent (Dakora, 2003).

Materials and Methods

Pigeonpea Varieties Selected for the Study

The following varieties of Pigeonpea (*Cajanus cajan* (L.) Millsp.) were selected for the study PUSA-992, BDN-2, BDN-708, BSMR-853, BSMR-736, BSMR-175, ICP- 8863, ICPL-87119, ICP-2376 and AKT-9913.

Collection of Rhizosphere Soil Samples

The rhizosphere and non rhizosphere soil samples of the above varieties were collected by shaking up-rooted plants at non-flowering, flowering and fruiting stages of plant growth in sterile paper bags from Research Centre for Pulses at Badnapur, Dist. Jalna (M.S).

Isolation of Rhizosphere Mycoflora

Isolation of rhizosphere and non rhizosphere mycoflora was done by dilution plate (Waksman, 1927), soil plate (Warcup, 1950) methods at different stages of plant growth (non-flowering, flowering and fruiting stages). The developed colonies were identified (Barnett and Hunter, 1972; Gilman, 1957; Subramanian, 1971; Mukadam, 1997). Unidentified cultures were sent to Agharkar Research institute for identification. The pure cultures were maintained on PDA slants. Number of colonies of each species as well as total number of colonies in each plate was recorded (Jalander and Gachande, 2011; 2013)

Collection of Root Exudates from of Pigeonpea

Root exudates were collected by growing the plants aseptically in sand culture (Shanker, 1973). Seeds were surface sterilized with 0.1 per cent mercuric chloride, and aseptically transferred to 250-ml flasks containing acid washed sand wetted with sterile distilled water. The sterility of the seeds was checked, before transferring to the sterile culture flasks, by plating a representative sample of 10 seeds on moist blotter paper. Ten seeds were transferred to each culture flask and 6 replicates were maintained for each variety. After 4 weeks of growth under diffused light, root washings were collected freeing the roots carefully from the sand. The washings from the sand were also collected. All the washings were bulked and concentrated in a water bath at 60°C to a final volume of 10 ml for each variety separately.

Effect of Root Exudates on the Growth of Rhizosphere Fungi

The effects on fungal growth were assessed by the linear growth rate method using Czapek's dox agar (CDA) medium at pH 6.0. Root exudates were filtered through seitz filter and 1 ml of the concentrated sample was incorporated into each petri dish along with 20 ml of the medium, hyphal tips of the test fungi were used as inoculums. The petri plates were incubated at 27±2°C and growth was measured after 7days. For each treatment 3 replicate maintained.

Results and Discussion

Isolation of Rhizosphere Mycoflora

Totally 58 species of fungi belonging to 23 genera were recorded from rhizosphere soil of different varieties of pigeonpea. Out of 58 species, 3 belongs to phycomycetes, 36 from ascomycetes, 14 from deuteromycetes and 5 are of mycelia sterilia (Table 53.1). More number of species was recorded at flowering stage of the plant growth. Out of 58 fungal species, isolated from rhizosphere of ten different varieties of pigeonpea, few of them were common at three stages of plant growth.

Table 53.1: Effect of Root Exudates of different Varieties Pigeonpea on Growth of Rhizosphere Fungi

Fungal Species	*Average Colony Diameter (mm) after 7 Days of Incubation*											*C.D. (p=0.05)*
	PUSA-992	*BDN-2*	*BDN-708*	*BSMR-853*	*BSMR-736*	*BSMR-175*	*ICP-8863*	*ICPL-87119*	*AKT-9913*	*ICP-2376*	*Con.*	
Alternaria alternata	26.20	28.00	30.20	33.60	28.60	29.60	30.30	33.20	30.20	36.00	25.20	3.4618
A. solani	30.00	32.00	31.30	28.00	27.60	28.30	32.00	30.20	38.20	34.60	28.30	4.0597
Alternaria sp.	28.00	30.20	30.60	28.200	24.20	27.60	27.50	26.50	26.60	36.50	26.20	3.4919
Aspergillus candidus	32.00	33.30	30.20	36.20	35.80	33.20	33.60	35.60	38.50	30.20	30.00	3.0225
A. flavus	80.20	83.40	82.20	80.60	86.50	87.30	86.50	85.60	86.50	70.50	68.00	7.1206
A. fumigatus	65.60	78.00	64.30	68.60	65.50	65.50	70.10	72.60	68.50	65.50	60.00	5.11446
A. niduans	32.40	38.00	33.40	32.00	36.20	34.00	38.50	28.50	27.50	27.20	25.00	4.9046
A. nidulans (1)	32.00	38.60	36.80	36.20	38.80	32.60	32.50	32.20	32.00	30.30	28.00	3.7482
A. niger	85.20	88.60	86.30	88.60	82.50	78.20	84.20	87.60	84.20	70.50	65.20	8.2626
A. niger (1)	75.60	76.60	78.50	68.40	68.00	72.00	74.60	75.50	68.20	62.20	40.10	11.6135
A. niger (2)	34.20	36.00	38.20	36.20	38.50	38.50	34.20	36.50	31.20	32.60	30.00	3.1618
A. parasiticus	25.40	25.00	28.10	28.60	26.60	28.60	27.50	27.50	24.50	24.60	22.00	2.2600
A.terrus	36.02	35.50	38.00	40.20	41.50	42.60	38.20	40.80	36.20	32.20	31.00	4.0203
A. terrus (1)	34.20	28.40	39.10	28.20	24.60	29.60	28.60	26.50	27.60	25.50	25.00	4.6575
A.terrus (2)	55.30	42.60	60.80	45.60	48.50	46.50	48.50	56.50	52.60	50.20	40.00	6.6874
Aspergillus sp.1	52.80	50.20	50.60	50.50	56.20	54.60	58.20	55.00	52.30	50.20	43.20	4.2673
Aspergillus sp.2	27.60	32.20	30.20	32.00	25.60	28.60	30.20	24.60	24.60	26.80	20.50	3.8628
Aspergillus sp.3	73.40	66.20	70.4	72.40	68.20	76.30	66.80	76.50	74.00	74.50	60.50	5.3556
Aspergillus sp.4	17.50	20.00	20.20	10.50	12.60	18.60	15.50	16.20	21.00	20.00	05.20	5.3127
Aspergillus sp.5	12.50	13.60	15.20	16.20	15.00	12.20	18.60	20.20	10.20	20.50	09.00	4.1313
Aspergillus sp.6	25.20	22.30	25.60	30.50	28.00	26.20	24.60	24.50	24.50	22.30	20.00	3.1070

Contd...

Table 53.1–*Contd...*

Fungal Species	*Average Colony Diameter (mm) after 7 Days of Incubation*											*C.D. (p=0.05)*
	PUSA-992	*BDN-2*	*BDN-708*	*BSMR-853*	*BSMR-736*	*BSMR-175*	*ICP-8863*	*ICPL-87119*	*AKT-9913*	*ICP-2376*	*Con.*	
Aspergillus sp.	25.80	15.20	20.20	24.00	20.00	21.60	15.60	18.60	22.60	20.40	12.00	4.3998
Cheatomium sp.	42.60	40.20	38.60	36.20	40.60	41.00	38.00	36.50	38.00	37.50	32.00	3.1085
Cladosporium herbarum	34.20	33.6	29.5	30.2	38.60	36.5	30.2	32.0	28.00	30.50	24.00	4.3747
Cladosporium sp.	32.60	38.0	35.20	32.50	36.00	38.50	34.20	33.30	32.00	28.60	25.00	4.2458
Curvularia lunata	33.60	30.30	32.30	28.60	27.50	25.50	29.20	24.60	28.00	28.80	24.00	3.0906
C. paliscence	22.00	22.50	22.50	23.00	21.50	23.20	20.00	22.50	20.50	18.00	12.00	3.5323
Curvularia sp.	24.00	26.20	18.50	20.20	15.60	18.50	22.60	18.60	18.50	15.00	13.00	4.2566
Drachslera tetramera	42.60	36.50	38.60	40.50	42.00	26.20	30.20	35.50	35.50	25.60	23.60	7.3640
Fusarium oxysporum	35.00	32.30	34.50	30.20	36.50	34.60	35.20	38.60	40.20	45.60	43.40	4.3246
Fusarium sp.	28.00	28.50	26.80	28.60	22.00	24.50	28.20	25.50	28.00	36.80	30.00	4.1563
Helminthosporium sp.	50.20	48.20	52.00	50.60	48.50	52.80	46.80	46.50	50.20	45.60	42.00	3.3888
Isaria felina	30.20	32.00	30.30	34.50	32.60	32.50	33.30	30.20	32.00	30.00	30.80	1.6077
Mucor sp.	36.00	42.02	35.50	38.20	34.50	32.00	30.50	38.60	30.00	28.20	26.00	5.3091
Myrothecium roridium	42.00	38.20	34.80	33.20	35.20	38.60	40.20	35.50	35.00	32.60	30.00	3.7983
Nigrospora sp.	28.40	25.60	24.20	27.50	30.50	28.60	22.50	26.40	24.80	20.50	18.00	4.0275
Paecilomyces variotii	78.50	82.50	78.00	85.60	82.00	80.50	75.50	78.40	76.20	75.60	70.00	4.5501
Penicillium adametzi	18.20	21.00	21.50	20.50	18.40	18.50	19.50	19.50	18.00	17.20	15.00	1.9969
P. citrinum	54.20	58.60	54.80	50.20	54.60	51.60	52.00	52.60	50.20	48.20	42.00	4.6253
P. citrinum (1)	13.20	12.50	16.60	15.30	12.50	15.60	15.20	12.50	12.50	10.00	10.00	2.3312
P.decumbens	22.00	22.50	24.50	24.50	20.50	20.60	22.20	20.20	22.40	18.60	14.00	3.1815

Contd...

Table 53.1–*Contd...*

Fungal Species	*Average Colony Diameter (mm) after 7 Days of Incubation*											*C.D. (p=0.05)*
	PUSA-992	*BDN-2*	*BDN-708*	*BSMR-853*	*BSMR-736*	*BSMR-175*	*ICP-8863*	*ICPL-87119*	*AKT-9913*	*ICP-2376*	*Con.*	
P. islandicum	22.50	25.50	24.50	25.40	22.30	20.50	20.50	21.50	24.50	18.00	12.00	4.2530
Penicillium sp.1	28.20	25.60	24.00	20.60	22.00	20.50	24.60	20.50	17.50	15.60	13.00	4.8509
Penicillium sp.2	38.00	30.50	42.60	41.10	35.60	32.50	36.20	38.60	32.50	30.00	23.00	6.0931
Penicillium sp.3	42.60	36.20	40.60	38.60	36.50	42.00	28.60	38.40	28.60	25.50	22.00	7.7686
Penicillium sp.4	35.20	36.50	41.60	42.60	35.60	38.60	40.50	36.50	34.60	32.50	30.00	4.1420
Penicillium sp.5	42.40	38.20	40.20	39.50	37.50	38.60	40.50	38.60	39.40	32.00	30.00	3.9738
Penicillium sp.6	43.40	43.50	40.20	42.50	40.50	42.60	42.50	41.50	33.50	38.20	32.00	4.2494
Pythium sp.	25.60	22.50	24.00	20.60	21.40	22.50	21.60	22.30	17.50	20.50	15.00	3.1464
Rhizoctonia solani	56.20	54.80	45.50	38.60	50.60	48.50	52.40	51.40	35.20	54.20	32.00	9.0144
Rhizopus stolonifer	54.00	58.60	60.40	54.80	55.500	52.60	62.00	54.80	50.40	51.20	40.00	6.3724
Talaromyces sp.	42.50	42.60	44.50	44.60	40.50	38.60	40.10	38.50	35.20	36.40	34.00	3.8950
Trichoderma viride	86.20	85.50	82.60	84.30	85.50	84.60	83.60	78.60	72.50	80.20	71.00	5.6922
Trichoderma sp.	68.40	70.50	68.40	66.50	66.50	68.50	64.80	62.30	58.20	62.50	56.00	4.9511
White sterile mycelium-1	25.00	20.60	21.00	18.60	20.60	24.000	18.60	19.60	18.60	24.00	16.00	2.9914
White sterile mycelium-2	22.00	18.60	20.50	21.50	28.40	20.50	15.80	20.40	22.30	18.60	14.00	4.0346
Brown sterile mycelium-1	20.00	16.80	18.20	20.40	21.60	24.50	20.40	18.60	18.40	20.50	15.00	2.7208
Brown sterile mycelium-2	26.00	24.50	20.60	18.60	22.60	22.30	20.30	21.50	20.40	18.60	16.00	3.0390

Effect of Root Exudates on the Growth of Rhizosphere Fungi

It is generally accepted that the main factors for the stimulation of micro-organisms in the rhizosphere are the excretion of organic substances and the sloughing off of root hairs and epidermal cells. Investigations on the nature of root exudates show that a wide range of substances are secreted. Root exudates play an important role in plant microbe's interaction in soil. Due to their different compounds like sugars, sugar alcohols, organic, amino acids and secondary metabolites. The various factors in root exudates have been shown to stimulate or inhibit spore germination in fungi, stimulate or inhibit mycelia growth and to attract zoospores.

Totally 58 rhizosphere fungi were grown on CDA media having 0.5 per cent root exudates of pigeonpea varieties. Without root exudates served as control. Root exudates from resistant varieties showed a stimulatory effect on the linear growth of 56 rhizosphere fungi and inhibitory to *Fusarium oxysporum* and *Fusarium* sp., but in case of susceptible variety ICP-2376 to wilt, it was different. The root exudates from var. ICP-2376 showed very low level of stimulation, than the resistant varieties but enhanced the growth of *Fusarium oxysporum* and *Fusarium* sp. (Table 53.1).

Root exudates obtained from wilt resistant varieties of pigeonpea such as PUSA-992, BDN-2, BDN-708, BSMR-87119, BSMR-736, BSMR-175, ICPL-87119, ICP-8863 AND AKT-9913 enhances the growth of rhizosphere fungi than susceptible variety ICP-2376. Only *Fusarium oxysporum* and *Fusarium* sp. was inhibited by the root exudates of resistant varieties and enhanced by the root exudates of susceptible variety. The similar results were also reported by Naqvi and Chauhan (1980) in case of three different cultivars of chilli (*Capsicum annum*). They found the good germination of rhizosphere fungal spores in root exudates of resistant cultivars than susceptible one. Bharathi *et al.* (2008) studied the effect of root exudates of different crops on pigeonpea wilt pathogen (*Fusarium udum*) and its antagonistic fungi. They reported that, the growth of antagonistic fungi was better than the pathogen. Sullia (1973) isolated the rhizosphere fungi from *Cassia tora* and *Crotalaria medicagina* and found that, root exudates caused only a marginal effect which was positive on all the tested fungi.Similar results were also obtained by Schroth *et al.* (1963), in case of bean root exudates, Clauduis and Meharothra (1973) in case of lentil root exudates, Kraft (1974) and Whalley and Taylar (1970) in case of pea root exudates.

Acknowledgement

The authors thank Principal, Dr. G.M. Kalamse, N.E.S. Science College, Nanded for permission to use Laboratory and Library facilities and Dr. D.U.Gawai, Head, Botany Dept. for permission to use Laboratory facilities. The first author also thanks to the authorities of U.G.C., New Delhi for the award of Rajiv Gandhi National Fellowship.

References

Bais, H.P. (2004). How plants communicate using the underground information *Trends in* superhighway? *Plants Science.*, **9** (1): 26-32.

Barnett, H.L. and B.B. Hunter (1972). *Illustrated genera of imperfect fungi.* Third Edition Burgess publishing company, Minneapolis, Minnesota.

Bharathi, V., K. C. Rao and L. Kishan Reddy (2008). Studies on the effect of root exudates of different crops on pigeonpea wilt pathogen (*Fusarium udum* Butl.) and its antagonistic fungi. *Internat. J. Agric. Sci.*, **4**(4): 670-673.

Clauduis, G. R. and R.S. Mehrotra (1973). Root exudates from lentil (*Lens culinaris* Medic) seeding in relation to wilt disease. *Plant and Soil*, **38**: 315- 320.

Dakora, F.D. (2003). Defining new roles for plant and rhizobial molecules in sole andmixed plant cultures involving symbiotic legumes. *New Phytologist*, **158**: 39-49.

Gilmam, J.C. (1957). *A manual of soil fungi*. Sec. Eds. Iowa State Univ. Oxford IBM Publ. Co. Calcutta press. pp. 410.

Hiltner, L. (1904). Uber neuere Erfahrungen und Probleme anf dem Gebiet der Bodenbakteriologie und unter besonderer Berficksichtigung der Grundung und 3rache. *Arlo. dtsch. Landw Ger.*, **98**: 59-78.

Jalander, V. and B.D. Gachande (2011). Rhizosphere and non-rhizosphere mycoflora of different varieties of pigeonpea (*Cajanus cajan* (L.) Millsp.). *Geobios*, **38**(1): 37-40.

Jalander, V.and Gachande, B.D.(2013).Studies on behavior of rhizosphere fungi in different varieties of pigeonpea (*Cajanus cajan* (L.) Millsp.). Frontiers in Life Sciences, (Eds Dr. Sayyed Iliyas Usman, Science impact Pub. Ahmedpur), pp.107-117.

Kraft, J. M. (1974) Influence of seedling exudates on the resistant of pea to *Fusarium* and *Pythium* root rot. *Phytopath.*, **64**: 190-193.

Mukadam, D.S. (1997). *The illustrated kingdom of fungi.* Akshar Ganga Prakashan, Aurangabad (MS) India.

Naqvi, S.M.A. and S.K. Chauhan (1980). Effect of root exudates on the spore germination of rhizosphere and rhizosphere fungi of chilli (*Capsicum annuuam* L.) cultivars. *Plant and Soil*, **55**: 397-402.

Sckroth, M. N., T.A. Toussoum and W.C. Snyder (1963). Effect of certain constituants of bean exudates on germination of chlamydospores of *Fusarium solani f. phaseoli* in soil. *Phytopathology*, **53**: 809-812. (in text missed)

Shanker, B. S. (1973). Effect of root exudates and extracts on rhizosphere fungi. *Plant and Soil*, **39**: 197-200.

Subramanian, C.V. (1971). *Hyphomycetes.* "An account of Indian species, except *Cercosporae.*" ICAR publ. New Delhi. Pp. 930.

Sullia, S.B. (1973). Soil moisture content and root exudates in relation to rhizosphere effect in leguminous weeds. *Proc. Plant Sciences*, **77**(6): 264-275.

Waksman, S.A. (1927). *Principals of soil microbiology.* Bailliere Tindal and Co., London.

Warcup, J.H. (1950). The soil plate method for isolation of fungi from soil. *Nature*, pp.116-117.

Whalley, W. M. and G.S. Tayler (1976). Germination of chlamydospores in physiologic races of *Fusarium oxysporum f. pisi* adjacent to susceptible and resistant pea culture. *Trans. Br. Mycol. Soc.,* **66**: 7-13.

Whipps, J. M. (2001). Microbial interactions and biocontrol in the rhizosphere. *J. Experimental Botany,* **52**: 487-511.

2015, Major Constrains and Verdict of Crop Productivity *Pages 678–686*
Editor: Dr. U.N. Bhale
Published by: ASTRAL INTERNATIONAL (P) LIMITED, NEW DELHI

Chapter 54

Effects of Chemical Pesticides on Plants with Special Reference to Grapes (*Vitis vinifera* L.): A Review

Vishal V. Naik[1]*, Vishal T. Aparadh[2] and S.P. Bartakke[3]

[1]Department of Botany, Shivaji University, Kolhapur – 416 004, M.S., India
[2]S.P.K. Mahavidyalaya, Sawantwadi, Dist. Sindhudurg – 416 510, M.S., India
[3]Krishna Mahavidyalaya, Rethare Bk., Tal. Karad, Dist. Satara – 415 108, M.S., India

ABSTRACT

Plants are easily attacked by pathogens. To defend these pathogens, use of chemical pesticides is extensive. The harmful chemicals used in pesticides can cause adverse effect on nutritional value of crops. The important nutrients like protein; carbohydrates etc. are combined with the chemicals used in pesticides. In present review article, side-effects of chemical pesticides are collectively discussed with special reference to grapes (Vitis vinifera L.).

Keywords: *Grapes, Pesticides, Nutrition.*

Introduction

The population of the world has been increasing threateningly. The world's population is expected to reach nine million in 2050 (Tort *et al.*, 2006). The continuous decline in soil and water sources and increasing population of the world, several

* *Corresponding Author.* E-mail: vishu1415@gmail.com

methods are used to increase agricultural production. But various factors such as climatic condition, diseases and pests cause severe crop losses (Oerke, 1994). It is estimated that crop losses caused by diseases is about 12 per cent of the total production, which is worth 42- billion dollars (Kazan and Gurel, 2001). One of the methods used to deal with agricultural problems is the use of pesticides.

In spite of unquestionable benefits of pesticides, their uncontrolled overuse negatively affects other untargeted living organisms, including human beings (Comelekolu and Mazamnc, 2001). The side-effects of DDT on environment were first described by Rachel Carson in 1962 in "Silent Spring". It is a known fact that widespread use of pesticides lead to a series of toxicological and environmental problems such as prolonged existence of some pesticides in the environments without breaking down and the evolution of more resistant species of harmful organisms (Suwalsky *et al.*, 2000).

Hazards of Chemical Pesticides

Pesticides used in appropriate doses recommended by the producers, do not normally lead to negative effect, but they may have some positive effect on the development of plants. But if they are given in overdoses cause phytotoxic effects and lead to anatomical and morphological deformities causes decrease in productivity. Pesticides cause biotic stress in plants (Lewitt, 1980).

Field crops grown in pesticide treated soils may contain measurable amounts of pesticides in the top portion of the plant. Translocation of chemical hydrocarbons insecticides in food crops has been studies extensively (Lichtenstein, 1650 and 1960; Lichtenstein and Schulz, 1960, 1965; Lichtenstein *et al.*, 1965). Most such studies were conducted with cucumbers (*Cucumis sativus*) (Lichtenstein, 1960), lettuce (*Lectuca sativa*) (Licthenstiein 1960), alfalfa (*Medicago sativa* L.) (King *et al.*, 1966; Zins *et al.*, 1991), canola (*Brassica napus* L.) (Dupont and Khan, 1993), and soybeans (*Glycine max*) (Bruce and Decker, 1966).

Toxicants first affect the most sensitive photosynthetic parts of a plant and then the enzymatic systems (Korte *et al.*, 2000). Azinphosmethyl and organophosphorus pesticides, was used on *Medicago sativa* L. plant and it was seen that 24 hours after the application, chlorophyll constituents was decrease more as compared to the control group (Flocco *et al.*, 2003). The application of high concentrated chemical preparations leads to stresses is the increase in the synthesis of specific peptides and proteins (Przymusiski, 2004). Most plant produce stress proteins in order to protect themselves from stress conditions. There are lot of studies suggesting that pesticide application increase protein synthesis and lead to the formation of stress proteins (Scarponi *et al.*, 2001; Saladin *et al.*, 2003).

The residue of $_{14}$C-DDT was reported by Nash *et al.* (1970) in soybeans and cotton (*Gossypium hirsutam* L.). The effects of DDT on the germination and growth of various plants were studied by Mitra and Raghu (1989). They found that, oil rich seeds of plants, such as pea nut (*Arachis hypogaea*) and mustard (*Brassica juncea*), were more prone to DDT induced inhibition of germination and subsequent plant growth than cereals, pulses and fibre crops, like rice (*Oryza sativa*), mungbean (*Vigna radiata*) and cotton (*Gossypium hirsutum*). It is suggested that lipids of the plant cell solubilize and disperse DDT in the cytoplasm, which in turn, affects normal metabolism within the cell (Mitra and Raghu, 1989). Verma and Pillai (1991) studied bioavailability of soil-bound residues of DDT and HCH to green gram (*Cicer arietinum*), maize (*Zea mays*) and dry land rice (*Oryza sativa*). They found that binding of DDT was more in the shoots and that of HCH in roots. Residue of DDT was reported in cowpea (*Vigna* sp.) by Kiflom *et al.* (1999).

Mani and Jayraj (1976) reported that the insecticides monocrotophos and phosphamidon likewise increased concentrations of nitrogen and phosphorous in rice plants; these changes were thought to contribute to a resurgence in numbers of the rice blue leafhopper. Pesticide can alter the chemical composition of plants. The changes that occur appear to be specific for both the plant and the pesticides involved. For example, certain organochlorine insecticides have increased the amounts of some macro- and micro-element constituents (Al, B, Ca, Cu, Fe, K, Mg, Mn, N,P, Sr, and Zn) of corn and beans, decreased the amounts of others (Cole *et al.*, 1968). In another study, DDT, aldrin, endrin and lindane were found to stimulate synthesis of the important amino acids arginine, histidine, leucine, lysine, praline and tyrosine in crop but decreased the content of tryptophan (Thakre and Saxena, 1972). According to Rouchaud *et al.* (1983) soil treatment to carrot with insecticides Nexion, Birlane and Dyfonate increased the concentration of free sugars in roots.

There are many reports of adverse effect of fungicides on plants. Now a day's many modern systematic fungicides are extensively used in agriculture. According Reyes (1975), the application of benlate, a fungicide, produced chlorosis and irregular depression at the central and marginal portion of saffron leaves. Alcholor metalaxyl induced sharp decreases in cell division (Coman *et al.*, 1975). John *et al.* (1975) reported that Triarimol inhibits the seedling growth of pea. Siddiqui and Ahmed (2002) studied effects of systematic fungicides on protein, carbohydrate, amino acids and phenolic contents of susceptible (Maxipak) and resistant (Povan) varieties of *Triticum aestivum* L. They found that application of systematic fungicide caused a significant decrease in total protein and carbohydrate content compared to the control. An osmotic shock effect of systematic fungicides results in the release of protein and loss of membrane transport ability in the leaf cells (Amar and Reinhold, 1973). It has been suggested that the toxicant produced by the application of systematic fungicides inhibits protein synthesis by the binding the larger ribosomal subunits inducing change enzyme system (Person *et al.*, 1975) ceasing ATP and NADP formation (Mishra and Waywood, 1968; Siddiqui, 1997). It has been reported that plants treated with fungicides suffer from chemical stress (Siddiqui *et al.*, 1997). The toxicant produced by the application of systematic fungicide inhibits respiration photosynthesis and protein synthesis by inhibiting activity NADH cytochrome "c" oxidases in the respiratory chain and accumulation of succinate and blocking of alternative pathway of respiration (Berger and Cwick, 1990; Pillonel, 1993). The extensive use of fungicides on pear, apple and raspberry trees during their flowering periods has negative effects on pollen germination and fruit formation (Redalen, 1980; Fell *et al.*, 1983; Marcucci and Filiti, 1984). Higher concentrations of Topsin M systematic fungicide have harmful effects on *Vigna radiata* L. Wilezek and *Pennisetum americanum* L. It is more effective on root growth than on stem growth (Siddiqui *et al.*, 1999b; Siddiqui and Khan, 2001). According to Siddiqui *et al.* (1999a), Beniate, a systematic fungicide, blocked root growth. It has also seen that root dry weights decreased especially after higher concentrations of phosphate fungicide were applied to *Cormbia calophylla* and *Banksia brownii* plants (Barrett *et al.*, 2003).

Tort and Turkyilmazi (2003) found that recommended dose and two-fold of recommended dose of Captan fungicide increased the amounts of proline in pepper. Saladin *et al.* (2003) reported that the application of fludioxionil and primethanil fungicides under chemical control of *Botrytis cinerea* to a vine plant (*Vitis vinifera* L.) encouraged protein synthesis. It was also reported that as a result of breakdown of structural proteins, fungicide applications might result in a decrease in the amounts of protein (Goicoechea *et al.*, 2000; Pinheiro *et al.*, 2001). Saladin *et al.* (2003) studied effects of fludioxonil and pyrimethanil, used against *Botrytis cinerea*, on carbohydrate physiology in *Vitis vinifera* L. They found that *in vitro* both fungicides decreased gas exchanges, photosynthetic pigment and starch concentration in the leaves, whereas soluble carbohydrates transiently accumulated, suggesting that plantlets mobilized starch in response to photosynthesis inhibition caused by fungicides.

Tort *et al.* (2006) studied morphological and physiological effects of a fungicide with a thiram agent on some corn culture forms. They found that application of fungicide cause decrease in root coleoptile length, root-stem length, fresh and dry weights as compared to the controls. Photosynthetic pigments substances decreased in cin cultivar but increased in set cultivar of corn. According to them, high concentration of fungicides affects plant metabolism of both cultivars negatively by causing biotic stress.

Spray adjuvants assist in the penetration of pesticides through plant surface. Spray adjuvants allow better coverage and penetration of pesticides and are used extensively in viticultural spray programs. However, spray adjuvant disrupt the cuticular layer that covers the plant surface. Mlikota Gabler *et al.* (2003) reported that fungal infection is more likely to occur with decreased cuticular thickness, or with an increased number of entry points through the cuticle. Therefore, even though adjuvant is used to increase the efficacy of a fungicide, it increases the susceptibility to other pathogens that were not a target of that fungicide (Rogiers *et al.*, in press). Adjuvants also alter the natural microflora on plant tissues through changes in the immediate physical and chemical environment. Some of the indigenous yeast, fungal and bacterial species are beneficial in that they deter the growth of pathogens (Blakeman and Fokkema, 1982).

Extensive Use of Pesticides Affects Grape Nutrition Quality

In developing countries, like India the area under grape cultivation increasing rapidly. Due to humid conditions, grape plants are easily attacked by pathogens. Use of chemical pesticides in grape cultivation is higher than any other crop. Pesticides exposure represents a major health hazard for sprayers in grape garden. Major chemical groups of pesticide used in grape gardens are organophosphate, carbonate, organochlorines, nitro and chlorophenols and pyridyl derivatives (Dave, 1998) It has been observed that maximum exposures occur while mixing and handling of the concentrated solutions. The biochemical effects produced by certain pesticides can be enzyme induction or enzyme inhibition organophosphorus and carbamate are widely used pesticides in grape gardens. These are inhibitor of enzyme cholinesterase (WHO, 1992 and 1993). According to WHO, the workers exposed to the organophosphorus pesticide alone or in combination with organochlorine, altered liver enzyme activities observed in them. Patil *et al.* (2003) studied biochemical effects of various pesticides on sprayers of grape gardens they found that the long term exposure of various pesticides on sprayers of grape garden affect liver, heme biosynthesis and decrease in serum cholinesterase. Rogier *et al.* (2005) found that spray adjuvant disrupt wax plate lets of grape berry and increases the susceptibility of *Botrytis cinerea*. They also found that spray adjuvant in some circumstances reduce the indigenous microflora on the berry's surface.

Conclusion

The above information gives idea regarding role of chemical pesticides and other substances creating the side effects on consumers as well as changing quality, taste, durability and the food value of fruits and other edible parts of different cultivated crops.

References

Amar, L. and Reinhold, L. (1973): Loss of membrane transportability in leaf cell and release of protein as a result of osmotic shock. *Plant Physiol.*, **51**: 620-625.

Barrett, S. R., Shearer, B. L. and Hardy, G. E. St. J. (2003): Root and Shoot development in *Corymbia calophylla* and *Banksia brownii* after the application of the fungicide phosphite. *Australian Journal of Botany*, **50** (2): 155-161.

Berger, S. and Cwick, K. (1990): Selected aspect of adverse nutritional effect of pesticides. *Ernahrun1.*, **4**: 411-415.

Blakeman J. P., Fokkema N. J. (1982) Potential for biological control of plant diseases on the phylloplane. *Annual Review of Phytopathology*, **20**: 167-192.

Bruce, W, N. and Decker, G. C. (1966): Insecticide residues in soybeans grown in soil containing various concentrations of aldrin, dieldrin, heptachlor, and heptachlor epoxide. *Journal of Agricultural and Food Chemistry*, **14**: 395-398.

Cole, H., Mackenzie, D., Smith, C. B. and Bergman, E. L. (1968): Influence of various persistent chlorinated insecticides on the macro and micro element constituents of *Zea mays* and *Phaseolus vulgaris* growing in soil containing various amounts of these materials. *Bull. Environ. Contam. Toxicol.*, **3**: 141-153.

Coman, N., Dordea, M. and Perseca, T. (1990): The estimation of alcholor and metaxyl on cell division. *Biologia*, **35**: 55-60.

Comelekoglu, U., Mazmanci, B., and Arpaci A. (2000): Investigation of the Liver Functions in Agricultural Workers Chronically Exposed to Pesticides, *Turk J. Biol.*, **24**: 461-466.

Dave, S. K. (1998): Occupational health services for agriculture workers. *Indian J. Occup. Environ. Med.* **2**: 96.

Dupont, S. and Khan, S. U. (1993): Bound and extractable $_{14}$C residues in canola plants treated with radio labeled atrazine. *Weed Research*, **33**(1): 9-16.

Fell, R. D., Rajotte, E.G. and Yader, K. S. (1983): Effects of fungicides sprays during apple Voom on polen viability and honey bee foraging. *Environ. Entomol.*, **12**: 1572-1575.

Flocco, C. G., Carranza, M. P., Carvajal, L. G., Loewy, R. M., Pechén de D'Angelo, A. M. and Giulietti, A. M. (2004): Removal of azinphos methyl by alfalfa plants (*Medicago sativa* L.) in a soil-free system. *Science of the Total Environment*, **327**(1–3): 31-39.

Goicoechea, N., Aguirreola, J., Cenoz, S. and Garcia-Mina, J. M. (2000): *Verticillum dahliae* modifies the concentrations of proline, soluble sugars, starch, soluble proteins and abscisic acid in pepper plants. *Eyr. J. Plant Pathol.*, **106**: 19-25

John, B., Shive Jr., and Sisler, H. D. (1975): Effects of ancymidol (a growth retardant) and Triarimol (a fungicide) on the growth, sterols and gibberellins of *Phaseolus vulgaris. Plant Physiol.*, **57**: 640-644.

Kazan, K. and Gurel, E. (2001): Increasing resistance to diseases in plants. In: Ozcan S, Gurel E, Babaoglu M (editors), Plant Biotechnology, Volume II, Genetic Engineering and Its Applications, page 261-287, Selcuk University Press, Konya (*in Turkish*).

Kiflom, W. G., Wandiga, S. O., Nganga, P. K. and Kamau, G. N. (1999): Variation of plant p, p2-DDT uptake with age and soil type and dependence of dissipation on temperature. *Environment International*, **25**(4): 479-487.

King, R. L., Clark, N. A. and Hemken, R. W. (1966): Distribution, movement, and persistence of heptachlor and its epoxide in alfalfa plants and soil. *Journal of Agricultural and Food Chemistry*, **14**: 62-65.

Korte, F., Kvesitadze, G., Ugrekhelidze, D., Gordezian, M., Khatisashvili, G., Buodze, O., Zoalishvili, G. and Coulston, F. (2000): Organic toxicants and plants. *Ecotoxical Environ. Safe*, **47**: 1-26.

Lewitt, J. (1980): Responses of plants to environmental stresses. Vol: 1, pp. 3-18 (Academic press, New York, U.S.A).

Lichtenstein, E. P. (1959): Absorption of some chlorinated hydrocarbon insecticides from soils into various crops. *Journal of Agricultural and Food Chemistry*, **7**: 430-433.

Lichtenstein, E. P. (1960): Insecticidal residues in various crops grown in soils treated with abnormal rates of aldrin and heptachlor. *Journal of Agricultural and Food Chemistry*, **8**: 448-451.

Lichtenstein, E. P. and Schulz, K. R. (1960): Translocation of some chlorinated hydrocarbon insecticides into the aerial parts of pea plants. *Journal of Agricultural and Food Chemistry*, **8**: 452-456.

Lichtenstein, E. P. and Schulz, K. R. (1965): Residues of aldrin and heptachlor in soils and their translocation into various crops. *Journal of Agricultural and Food Chemistry*, **13**: 57-63.

Lichtenstein, E. P., Myrdal, G. R. and Schulz, K. R. (1965): Absorption of insecticidal residues from contaminated soils into five carrot varieties. *Journal of Agricultural and Food Chemistry*, **13**: 126-131.

Mani, M. and Jayraj, S. (1976): Biochemical investigations on the resurgence of rice blue leaf hopper *Zygina maculifrons* (Motch.). *Indian J. Exp. Biol.,* **14***:* 636-637.

Marcucci, M. C. and Filiti, N. (1984): Germination of pear and apple pollen as influenced by fungicides. *Gartenbauwiss ecschaft.*, **49**, 28-32.

Mishra, D. and Waywood, E. R. (1968): The effect of BZI and kinetin on the nicotinamide and nucleotide content of senescence wheat leaf. *Can. J. Bot.*, **46**: 167-178.

Mitra, J. and Raghu, K. (1989): Effects of DDT on the growth of crop plants. *Environmental Pollution*, **61**: 157-170.

Mlikota, G. F., Smilanick, J. L, Mansour, M., Ramming, D. W. and Makcey, B. E. (2003): Correlations of morphological, anatomical, and chemical features of grape berries with resistance to *Botrytis cinerea*. *Phytopathology*, **93**: 1263-1273.

Nash, R. G., Beall Jr., M. L. and Woolson E. A. (1970): Plant uptake of chlorinated insecticides from soils. *Agronomy Journal*, **62**: 369-372.

Oerke, E. C. (1994): Estimated crops losses due to pathogens, animal pests and weeds. In: Oerke E. C., Dehne H. W., Schönbeck, F., Weber, A. (eds.), Crop Production and Crop Losses: Estimated Losses in major foods and Cash Crops, pp. 72-88. Elsevier, Amsterdam.

Patil, J. A., Patil, A. J. and Govindwar, S. P. (2003) Biochemical effects of various pesticides on sprayers of grape gardens. *Indian Journal of Clinical Biochemistry,* **18** (2): 16-22.

Person, C. D., Sambroski, J. and Forysth F. R. (1957): The effect of BZI on detached wheat leaves. *Can. J. Bot.* **180**: 1294-1295.

Pillonel, C. (1993): Interaction of benzimidazol-N-sulfonamide with the cytochrome b and b/c complex in *Phythium alphanidermatum*. *Pesticide Sci.* **43**: 107-113.

Pinheiro, C., Chaves, M. M. and Ricardo, C. P. (2001): Alterations in carbon and nitrogen methabolism induced by water deficit in the stems and leaves of *Lupinus albus* L. *J. Exp. Bot.* **52**: 1063-1070.

Przymusiski, R., Ruciska, R. and Gwód, E. A. (2004): Increased accumulation of pathogenesis-related proteins in response of lupine roots to various abiotic stress, *Environmental and Experimental Botany*, **52**: 53-61.

Redalen, G. (1980): Effect of fungicides on polen germination and fruit set in raspberries. *Gartenbauwiss Ecschaft*. **45**: 248-251.

Reyes, A. A. (1975): Phytotoxicity of benomyl to saffron. *Phytopath.,* **65**: 1-6.

Rogiers. S. Y., Whitelaw-Weckert, M., Radovanovic-Tesic, M., Greer, L.A, White, R. G. and Steel, C. C. (2005): Effects of spray adjuvants on grape (*Vitis vinifera*) berry microflora, epicuticular wax and susceptibility to infection by *Botrytis cinerea*. *Australasian Plant Pathology*, **34** (2): 221-228.

Rouchaud, J., Moons, C. and Meyer, J. A. (1983): Effects of selected insecticides and herbicides on free sugar contents of carrots. *J. Agric. Food Chem.,* **31**: 206-210.

Saladin, G., Magne, C. and Clement, C. (2003): Physiological stress responses of *Vitis vinifera* L. to the fungicides fludioxonil and pyrimethonil. *Pesticide biochemistry and Physiology,* **77**: 125-137

Saladin, G., Magne, C. and Clement, C. (2003): Physiological stress responses of *Vitis vinifera* L. to the fungicides fludioxonil and pyrimethanil. *Pesticide biochemistry and Physiology,* **77**: 125-137.

Scarponi, L., Vischetti, C. and del Buono, D. (2001): Imazamox in maize, uptake, persistence and interference on protein and carbohydrate formation. *Ital. J. Food Sci.,* **13**(2): 213-220.

Siddiqui, Z. S. (1997): Effect of systemic fungicides on total protein, carbohydrate and phenolic contents of *Solanum melongena* and *Avena sativa*. *Appl. Ent. Phytopath.,* **64**: 17-22.

Siddiqui, Z. S. and Khan, S. (2001): Effect of Systemic Fungicides and Insecticides on Absorption Spectra, Chlorophyll and Phenolic constituents of *Vigna radiata* (L.) Wilczek. *Pakistan Journal of Biological Sciences,* **4** (7): 812-814.

Siddiqui, Z. S. and S. Ahmed (2002): Effects of systemtic fungicides on protein, carbohydrates, aminoacids and phenolic contents of susceptible (Maxipak) and resistant (Povan) varities of *Triticum aestivum* L. *Turk. J. Bot.,* **26**: 127-130.

Siddiqui, Z. S., Ahmed, S. and Gulzar, S. (1997): Effect of topsin-M (Methylthiophenate) and Bayleton (Triademifon) on seedling growth, biomass, nodulation and phenolic content of *Sesbania sesban*. *Bangl. J. Bot.* **26**(2): 127-130.

Siddiqui, Z. S., Ahmed, S. and Shaukat, S. S. (1999a): Effect of Systemic Fungicide (Topsin-M) and Insecticide (Dimecron) on germination, seedling growth, Phenolic constituent of *Pennisetum americanum* L., *Pakistan Journal of Biological Sciences,* **2** (1): 182-184

Siddiqui, Z. S., Ahmed, S. and Stres, A. (1999b): Effects of Systemic fungicides on germination, seedling growth, DNA, RNA and Phenolic constituent of *Brassica compestris* L. *Pakistan Journal of Biological Sciences,* **2** (4): 1350-1351.

Suwalsky, M., Benites, M., Narris, B. and Sotomayor, P. (2000): Toxic effects of the fungicide benomyl on cell membranes. Comparative Biochemistry and Physiology Part C: *Pharmacology, Toxicology and Endocrinology,* **125** (1): 111-119

Thakre, S. K. and Saxena, S. N. (1972): Effect of soil application of chlorinated insecticides on amino acid composition of maize (*Zea mays*). *Plant Soil,* **37**: 415-418.

Tort, N. and Türkylmaz, B. (2003): Physiological effects of Captan fungicide on Pepper (*Capsicum annuum* L.) plant. *Pakistan Journal of Biological Sciences,* **6** (24): 2026-2029.

Tort, N., Dereboylu, A. E. and Turkyilmaz, B. (2006): Morphological and physiological effects of a fungicide with a thiram agent on some corn culture forms. *Journal of the Faculty of Science* **29**: 67–79.

Verma, A. and Pillai, M. K. K. (1991): Bioavailability of soil-bound residues of DDT and HCH to certain plants. *Soil Biology and Biochemistry*, **23**(4): 347-351.

WHO (1992 and 1993). World Health Organisation. The ICD-10 classification of mental and behavioural disorders. *Diagnostic criteria for research*. Geneva. www.who.int/classifications/icd/en/GRNBOOK.pdf

Zins, A. B., Wyse, D. L. and Koskinen, W. C. (1991): Effect of alfalfa roots on movement of atrazine and alachlor through soil. *Weed Science*, **39**: 262-269.

Index

C

D

G

H

N

O

P

Q

R

S

Figure 27.1 (P. 377)

Kunapajala

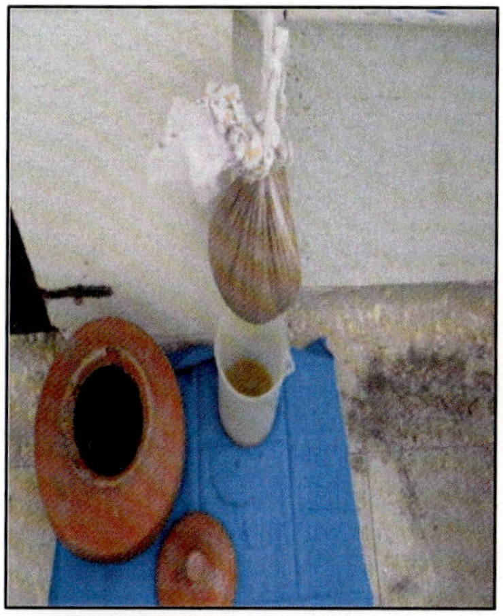

Filtration of *Kunapajala* (I)

Filtration of *Kunapajala* (II)

Figure 28.1 (P. 388)

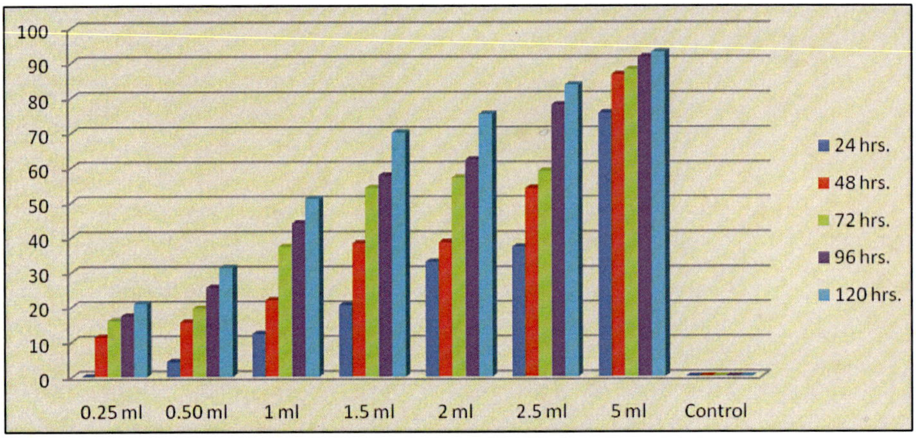

Figure 30.1 (P. 416)

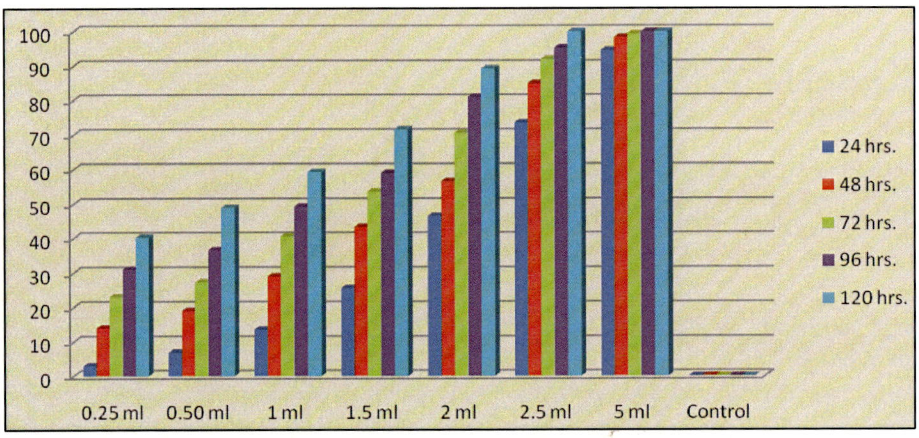

Figure 30.2 (P. 417)

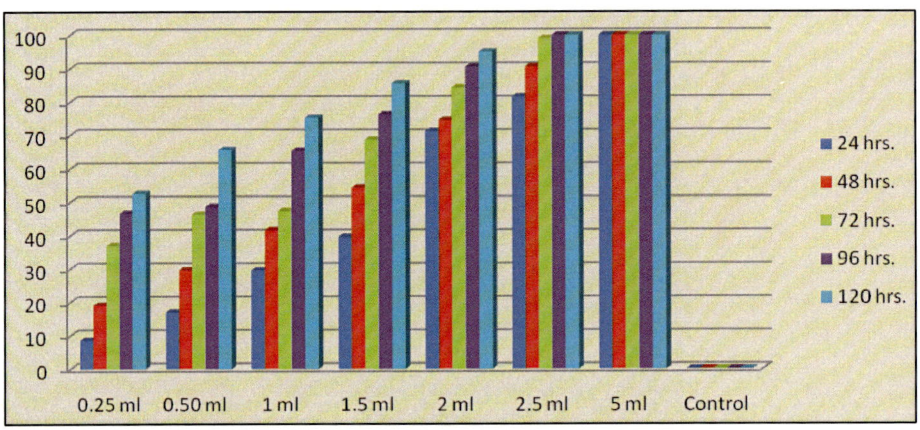

Figure 30.3 (P. 418)

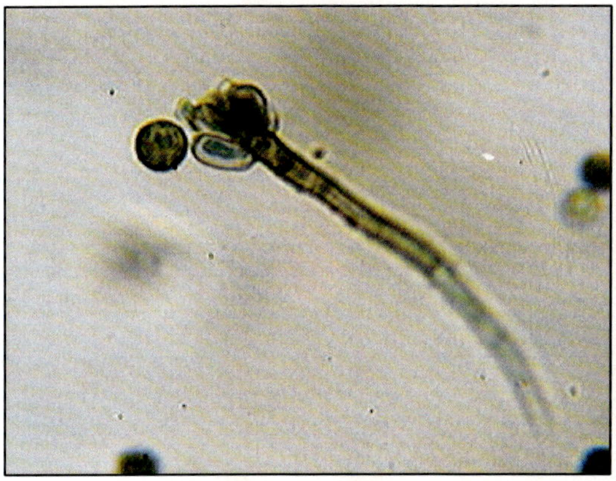

Figure 34.1 (P. 447)

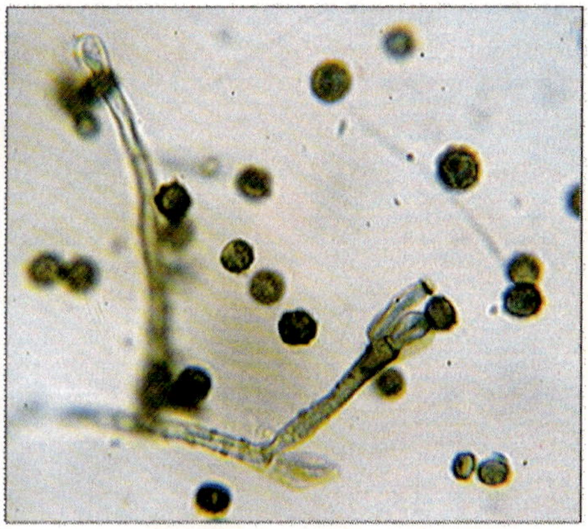

Figure 34.2 (P. 448)

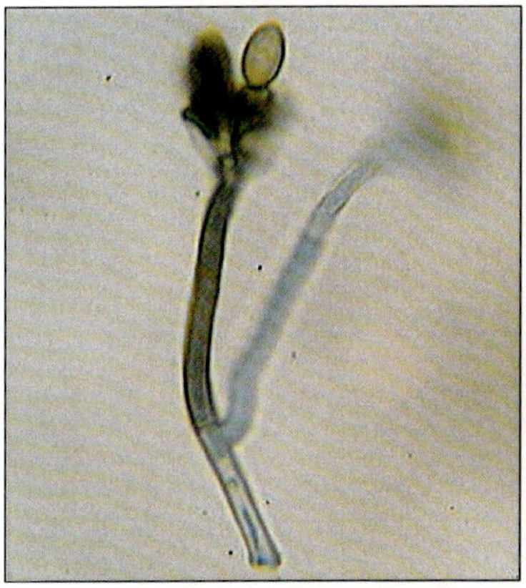

Figure 34.3 (P. 449)

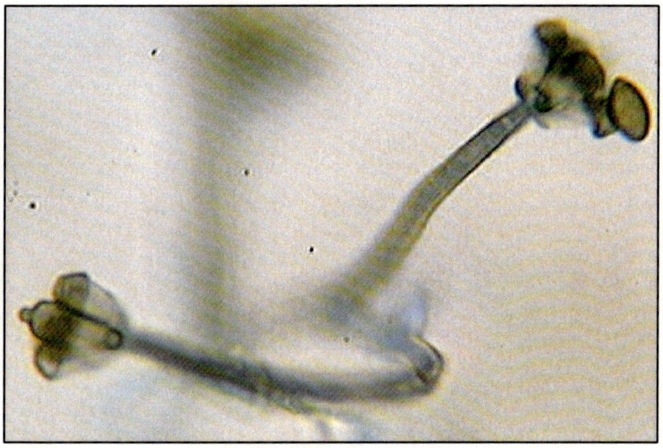

Figure 34.4 (P. 449)

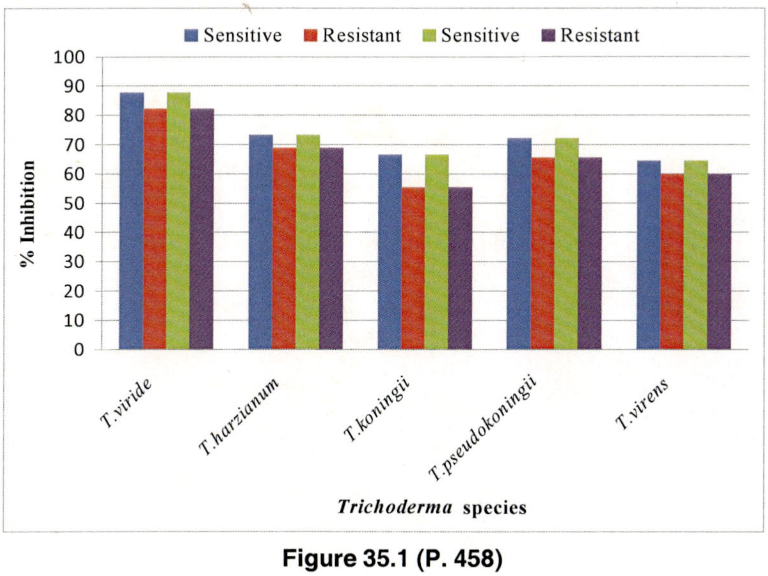

Figure 35.1 (P. 458)

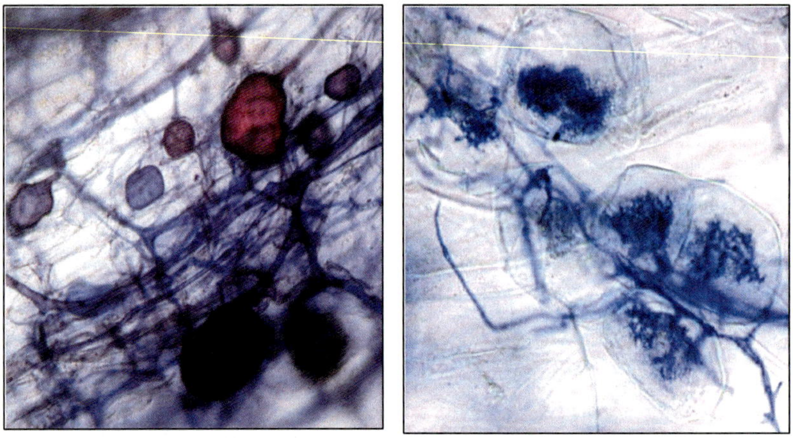

Intraradical-Vesicles **Hyphal and Arbuscular**

Figure 44.2 (P. 555)

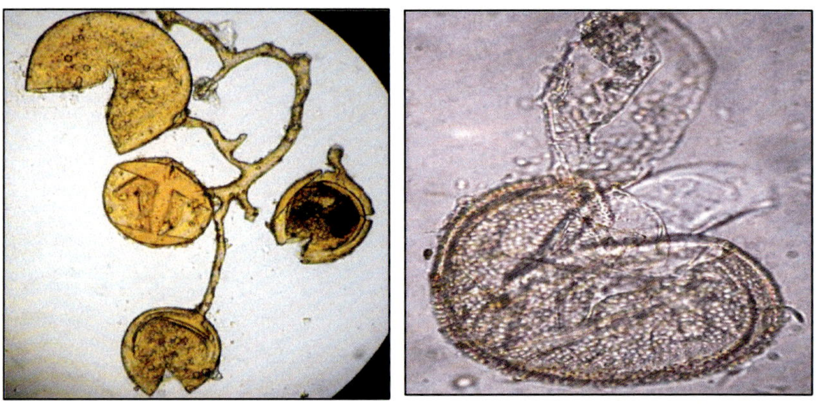

Glomus aggrigatum ***Acaulospora scrobiculata***

Figure 44.3 (P. 555)

Plate 45.1 (P. 562)

Figure 46.5 (P. 584)